Atomic Force Microscopy for Biologists

Atomic Force Microscopy for Biologists

V J Morris

A R Kirby

A P Gunning

Institute of Food Research
Norwich, UK

Imperial College Press

Published by

Imperial College Press
57 Shelton Street
Covent Garden
London WC2H 9HE

Distributed by

World Scientific Publishing Co. Pte. Ltd.
P O Box 128, Farrer Road, Singapore 912805
USA office: Suite 1B, 1060 Main Street, River Edge, NJ 07661
UK office: 57 Shelton Street, Covent Garden, London WC2H 9HE

British Library Cataloguing-in-Publication Data
A catalogue record for this book is available from the British Library.

ISBN   1-86094-199-0

Printed in Singapore by Regal Press (S) Pte. Ltd.

This book is dedicated to Martin Murrell and Mark Welland who introduced us to probe microscopy, with whom we have subsequently spent many happy hours delving into the workings of our AFM, and to Christina, Gloria and Yvonne for their patience and understanding during the writing of this book.

ACKNOWLEDGEMENTS

The authors wish to acknowledge the kind co-operation of those who provided figures for inclusion in this book; individual accreditation is given in the relevant figure captions. The beauty of the images stands not only as testament to the dedication of the researchers, but also to the power and rapid progress of the technique. Alan Mackie and Pete Wilde gave helpful advice on the topics discussed in chapter 5 and provided some of the schematic figures used. We wish to thank Imperial College Press for giving us the opportunity of writing this book, and in particular Geetha and Yuga on the editorial staff for answering our many questions. Finally, we acknowledge the Biotechnology and Biological Sciences Research Council for providing funding for our research through its core grant to the Institute of Food Research.

CONTENTS

CHAPTER 3 **BASIC PRINCIPLES** 44

CHAPTER 4 **MACROMOLECULES** 76

CHAPTER 5 **INTERFACIAL SYSTEMS** 160

CHAPTER 1

AN INTRODUCTION

The atomic force microscope (AFM) is perhaps the most versatile member of a family of microscopes known as scanning probe microscopes (SPMs). These instruments generate images by 'feeling' rather than 'looking' at specimens. This novel mode of imaging results in a magnification range spanning that associated with both the light and the electron microscope, but under the 'natural' imaging conditions normally associated with just the light microscope. The potential to image biological systems in real time, under natural conditions, with molecular, or even submolecular resolution is clearly of interest to biologists. Thus, since their inception, SPMs have had an obvious appeal to biologists and biophysicists. Since these first studies in the early 1980s publications describing biological applications of SPM have grown extremely rapidly. One of the aims of this book is to look at what impact these techniques have made in the biological sciences, and to try to assess their future potential.

Scanning probe microscopy began in the early 1980s when Binnig and Rohrer revolutionised microscopy through the invention of the scanning tunnelling microscope (STM). The importance of this discovery was recognised through the award of the Nobel prize in Physics in 1986. The STM is the first of this large and growing family of probe microscopes, which sense the structure of a surface by scanning it with a sharp probe and measuring some form of interaction between the surface and the probe. The development of the STM arose from an interest in the study of the electrical properties of thin insulating layers. This led to an apparatus in which the probe-surface separation was monitored by measuring electron tunnelling between a conducting surface and a conducting probe. A few years later Binnig and colleagues (1986) announced the birth of the second member of the SPM family - the atomic force microscope, also known at the scanning force microscope (SFM). In the late 1980s commercial STMs became available. Commercial AFMs began to appear in the early 1990s and have evolved through several generations. Refinements and new types of SPMs have appeared and will undoubtedly continue to be developed in the future. Particular developments of importance in biological research are combinations of probe microscopes with light or electron microscopes, cryo AFMs and scanning near field optical microscopes (SNOMs).

SPMs are not strictly microscopes: they visualise surfaces by 'feeling' them with a sharp probe. Conventional (far field) micoscopes image by collecting radiation transmitted through, or reflected from the sample. The ultimate resolution is diffraction limited and depends on the wavelength of the radiation.

Thus light microscopes are limited to a resolution of ≈ 200 nm. Higher resolution images of biological materials can be obtained using high energy electrons in the electron microscope (EM). Despite recent advances in the development of environmental EMs even molecular resolution still requires that specimens are examined under vacuum. Electron microscopists have developed many elegant preparative methods to preserve the 'native' structure of biological materials. SPMs image by a different mechanism (Fig. 1.1) and different criteria determine their resolving power.

Figure 1.1 Scanning electron microscope image of an AFM tip used to probe the structure of the sample surface. Magnification approximately x 10,000. The probe 'feels' the sample surface. Image courtesy of P.A. Gunning.

In SPMs images are obtained by measuring changes in the magnitude of the interaction between the probe and the specimen surface as the surface is scanned beneath the probe. Hence the resolution will depend on the sharpness, or apparent sharpness of the probe tip, and the accuracy with which the sample can be positioned relative to the probe. SPMs are capable of 'atomic' resolution of flat surfaces and such resolution can be achieved in gaseous or liquid environments. For macromolecules atomic resolution is only possible for simple molecules in which each atom is in intimate contact with the surface of a flat substrate. However SPMs do allow sub-molecular resolution on most biopolymers under 'natural' conditions. Thus the attractive potential of SPMs for biologists is the ability to visualise molecular processes under natural or physiological conditions. They offer the resolution of most commercial electron microscopes but under the experimental conditions familiar to the light microscopist.

The first biological studies were made using STM. The tunnelling current decays exponentially with increasing separation between the surface and the probe. A change in probe-sample separation of atomic dimensions will lead to an order of magnitude change in tunnelling current. This means that tunnelling effectively occurs from the atom on the tip nearest the surface, and the probes behave as if they are atomically sharp. Thus STMs offer the highest resolution obtainable by SPM. However, the rapid decay in the tunnelling current basically restricts investigations to the study of thin interfaces, or individual biopolymers. For larger biological systems the probe-sample separation would become too large and any tunnelling current would be expected to be too small to detect. Furthermore, the sample surface needs to be conducting and this usually means coating the biological sample, offsetting the main advantages of the SPM method. With the AFM there are no such restrictions on the size of the specimen that can be examined, and biological samples ranging in size from individual molecules to cells or tissues can be, and have been, imaged in their native state. AFMs, and refinements such as cryo-AFM, have become the preferred SPM methods in biology. It was originally believed that SPMs were non-invasive techniques. In practice sample damage and displacement plagued the early uses of STM and AFM. Understanding and overcoming these problems has led to reliable and reproducible methods of imaging. Emphasis has passed from validation of the microscopes to their use to study biological problems.

Other types of SPM will also become important in biology. A likely candidate, now available commercially, is the scanning near-field optical microscope (SNOM). This is a near field optical microscope which breaks the diffraction limit to spatial resolution by placing the sample within a fraction of a wavelength of a narrow optical aperture. The basic principles of SNOMs and appropriate biological applications will be discussed. There is a growing appreciation of the value of combining AFMs or SNOMs with conventional optical or confocal microscopes, or of using them in conjunction with surface techniques such as surface plasmon resonance, and the basis of some of these combined microscopes will also be described

This book will concentrate on biological applications of AFM. The advantages and limitations of the technique will be assessed. The literature in this area is vast and it is not possible to reference all of the published papers on a particular topic. Rather we have tried to cite books, recent reviews and selected research papers. The papers have been chosen to emphasise a point, or to provide a route to the literature in this area. The choice is not meant to imply priority and omission of papers is purely the result of the limitations of space and is not necessarily a reflection on the quality of the publication.

What do we wish to achieve in writing this book? One aim is to introduce the AFM, to describe the type of apparatus available and how it is used. A second aim is to look at the types of biological samples which have been studied, to look at how successful these studies have been, and to assess whether the use of AFM has generated new knowledge or understanding in these areas. In general terms we wish to look at what can and what cannot be done, if it is possible to do it then how is it done, and at what has been done and where things may go in the future. Who do we hope will benefit from reading this book? We hope that the book will provide a good resource base for literature on biological applications of AFM. The information presented should allow the reader to critically evaluate published present and future AFM data on biological systems, to decide whether AFM is likely to be useful in their areas of interest and, for new recruits to AFM, provide a basis for deciding what sort of technique to invest in, its inherent limitations, and how to optimise its use.

CHAPTER 2

APPARATUS

2.1. The atomic force microscope

Despite its rather grandiose title the atomic force microscope is probably one of the easiest forms of microscopy to understand at a rudimentary level. However, the first thing to consider when thinking about how an AFM works is that all notions of conventional microscope design need to be disregarded, since it has no lenses of any kind. In fact the AFM images samples by 'feeling' rather than by 'looking'. A good analogy is a blind person feeling objects with their fingers and then building up a mental image of what they touch. Like the blind persons' fingers this method of imaging can produce an exquisitely detailed picture, not just of the topography of the surface being touched but also of its' texture or material characteristics, soft or hard, springy or compliant, sticky or slippery. This latter aspect of AFM will be discussed in more detail elsewhere (sections 3.2 and 7.1.2) For now let us concern ourselves only with the topographical information that the AFM provides and how this is achieved optimally in modern instruments. The schematic in Fig. 2.1 illustrates the main features of an atomic force microscope.

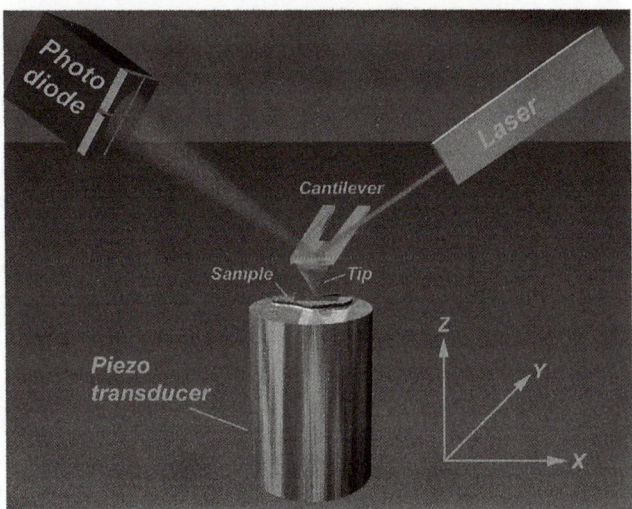

Figure 2.1. Schematic representation of the atomic force microscope.

The first and arguably most important part of the AFM is the stylus, or tip, which does the 'feeling', and an actual AFM tip is shown in the electron micrograph in Fig. 2.2.

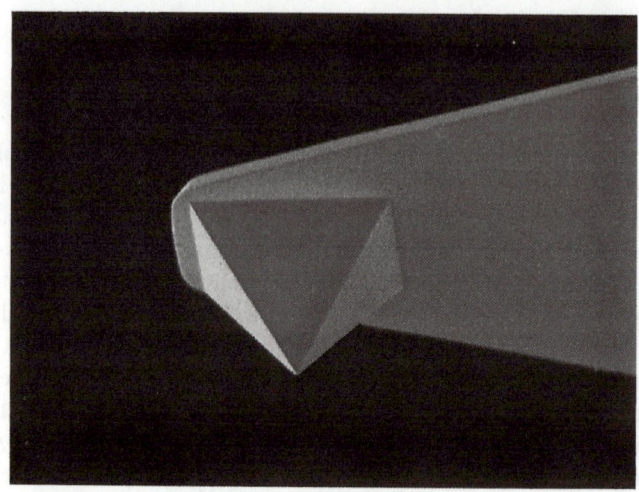

Figure 2.2. Scanning electron micrograph of an AFM tip and cantilever. To give some sense of scale the base length of the pyramid is 4µm. Image courtesy of P.A. Gunning.

The tip consists of a micro-fabricated, extremely sharp spike mounted on the end of a cantilever. This tiny assembly is bonded to a glass chip to allow easy handling. The sharpness of the spike, or tip as it is more usually called, determines the resolving power of the instrument. The cantilever on which it is mounted allows the tip to move up and down as it tracks the sample, in the same way that a record player stylus tracks a record (or rather used to in the days before CDs!). Furthermore, the cantilever typically has a very low force constant (also referred to in physics text books as spring constant) enabling the AFM to control the force between the tip and the sample with great precision. The cantilever-tip assembly is generally made of silicon or silicon nitride, these materials being both hard, and so wear resistant, and ideally suited to micro-fabrication.

The second crucial feature of the AFM is the scanning mechanism. Simply having a very sharp tip is of no use whatsoever without a means of accurately positioning it relative to the sample surface. This is done by means of a piezoelectric transducer (Fig.2.3). The principle is the same as in a piezoelectric gas-lighter - namely that when a crystal of piezoelectric ceramic is squeezed it produces a potential difference (i.e. a bias voltage) large enough to generate a spark. If this process is reversed and a potential difference is applied to the piezoelectric ceramic, it will expand. This motion is incredibly reproducible and sensitive such that, with a clean enough electrical signal, the piezoelectric ceramic can be made to move with an accuracy of atomic dimensions. This provides the AFM, and indeed all probe microscopes, with the accuracy of positioning sample or tip that they require. There are many different geometries which will be discussed later (section 2.2) but a generic layout is shown in Fig. 2.1 in which the sample is mounted on top of the piezoelectric transducer. The motion of the sample can be controlled in the three orthogonal directions x, y and z and these are assigned to three channels in the instruments' control electronics. The sample can now be positioned in extremely close proximity to the AFM tip (often actually in contact) using the z channel, and then raster-scanned (i.e. line by line) using the x and y channels in order to build up an image of a selected area on the sample surface.

The final feature of the instrument is the detection mechanism. The motion of the tip as it traverses the sample must be monitored. Again there are several different ways of doing this which are described in detail later (section 2.5) but, for the sake of simplicity, the most common method, the optical lever system, is illustrated in the schematic in Fig. 2.1. A laser beam is focused onto the end of the cantilever, preferably directly over the tip, and then reflected off onto a photodiode detector. In modern instruments the photodiode is split into four segments. As the tip moves in response to the sample topography during scanning, the angle of the reflected laser beam changes, and so the laser spot falling onto the photodiode moves, producing changes in intensity in each of its quadrants. This surprisingly simple system, which is in effect a mechanical amplifier, is sensitive enough to detect atomic scale movement of the tip as it traverses the sample. The difference in laser intensity between the top two segments and the bottom two segments produces an electrical signal which quantifies the normal (up and down) motion of the tip, and the difference between the laser intensity in the left and right pairs of segments quantifies any lateral or twisting motion of the tip. Thus frictional information can be distinguished from topographical information.

When the sample is scanned the topography of the sample surface causes the cantilever to deflect as the force between the tip and sample changes. In the simplest operating mode (for more details on modes of operation see section 3.3) the cantilever deflection is maintained at a constant predefined level by a control loop which moves the sample or tip in the appropriate direction at each imaging point. In this mode of operation this feedback mechanism is crucial to generating an image. The x,y and z displacements of the piezoelectric scanner are recorded and displayed to produce an image of the sample surface.

2.2 Piezoelectric scanners

Modern AFMs use one of two basic types of scanning mechanism: there are some that scan the sample, and others that scan the tip. Both however, rely upon piezoelectric transducers. The *piezoelectric effect* is the generation of a potential difference across the opposite faces of certain non-conducting crystals (piezoelectric crystals) as a result of the application of stress and is highly sensitive. The electrical polarization produced is proportional to the stress and the direction of the polarization changes if the stress changes from compression to tension. The *reverse piezoelectric effect* is the opposite phenomenom and the one that AFM scanners employ. If the opposite faces of a piezoelectric crystal are subjected to a potential difference, the crystal changes shape. The piezoelectric materials used in AFMs are ceramics, generally of the so-called PZT type (lead zirconate titanates). In the early days of scanning probe microscopy the geometry of the scanner had three blocks of piezoelectric ceramic arranged in a tripod, moving the tip or sample in the three orthogonal directions x, y and z. These days piezoelectric tripods have been superseded by tubes of piezoelectric ceramic materials (or in some cases a combination of tripods and tubes) which were first introduced by Binnig and Smith in 1986 (Binnig and Smith, 1986). This geometry has many advantages over the tripod arrangement, but the principle one is that larger scan ranges are possible with what is a far more compact, and symmetrical geometry of the scanner.

The scanner tube consists of a thin-walled hard piezoelectric ceramic which is polarised radially. Electrodes are stuck to the internal and external faces of the tube, and the external face of the tube is split into quarters parallel to the axis, as shown in the schematic in Fig. 2.3. By applying a bias voltage between the inner and all the outer electrodes the tube will expand or contract ie. move in the z direction. If a bias voltage is applied just to one of the outer electrodes the tube will bend ie. move in the x and y directions. To make this bending even more pronounced, in order to improve the scan range, the outer electrodes are arranged opposite each other in terms of direction ie. $+x$ opposite $-x$ meaning that if the tube is biased with $+n$ volts on one electrode and $-n$ volts on the opposite electrode, the resultant bending of the tube will be twice as much as by simply biasing one electrode. A detailed mathematical study of the dynamics of tube scanners has been given elsewhere (Taylor, 1993). Tube scanners have two main disadvantages. The first is that the motion of the end of the tube, which drives the tip or sample in the x and y directions, traces out an arc rather than a straight line, leading to an effect known as 'eyeballing' when large scans are carried out. A flat surface appears as if it is part of the surface of a sphere - hence the term. This effect can be corrected with image processing software. The second problem is that of scan speed. Tubes cannot move quickly and, if very fast processes are to be studied, an alternative scanner arrangement consisting of small stacks of piezoelectric ceramics is used. Other problems relating to the non-linearity of piezoelectric scanners are discussed in the calibration section (section 2.8) in this chapter.

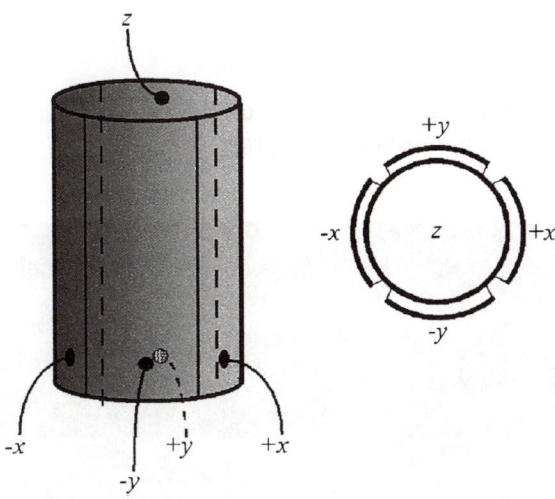

Figure 2.3. Schematic representation of a piezoelectric tube scanner. Although the tube moves in only the three orthogonal directions x,y,z there are actually five electrodes $+x,-x,+y,-y$ and z required to achieve this.

2.3. Probes and cantilevers

The heart of AFM is the tip since this is the part which interacts with the sample. Rather like fitting bald worn-out tyres to a Ferrari, fitting a blunt tip to even the most sophisticated AFM produces results which are at best disappointing and, at worst, worthless. Quality and consistency are the key to good AFM tips. This means that even since the very early days of AFM the need to have specially manufactured tips was recognised (Binnig *et al* 1987) and the home-built variety were quickly abandoned! Modern AFM tips and cantilevers are made by micro-fabrication, using many of the techniques that have been developed for integrated circuit manufacture, such as lithographic photo-masking, etching and vapour deposition. Cantilevers and tips are nearly always made from either silicon, silicon nitride, or diamond. They can be conducting or non-conducting, and are often coated with another material. If optical sensing methods are to be used to monitor cantilever deflection then they are coated with a thin gold layer to improve their reflectivity or, if magnetic sensitivity is required a ferromagnetic coating may be applied. Commercial AFM tips are usually supplied in 'wafers' containing a few hundred tips. Each tip needs to be separated according to the manufacturers instructions. A detailed description of the manufacture of tips and cantilevers is given by Boisen and co-workers (Boisen *et al* 1996).

2.3.1. Cantilever geometry

There are two basic geometries for the cantilever on which the tip is mounted and these are shown in the electron micrograph in Fig. 2.4.

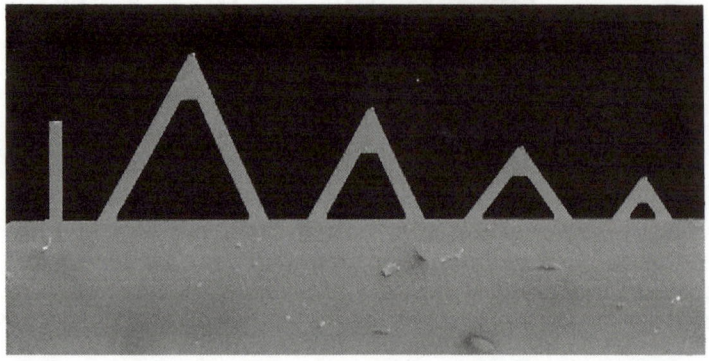

Figure 2.4. SEM image of AFM cantilevers. Magnification approximately 100x. Image courtesy of P.A. Gunning.

The triangular geometry seen in Fig. 2.4, usually referred to as a 'V' shaped lever, is designed to minimise torsional motions, or twisting of the cantilever whilst scanning a sample, and is the lever of choice for purely topographical imaging. If one wishes to measure the frictional properties of a sample then a simple beam or rectangular geometry, the more familiar notion of a cantilever, also shown in Fig. 2.4 is best, having a greater degree of rotational freedom making it sensitive to lateral forces. Irrespective of geometry, the force contribution (F) from the bending of the cantilever which acts on the sample is determined by the following simple equation, which is known as Hookes' Law:

$$F = -kd \tag{2.1}$$

where k is the force constant of the cantilever and d is its displacement. The minus sign indicates that the force acts in the opposite direction to the displacement of the cantilever. The force constant, k increases with lever thickness but decreases with cantilever length.

Q: " *Since longer cantilevers have lower force constants, are they the best ones to use for imaging delicate biological samples with low forces in order to get better resolution?"*
A: "Whilst they do allow lower forces to be achieved, in practice this factor alone often doesn't lead to better resolution due to their poorer sensitivity"

A general rule for both cantilever geometries is that longer cantilevers are less sensitive if the optical beam detection method is employed (which is the case for nearly all present day AFMs). The relationship between the angular displacement of the reflected laser beam, θ, and cantilever length, l, is:

$$\theta \propto \frac{1}{l} \tag{2.2}$$

i.e. a given displacement of the cantilever deflects the laser beam through a smaller angle than if the cantilever is shorter. Consequently the laser spot moves a smaller distance over the face of the photodetector producing a smaller output signal for the control loop. A practical application of this lack of sensitivity is that longer cantilevers are better for imaging rougher samples, because larger displacements of the tip produce less angular deflection of the laser beam, reducing the possibility that it will miss the photodiode detector, thus inactivating the feedback mechanism with the inevitable loss of control this would cause. However, very low force constant cantilevers have the disadvantage of low resonant frequencies (typically 7-9 kHz) which makes them less stable, and more difficult to use if scan conditions are not chosen carefully. The principle is that a cantilever should have a resonant frequency at least ten times higher than the fastest scan speed it will encounter during imaging, otherwise it may be excited into resonant oscillation causing loss of image quality (see section 3.6.2).

Since the cantilever is the spring in the system it determines the categories that AFM probes are divided into, contact, non-contact, Tapping and force modulation. These are imaging modes in their own right and described in detail later (Table

2.1 and section 3.3). However a brief description is given in table 1 at the end of this chapter which gives typical values of force constant, k, and resonant frequency, f_r, and explains why particular levers are used.

2.3.2. Tip shape

The actual shape of the tip used in AFM is an important consideration, and choice of tip shape is closely linked to the properties of the sample under study. There are all manner of different tips available commercially, each with a specific function. First of all they can simplify be divided into two important categories; high and low aspect ratio tips. In deciding which type is needed for any given sample a concept familiar in other forms of microscopy is relevant, namely depth of field. If a large depth of field is required, such as when a sample is rough, then high aspect ratio probes are needed, as illustrated in Fig. 2.5.

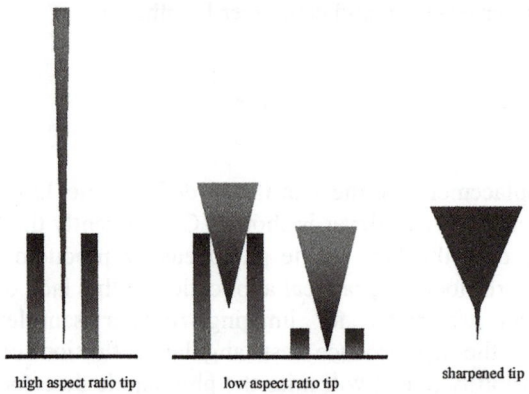

high aspect ratio tip   low aspect ratio tip   sharpened tip

Figure 2.5. Schematic representation of different aspect ratio AFM tips. Left shows a high aspect ratio tip suitable for rough samples, middle illustrates that low aspect ratio tips are limited to relatively flat specimens, and right details a sort of hybrid tip known as a 'sharpened' tip.

Having said this it should be born in mind that AFM is actually quite limited in its depth of field, and is not a suitable form of microscopy for very rough samples, irrespective of tip aspect ratio. It is obvious from Fig. 2.5 that when sample roughness approaches, or exceeds the height of the tip, proper imaging is impossible. If, however, the sample is flat then a low aspect ratio tip is suitable (Fig. 2.5). Some low aspect ratio tips, referred to as 'sharpened' tips, have high aspect ratio sections at the apex, in order to improve resolution on relatively flat samples with only modest increases in cost (Fig. 2.5). The three most common tip shapes are shown in Fig. 2.6 , and these are; pyramidal (left), isotropic (middle) and 'rocket-tip' (right).

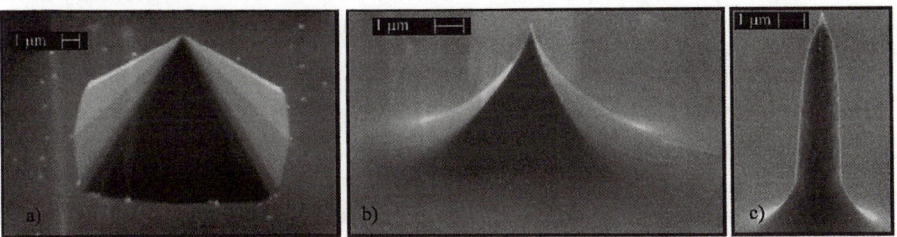

Figure 2.6. SEM image of common tip shapes. Reproduced with permission of the authors and publisher, Institute of Physics publishing, from Boisen *et al* 1996.

From a practical point of view the importance of different tip shapes can be defined by the 'opening angle' of tips: which is a way of defining the aspect ratio at the business end of the tip. Put simply this factor determines to what extent the AFM images of a sample are 'softened' or blurred by tip shape, low opening angles giving sharper images (this effect known as probe-broadening is discussed in more detail in section 2.8.2). The rougher a specimen the lower the opening angle of the tip should be, in order to avoid excessive tip convolution in the images. Note that the opening angle should not be confused with tip sharpness, which is usually defined as the radius of curvature at the apex of an AFM tip - for example a pyramidal tip can be just as sharp at its actual apex as a rocket tip.

2.3.3. Tip functionality

Recently it has been demonstrated that AFM tips can be functionalised. By coating the tips with certain materials the force between tip and the sample can be varied. This can be used to investigate particular types of tip-sample interaction or to map selected sites on surfaces (see section 7.1.2.). Coating can be used to enable a particular chemical sensitivity to be obtained (Frisbie *et al* 1994). Recently in a significant refinement of this approach, chiral sensitivity was demonstrated using a functionalised AFM tip (Mckendry *et al* 1998). Both have obvious implications for the application of AFM to biology, and it is now possible to purchase tips with defined functionality from several specialist labs (Bioforce, NT-MDT). Many of the functional attributes are based on attaching biological molecules to tips. They range from fairly simple hydrophilic or hydrophobic coatings to more sophisticated functionality such as antibody or antigen coated tips, and ligand or receptor coated tips. Another recent development which shows promise is the attachment of carbon nanotubes to AFM levers (Dai *et al* 1996). This gives a flexible, high aspect ratio tip with an extremely well characterised geometry at its apex, i.e. a C_{60} molecule, which is amenable to further chemical functionalisation (Wong *et al* 1998).

2.4. Sample holders

For imaging in air the sample is simply mounted on a small metal disc. It is best to stick the sample to the disc using double sided conducting adhesive tape or silver Dag^{TM} in order to prevent movement during imaging and build up of static charge on the sample. However, imaging in air is not generally the best option for biological samples and most of the time a liquid cell is used.

2.4.1. Liquid cells

The feature which really sets the AFM apart from the electron microscope is its ability to operate under liquids at unparalleled resolution. In order to do this a liquid cell is needed. All liquid cells, irrespective of their design, basically perform three functions; contain the sample, contain the liquid, and provide a stable optical path for the laser beam which is reflected off the back of the cantilever. Obviously if the optical beam detection method is being used the beam cannot simply pass through a liquid-air interface since it will be refracted 'all over the place' due to movement of the liquid surface. The solution to this problem is to use a glass sighting plate which is submerged in the liquid. A general layout of a liquid cell is shown in Fig. 2.7. Some liquid cells are sealed by means of compressing an 'O-ring' between the sample holder and the tip holder, others are sealed with a latex membrane, and some are left open. Sealed cells are more 'fiddly' to set up but they prevent evaporation of the liquid and allow flow of liquids through the cell during an experiment. Open cells on the other hand make flow of liquid through the cell difficult, but partial exchange of liquid or addition of liquid is still possible. The reason for adding, exchanging or flowing liquids through the cell whilst imaging is to allow the study of dynamic events, for example enzymatic catalysis. It is best to begin with a stable static system before the reagent is added so that a clear starting point is defined. The liquid cell should be made of an inert material and glass, PTFE or stainless steel are used depending upon the nature of the sample and liquid being used. Another requirement is that the sample needs to fixed to the bottom of the cell. This is normally done using a push-fit washer for plastic and glass cells, but magnetic washers can also be used. Glue may be used to fix the sample to the bottom of the cell but it is the least satisfactory option, carrying the risk of sample contamination.

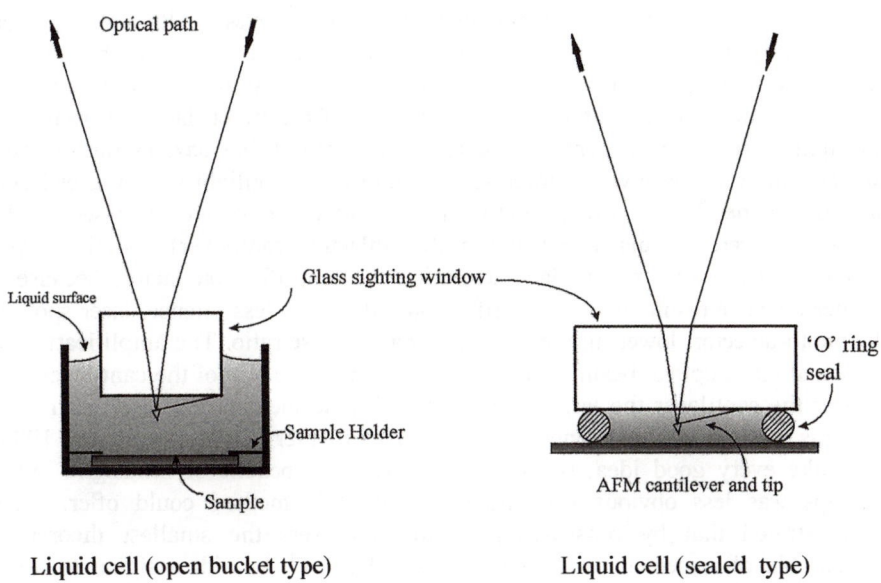

Liquid cell (open bucket type)   Liquid cell (sealed type)

Figure 2.7 Bucket type (left) and 'O-ring' type (right) liquid cells and optical detection.

2.5. Detection methods

Several different methods have been developed for detecting the motion of the AFM tip as it traverses the sample. The choice can be rather bewildering to the beginner, so a brief description of the most common methods used is given here. Broadly speaking these fall into two categories; optical and electrical methods.

2.5.1. Optical detectors: laser beam deflection

Laser beam deflection is the most common detection method used in modern commercial AFMs, and it is no coincidence that it is also the simplest and most versatile. A generic layout is shown in the schematic diagram in Fig. 2.8. The principle is familiar to every unruly schoolchild who, sitting at the back of a classroom on a sunny afternoon when they'd rather be somewhere else, have used their watch-face to redirect the sun into the teachers' eyes. Although distance equals safety for the transgressor it also increases the difficulty of hitting the target, and herein lies the crux of this method; namely mechanical amplification. Very small movements of the reflecting surface of the watch-face are translated into quite large displacements of the reflected sunlight that dazzles the teacher. Replace the watch-face with a micro-fabricated tip, the sunlight with a laser beam and the teachers' eyes with a photo-detector and you have an AFM sensor! In AFM, however, the distance between the reflector (cantilever) and the target (photo-detector) does not in fact determine the amplification factor, because a greater distance results in a more diffuse, and therefore less intense, laser spot on the photo-detector, lowering the overall signal to noise ratio. The amplification in an actual AFM optical beam set-up is determined by the size of the cantilever; the shorter the cantilever the larger the angular displacement of the laser beam will be. This mechanism was pioneered by Meyer and Amer (Meyer and Amer, 1988) and, like every good idea, seems obvious with the benefit of hindsight. What perhaps was less obvious was the sensitivity this method could offer. They demonstrated that by constructing small cantilevers the smallest theoretical measurable displacement was approximately 4×10^{-4} Å, with a signal to noise ratio of 1. In practice the displacement sensitivity is limited by random thermal excitation of the cantilever on which the tip is mounted but, with proper design, this is easily good enough to detect atomic scale displacement. The photo-detector is usually a simple photodiode, a semiconductor device which turns light falling on it into an electrical signal such that, as the incident light becomes brighter, the electrical signal increases. The photodiode is split into four sections enabling both normal and lateral motions of the tip to be differentiated. By comparing the relative intensity of the reflected laser light in each quadrant approximate quantification of tip displacement can be achieved. However for more precise measurement of tip displacement, such as is necessary if one is principally interested in performing force-distance spectroscopy on single molecules, then linear position sensitive detectors are a better choice of detection method (Pierce *et al* 1994).

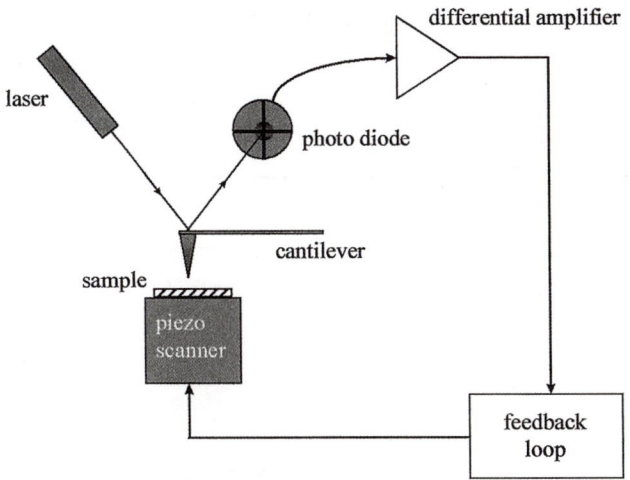

Figure 2.8. Laser beam deflection detection method.

2.5.2. *Optical detectors: Interferometry*

This was initially the obvious choice for detecting the very small displacements of the AFM tip, since it is a well-established technique for high precision measurement of small displacements invented as long ago as 1880 by Michelson (Hecht, 1987). Its implementation in AFM is shown schematically in Fig. 2.9. The phase change of the laser beam reflected from the back of the AFM cantilever is measured relative to a standard using an optical interferometer. This phase shift represents the relative displacements of the cantilever. It has advantages over optical beam deflection. This approach has the ability to cope with large deflections and has a superior signal to noise ratio. However, in practice it leads to an instrument which is harder to setup, requires an optical table to achieve sufficient vibration and acoustic isolation, and is susceptible to thermal drift and variation in the laser frequency. Despite these problems it was successfully implemented in AFM by Erlandsson and coworkers (Erlandsson *et al* 1988) and several essentially interferometric detection methods have been developed by others (Martin *et al* 1987; Schönenberger *et al* 1989; Rugar *et al* 1988;1989). The technique continues to be used in a few specialised research AFMs but has found greater application as an independent means of detecting motion of the piezoelectric scanner in so called 'closed-loop' feedback systems for commercial AFMs.

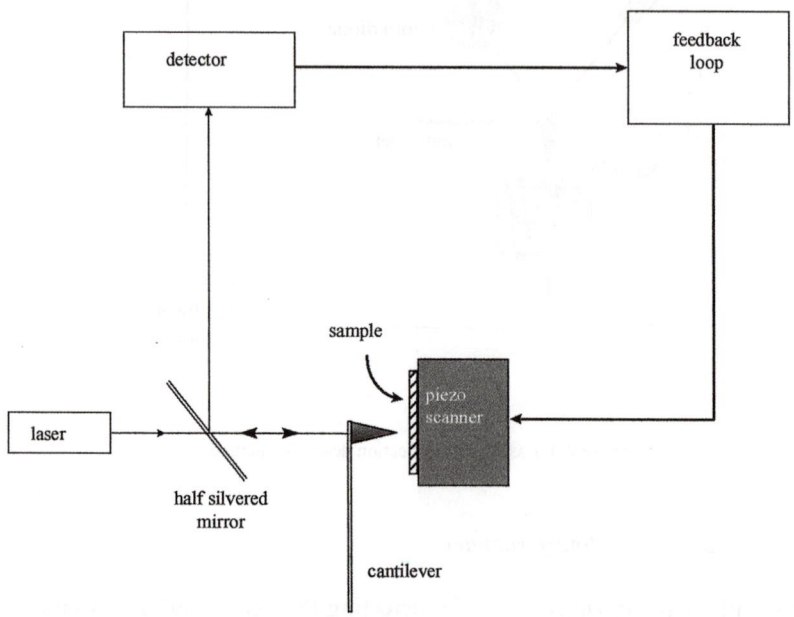

Figure 2.9. The Interferometer detection method.

2.5.3. *Electrical detectors: electron tunnelling*

An STM sensor is used to measure the displacement of the AFM tip and this is shown schematically in Fig. 2.10. The cantilever must be conducting, or coated with a conducting material (usually gold) and then forms the 'sample' of the STM. The tunnelling current varies exponentially with STM tip-cantilever separation and is used to monitor the motion of the AFM tip. Because of this high sensitivity to tip-'sample' separation it is necessary, in order to have any sort of useful working range, for the tunnelling current to be used in a feedback loop and to maintain constant tip-'sample' separation, i.e. between the STM tip and the AFM cantilever. The technique has two drawbacks, which ironically are both related to its extreme sensitivity. The first is that STM feedback control and vibration isolation are problematic. The second is that the STM sensor is sensitive to the roughness of the cantilever, so that if the tunnelling site on the cantilever changes due to lever flexure during imaging, then the image produced is a composite of the actual sample surface, plus the topography of the back of the cantilever. This was the method used in the first ever AFM which was built by Binnig, Quate and Gerber in 1986 (Binnig *et al* 1986) but is not used in any modern instrument.

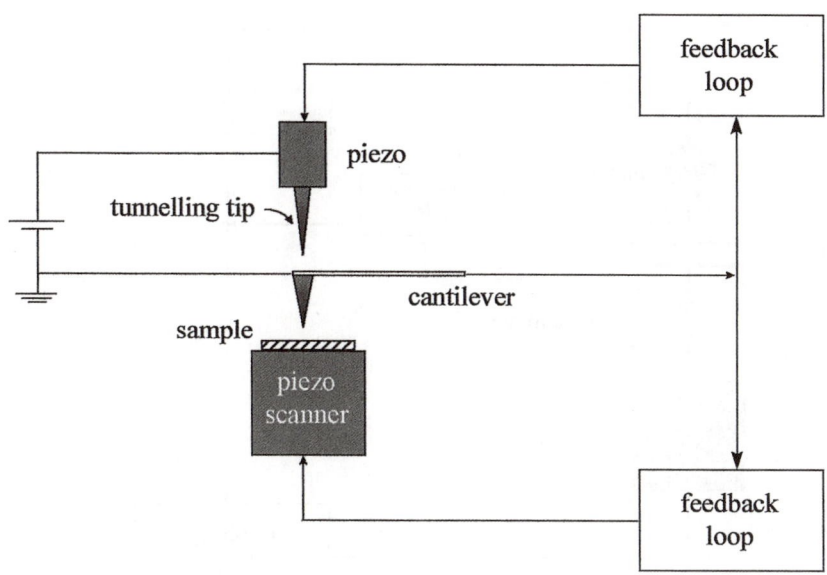

Figure 2.10. Electron tunnelling detector.

2.5.4. Electrical detectors: capacitance

A small metal plate, typically 300 x 300 µm, is attached to the back of an AFM cantilever, with a second plate attached to the end of a piezoelectric transducer (this is to give it a better dynamic z range) with a plate separation of about 1mm (Göddenhenrich *et al* 1990). A schematic layout of a capacitance force sensor is shown in Fig. 2.11. The capacitance depends on the separation between the plates and so provides a measure of cantilever displacement. It is very compact, UHV compatible and easy to configure, but is susceptible to temperature induced drift of the reference capacitor used in the measurement circuitry. Another disadvantage is that the force constant of the cantilever varies with capacitance. As for interferometric displacement sensing, capacitance sensing has found greater application for monitoring the displacement of piezoelectric scanners in closed loop feedback systems rather than as the primary means of sensing AFM tip motion.

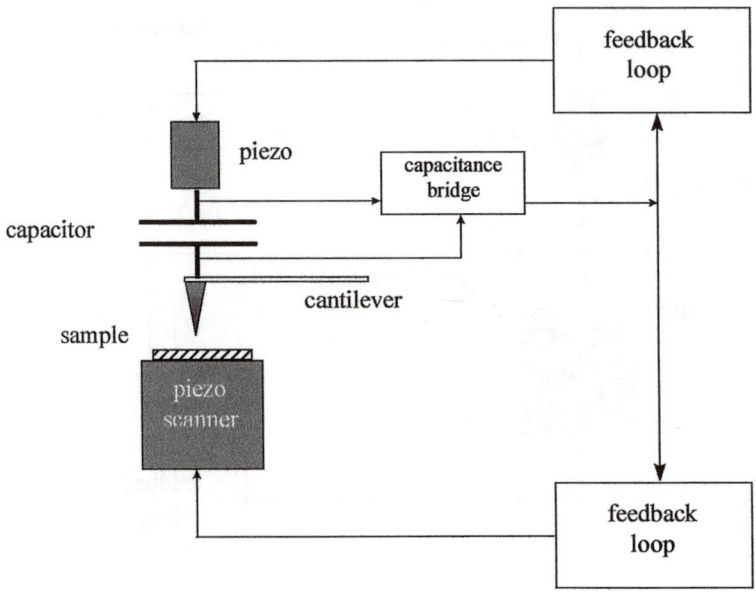

Figure 2.11. Capacitance detection method.

2.5.5. Electrical detectors: piezoelectric cantilevers

In this method the AFM cantilever is made from, or coated with, a piezoelectric material (Tansock and Williams, 1992). The principle is illustrated in the schematic diagram shown in Fig. 2.12.

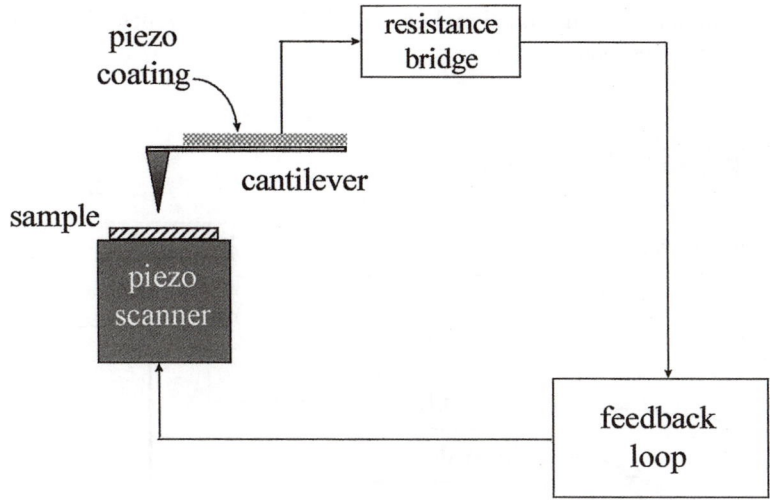

Figure 2.12. Piezoelectric cantilever detection method.

As the cantilever flexes in response to sample topography the bending of the piezoelectric material generates a potential difference which corresponds to the displacement of the lever. Flexure of the piezoelectric material also causes its resistance to change, which provides an alternative means of detection. This method has obvious benefits for examining light sensitive samples and this, combined with its reasonable sensitivity, makes it the most popular alternative to optical detection methods in modern commercially available AFMs. One disadvantage is the relatively high cost of the cantilevers, although this might reasonably be expected to come down as production increases.

2.6 Control systems

2.6.1. AFM electronics

Nearly every modern AFM has a digital control system (Wong and Welland, 1993; Baselt *et al* 1993). This consists of four elements which are illustrated in the schematic diagram shown in Fig. 2.13. The first is the digital control electronics, often present in the form of a digital signal processor (DSP) card, which resides in the image acquisition computer.

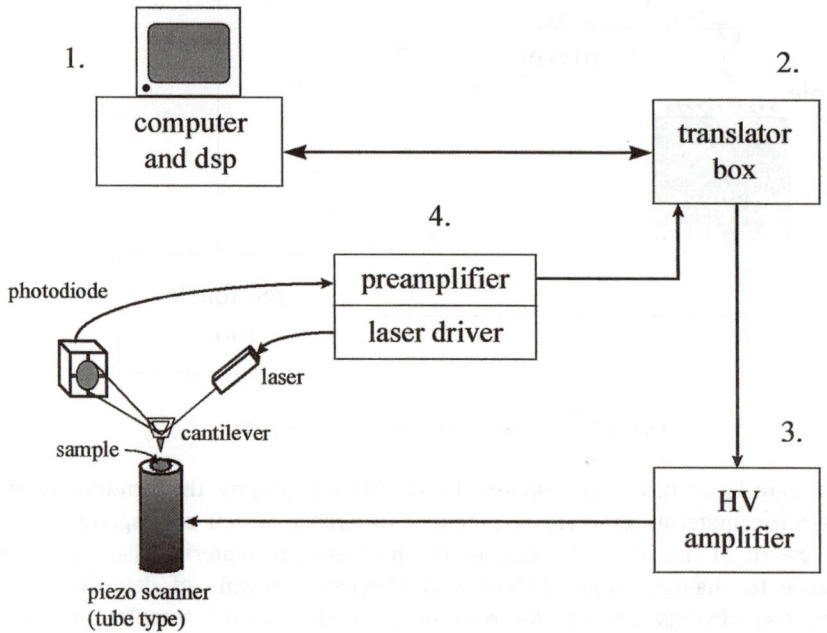

Figure 2.13. Typical AFM control system broken into main components.

The DSP performs all of the signal processing and calculations involved in operation of the AFM in real-time, this is the really clever bit. It is linked to the second element of the electronics which, for want of a better description, we will call a translator box. This translator box performs the conversion of the digital signals sent from the DSP card to an analogue form in order to run the microscopes' scanner mechanism and, unsurprisingly, incoming analogue data signals from the microscope head are converted to digital form for the DSP card. Basically the translator box is akin to the CD player in your living room, it is full of digital to analogue converters (DAC) and analogue to digital converters (ADC) and usually has an independent power supply for reasons of noise reduction.

The third part of the control electronics is the high voltage (HV) amplifier. The input of this amplifier is fed from the analogue signal which has come from the DACs in the translator box . It amplifies this low voltage signal to produce a high voltage signal, typically ± 150V, that drives the piezoelectric scanner described earlier. The final pieces of the jigsaw are the laser driver electronics (assuming the AFM uses an optical detection method) and the raw data signal preamplifier. These functions are usually combined in one box, or area of the control electronics, which is well shielded from the digital translator and high voltage amplifier electronics, both of which generate electrical noise. This noise (high speed switching noise from the digital translator box, and mains frequency noise from the high voltage amplifier) can be a real problem if it finds its way onto the incoming low voltage raw data signal. The laser driver circuitry provides power for the AFMs' laser and, because light output stability is an important requirement for the laser and detection circuitry, it has its own feedback loop which varies the laser power supply to maintain a constant laser intensity. The details of the raw data signal preamplifier obviously vary according to the type of detection system used but, whichever system, it amplifies the small output signal which comes direct from the AFM cantilever displacement sensor.

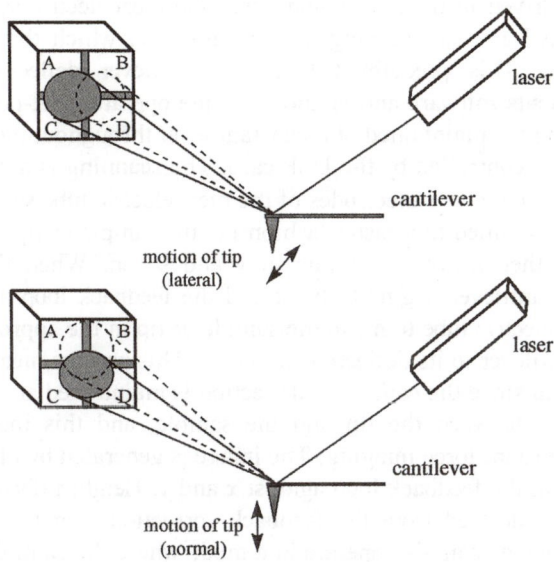

Figure 2.14. By splitting the photodetector into four segments, lateral motion or twisting of the cantilever (top) can be distinguished from normal or vertical motion (bottom).

For the optical beam method used by most AFMs, in addition to simple amplification, it also has a series of summing and difference amplifiers to produce an output signal for the translator box which defines either normal or lateral motion of the AFM tip (Fig 2.14).

2.6.2. Operation of the electronics

The function of each part of the AFMs' control electronics has been described in the last section. This section describes how they work together whilst the instrument is actually scanning a sample.

The whole system actually works in a feedback or control loop so that the tip can accurately track the surface of the sample in a controlled manner at all times. The most obvious demonstration of this is when the AFM is running in the so-called 'constant force mode' (section 3.3), and it is easiest to describe this in terms of contact mode imaging. In contact mode imaging (also referred to as dc mode) the AFM tip actually touches the surface of the sample just like a phonograph stylus playing a record. The tip is brought into contact with the sample surface and then driven in the z direction using the piezoelectric scanner mechanism by a predefined amount, causing the cantilever on which the tip is mounted to be deflected. This predefined level of cantilever deflection is determined in the instruments software and is known as the operating set-point of the instrument. The set-point is maintained at a constant level throughout the scan by a feedback loop which is controlled by the DSP card. The scanning is achieved by sending drive signals to the $\pm x, y$ electrodes of the piezoelectric tube such that the area to be examined is scanned in a raster fashion i.e. the sample or tip moves along one line in x and then moves up a line in y and so on. When the tip encounters an object the cantilever begins to bend and the feedback loop adjusts the z channel of the piezoelectric tube to move the sample or tip in the appropriate direction to return the cantilever to its deflection set-point. This process means for a homogeneous sample that since the cantilever deflection is maintained at a fixed value so also is the force between the tip and the sample, and this mode of operation has the name constant force imaging. The image is generated by plotting the z correction signal from the feedback loop against x and y. Height information in this mode is, therefore, derived from the feedback correction signal and not directly from the AFM tip. One can also operate in a mode where the cantilever is allowed to bend freely in response to sample topography, and this mode is known as the variable deflection mode. In this case the height data in the image comes directly from the AFM cantilever and tip displacement. In practice even this mode uses some feedback control but the gain of the loop is set to a very low value.

Whilst there are several other imaging modes, which are described in more detail in the next chapter, there is a third, in between state of affairs, known as 'error-signal' mode imaging. Error-signal mode imaging is described here because it is a special case in that it provides a way of overcoming a fundamental limitation of the feedback loop used in the AFM. In the error-signal mode the feedback loop is operational but at a reduced level, meaning that the deviation of the cantilever from its set-point (i.e. the 'error-signal') is relatively large on steep gradients, but negligible on small gradients, and so by recording the cantilever deflection directly it produces an image which enhances fine detail at the expense of coarse detail. This mode is particularly useful for rough samples such as whole cells, where the overall shape is of little interest but information on fine surface detail is required. The principle is that the feedback loop does just enough to remove the low frequency background (coarse detail) information from the image, leaving the high frequency (fine detail) information to be displayed.

2.6.3. Feedback control loops

The feedback control loop used in the AFM can be varied by the user through the gain setting of the instrument. The basic function of a controller is to maintain a predefined set-point, an everyday example is the thermostat unit in the central heating system in your house. The most basic form of control is so-called on-off control.

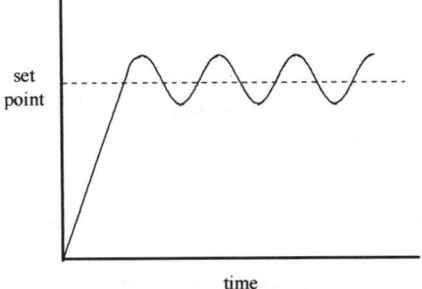

Figure 2.15. On-Off or 'bang-bang' control; the set-point is never achieved.

For the heating in your house say this would mean simply switching the heating off when the desired temperature is reached and then switching it on again when the temperature drops. This is shown graphically in Fig. 2.15. It is obvious from the graph that by using this method the set-point is never actually achieved, rather the signal (temperature in the case of a heating system) continually hunts around it, overshooting then undershooting. Clearly this would be a highly unsatisfactory way to try to control the motion of the AFM tip or scanner. The AFM control system should be as accurate as possible but also as fast as possible in order to give it a sufficient bandwidth. This is achieved by introducing mathematical terms to the control signal of which three types are commonly used; proportional, integral and derivative (PID).

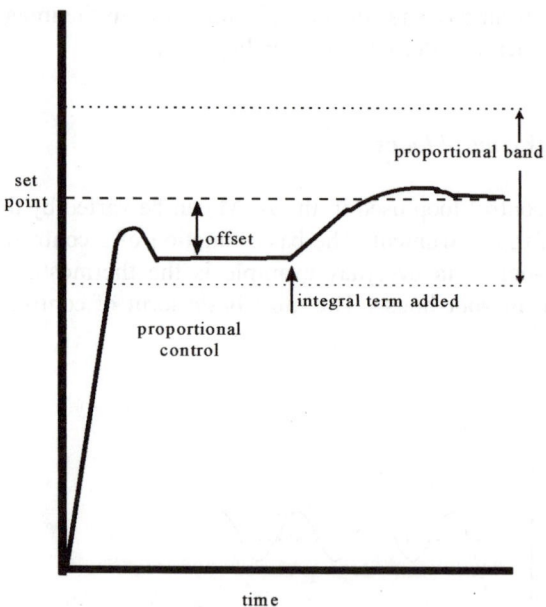

Figure 2.16. Response of a controller with proportional and integral terms added.

Proportional control simply amplifies the error between the set-point and measured value in order to establish the size of the correction signal required. It does this by defining a proportional band which is some percentage of the total span of the controller, the actual percentage being the gain of the system; for example if the proportional band is 25% then the proportional gain is 4, a band of

5% would translate to a gain **of 20 and so on**. High gain therefore translates to a narrow width for the proportional band. The width of the proportional band determines the magnitude of the response to an error signal. Proportional control leads to a response as shown in Fig. 2.16, where the signal settles to a point which is slightly lower (or higher) than the set-point, i.e. there is an offset. In theory the width of the proportional band could then be reduced by increasing the proportional gain to reduce the offset. In practice, however, this leads to a highly unstable system, because if the signal moves outside the new narrower band there will be an inevitable reversion to on-off control, i.e. oscillation.

A better way of removing the offset is by adding an extra term which integrates the deviation from the set point over a small time period (Fig. 2.16). It is important to remember though, that if this time period is made too small, then the control signal could shift faster than the piezoelectric scanner can actually respond, and so oscillation will occur. Finally a derivative term proportional to the rate of change of the control signal is sometimes applied to reduce the tendency of the control loop to erratic behaviour, since it will respond quickly to large deviations. Incidentally, derivative control is also the most beneficial for recovering from small perturbations from the set-point. For an AFM control loop the proportional gain responds quickly to small features on the sample surface, and the integral gain helps maintain an accurate set-point (it cannot respond to small features). Derivative gain reduces unwanted oscillations, but can exaggerate high frequency noise if set to excessive values. Some AFM systems may have only a single gain parameter that the user can adjust but the instruments' software will then implement this using an algorithm which varies all three gain controls. In many systems the user has access to only the first two forms of control (or gain) and a good rule of thumb for setting these is that proportional gain should be set 10-100 times higher than integral gain. Some systems however allow access to all three gain controls and so the following points should be kept in mind:

Too high a proportional gain gives control loop oscillation.

Too low a proportional gain gives sluggish control.

Too high an integral gain gives over eager response and instability.

Too low an integral gain slows down both the approach to set-point on start-up and return to set-point after a disturbance.

Although derivative action helps to achieve stability an excessive derivative gain may bring instability by over eager correction. It is probably the most difficult of the gain terms to set correctly, and is therefore usually set to zero except when scanning in non-contact (ac) mode of operation.

Finally the practical effects of instrument gain settings on actual AFM images is demonstrated in the next chapter (section 3.6.2)

2.6.4. Design limitations

Operation with a feedback control loop implies certain limitations on performance. These are both electronic and mechanical. In terms of electronic effects factors such as the bandwidth (frequency response) and stability of amplifiers, speed of analogue to digital conversion, or the sampling rate as it is normally called, all place limits on the operation of the AFM. In practice the bandwidth of a properly designed AFM system as a whole should be limited by the mechanics of the piezoelectric tube scanner rather than the electronics, because a tip traversing a very rough surface will require extremely fast correction signals from the feedback electronics to maintain a constant deflection, and there is a limit to how fast the piezoelectric scanner will respond to such high frequency signals. Bandwidth is an important consideration in any design of AFM since it places limits on the level of detail in an image. This is because fine detail in an image represents high frequency information.

Piezoelectric tubes have a relatively low resonant frequency, particularly in the x,y axis, typically around 5-10 kHz, so that they cannot be driven too quickly or resonant oscillation will occur with an inevitable loss of stability. Therefore they have limited bandwidth. The scan range of a piezoelectric tube increases with length, but its resonant frequency decreases with length, so we have a "Catch-22" situation. Longer tubes enable larger scans but have to be operated at slower speeds, which is not a problem in itself since when performing large scans it is wise to scan slowly to keep the tip velocity within reasonable bounds. However, a low resonant frequency makes them prone to unwanted resonant oscillation if the scan speed is increased. This means that even when scanning a smaller area, in order to produce a higher magnification image, the scan must be done slowly. Another factor which follows the same downward trend of piezoelectric tube stability versus length is their sensitivity; since long piezoelectric tubes deflect through a larger distance per volt, any noise on the driving signals causes a larger unwanted movement of the scanner for a longer piezoelectric tube. This causes a problem if scanning a very small sample such as a single molecule; the deviation of the AFM tip away from the set-point will be small, and so the z feedback voltage correction signal will be correspondingly small, meaning that the signal to noise ratio deteriorates. This implies that no one piezoelectric tube scanner is ideal for every sample, but rather the scanner should be chosen carefully to match the particular sample. So, for example, if one wishes to image large rough samples such as whole cells then a large scan range piezoelectric tube must be used but the ultimate resolution will not be as high, but, if one wishes to study single molecules on a flat surface such as mica then a smaller scan range tube is better, and will provide higher resolution and permit higher scan speeds. For this very reason many commercial AFMs have interchangeable piezoelectric-tube scanners.

2.6.5. Enhancing the performance of large scanners

Although one is forced to use a large scan range piezoelectric tube for large samples there are ways of improving their performance so that high magnification is still possible. After all what use is an AFM that can image a whole cell but then resolves little more detail than that obtainable from an optical microscope? Fortunately the signal to noise problem associated with long, large scan range piezoelectric tubes can be improved by scaling down unwanted headroom on the high voltage amplifier that drives the scanner when scanning small areas. This is done by scaling down the *input* signal to the high voltage amplifier using either potential dividers, to simply divide the output signal from the DACs in the translator box, or by reducing the voltage swing of the DACs with software control from the DSP card. This is easier and safer than trying to divide the output of the HV amplifier directly.

2.7. Vibration isolation: thermal and mechanical

Any high resolution measuring device requires a stable environment to operate at its optimum and the AFM is no exception to this rule. The two most important factors which must be controlled are temperature and vibration. AFM design tries to reduce the effect of thermal gradients through the careful selection of materials in its construction, i.e. ones with similar coefficients of thermal expansion, but nevertheless don't necessarily believe any salesperson who may claim that air-conditioning isn't needed! Indeed some AFM practitioners advocate allowing periods of 1-2 hours after inserting a sample into an AFM before commencing imaging so that instrumental drift is negligible. Fortunately control over temperature is relatively straight-forward to achieve with air-conditioning. A unit capable of controlling the room temperature to within 1°C is sufficient.

The AFM by its very nature is sensitive to very small displacement of the sample or tip. This of course means that vibration isolation is crucial to obtaining high resolution images. Of course AFM design strives to reduce the effect of mechanical vibration by making the microscope construction as compact and rigid as possible, in a similar way to that used for high performance optical microscopes. Just like the piezoelectric tube scanner and the AFM cantilevers, the desired effect is to move the resonant frequency of the microscope head itself as far as possible from the frequency of typical mechanical vibrations which may affect it, such as building vibrations (15-20 Hz). What is this 'resonance' stuff all about? A useful comparison here is how a musical instrument such as a guitar works. The sound produced by plucking a string is actually quite small, but the acoustic cabinet or body of the instrument works by having its resonant frequency very close to the frequency of the strings so that it oscillates sympathetically, increasing the amplitude of the oscillation, and hence amplifying the sound; an effect known as *resonance*. Anyone who has strummed an unplugged electric guitar and then compared this to strumming an acoustic guitar will know that the difference in sound levels is like night and day. The frequency of oscillation of the strings is identical in both cases, assuming that they are both in tune, but the resonant frequency of the solid electric guitar body is not matched to the strings and so no (mechanical) amplification takes place. So you can imagine that AFM design has the completely opposite objective to musical instrument design; external vibration should never be amplified. In addition to this, even without resonant amplification by the instrument, mechanical vibrations can still cause problems for the AFM. The solution to this is to mount the instrument on some form of mechanically-isolated platform. These range from optical tables isolated from the ground with air-bearings to heavy stone plinths hanging from bungee cords. Actually the second solution is simpler, cheaper and provides better isolation from sideways (shear) movement since it can swing freely, a factor which is particularly useful if the AFM is not sited in a ground floor laboratory. Having said all of the above most AFMs will work surprisingly well simply sitting on a sorbothane^{TM} mat on a sturdy table, but ultimately very high resolution may elude you.

2.8. Calibration

Whatever the sample being imaged with the AFM there always comes a time when one is required to make some quantitative measurement of the images. After all an AFM image provides us with a three dimensional map of the sample surface and so there is data aplenty other than the pleasing image which greets the eye. As the technique has moved from the stage of simply asking what can be imaged by AFM to using it as a tool for solving problems, so more is being asked from the data it gathers. This is generally a good thing but it does mean that now, more than ever, calibration of the instrument is very important. There are a growing number of papers concerned with using AFM in metrology (measurement science), and this area of research is helping to produce standards which can assess the reliability of such measurement (SPMet EU network).

There are potential problems with using AFM to make measurements of samples: some of the problems are scanner related and some are probe-related.

2.8.1. Piezoelectric scanner non-linearity

Although piezoelectric scanners can move with incredible precision to allow high resolution imaging they are inherently non-linear in almost every conceivable geometry. This non-linearity is negligible for small displacements but for large scans it can become a problem. Here large scans mean more than 70% of the full scale displacement of the scanner. The most common configuration of scanner, the piezoelectric tube, suffers from hysteresis in its forward and backward traces as shown in Fig. 2.17.

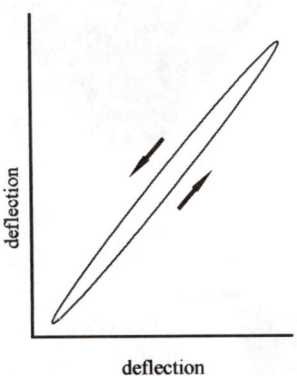

deflection

Figure 2.17. Hysteresis in the response of a piezoelectric tube scanner between forward and reverse directions.

Furthermore, the scanner can be non-linear with respect to scan speed, and the scanner moves faster in the x direction than the y scan direction, meaning that the calibration factors can be slightly different for each of these directions. The scanners are also susceptible to creep, which is a sort of relaxation which occurs under constant stress. Many modern commercial AFMs feature 'closed-loop' feedback systems, which independently monitor the motion of the scanner using interferometry, capacitance or strain-gauge sensors and then correct its motion for non-linearity. Thermal drift can also affect the scanner, but in this respect piezoelectric tubes are far superior to piezoelectric tripod scanners. Finally the electrical and physical characteristics of piezoelectric scanners change with age, particularly if the scanner is not used for long periods of time.

2.8.2. Tip related factors

AFM tips are not infinitely sharp and so every image will be a convolution of the actual surface topography and the shape of the tip. The degree to which tip shape imposes itself on an image is governed by the nature of the surface being examined, i.e. it is sample dependant. When a rough surface is being scanned the edges of the tall features will produce a mirror image of the side-walls of the tip rather than of the object itself as illustrated in Fig. 2.18.

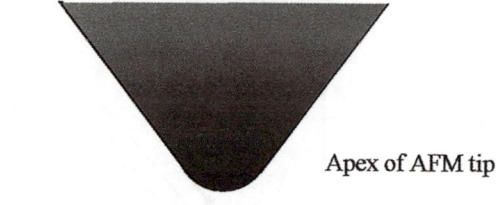

Apex of AFM tip

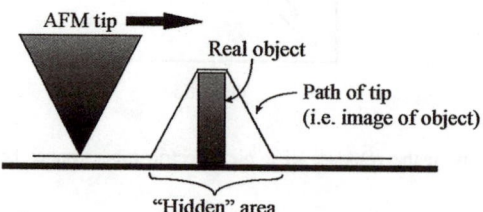

Figure 2.18. Tip convolution: because the AFM tip is not infinitely sharp (top) the image of an object is smeared out by the profile of the tip. The degree of this convolution is dependant upon both the tip shape and the height of the object being imaged (bottom). Note also that there are regions of the surface near a tall feature which are hidden from the tip.

For flat surfaces the degree of tip-related topography is greatly reduced since the tip should approximate to a hemisphere at its apex. Nevertheless, even small objects such as single molecules of DNA for example, which are about 2 nm in diameter, are subject to an effect known as 'probe-broadening'. This effect is just the same as that illustrated in Fig. 2.18, but, in this case, only the hemispherical region at the tip apex touches the molecule, so that the image of the molecule does not look angular in cross section as is seen for a tall feature, but the width of the molecule is greatly overestimated. Even with accurate calibration of the instrument probe-broadening will *always* occur, meaning that, unless one has access to good tip-deconvolution software (University of Nottingham SPM web page listed at the end of this chapter), it is better to use the heights of symmetrical features as a measure of their sizes, since the displacement of the tip in the z direction is unaffected by probe-broadening. A word of warning though, soft biological samples can be compressed by the tip (Weisenhorn *et al* 1993) and, if imaging is done in aqueous liquids, then electrostatic effects can alter apparent heights (Müller and Engel, 1997).

The geometry of the tip shape can effect the level of broadening or tip convolution, particularly for taller objects. This means that the scan angle of the tip with respect to the sample will have an effect on the level of probe-broadening seen in the image. Taking, for example, pyramidally-shaped tips, this is because opposite edges of the pyramidal tip are spaced further apart than opposite faces as shown in Fig. 2.19.

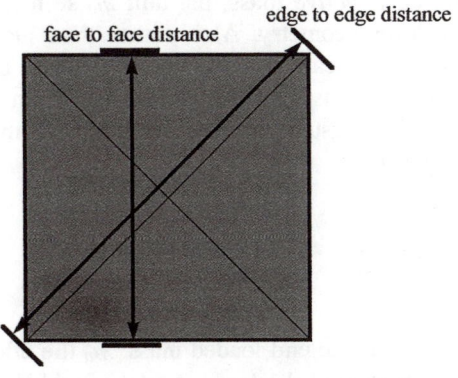

Figure 2.19. The distances face-to-face and edge-to-edge form the bases of two triangles, the apexes of which define the sharpness of the tip in the direction of scanning. Obviously the larger base length edge-to-edge means a larger angle at the apex and so the tip is effectively 'blunter' in this direction.

Things can be complicated even further by the fact that the AFM tip can change shape during imaging by being contaminated with the sample, or by being blunted due to wear. In both cases this will affect measurements to some degree because the degree of probe broadening or tip convolution will alter.

2.8.3. Determining cantilever force constants

For some applications of AFM in biology, notably single molecule force-distance spectroscopy, the force constant of the AFM cantilever must be determined experimentally, since the quoted value is only an approximation based on the as-manufactured length and thickness. The resonant frequency of a simple harmonic oscillator, for example a mass on the end of a spring, can be determined using the following equation, which can be found in any basic physics textbook.

$$f = \frac{1}{2\pi}\sqrt{\frac{k}{m}} \qquad (2.3)$$

where k is the force constant of the spring and m is the mass of the attached weight. But of course an AFM cantilever beam is not a simple point mass, but has its weight distributed along its length, so that the above equation needs to be modified slightly by using an *effective* mass, the unit m_0 seen in Eq. 2.4 below, which is governed by the lever geometry. Also, rather than just measuring the resonant frequency of an AFM cantilever alone, the force constant can be determined more accurately by measuring the changes in resonant frequency as small masses (in the form of tungsten spheres) are added, using the following equation (Cleveland *et al* 1993).

$$\omega^2 = \frac{k}{m^* + m_0} \qquad \text{since} \qquad f = \frac{\omega}{2\pi} \qquad (2.4)$$

where k is the force constant, m^* the end loaded mass, m_0 the effective cantilever mass, and ω the angular frequency of the lever. A plot of added mass, m^*, versus ω^{-2} has a slope equal to k, and an intercept equal to the effective cantilever mass m_0. By carefully performing the measurements described above Cleveland and co-workers (Cleveland *et al* 1993) derived the following equation which allows calculation of *k* with reasonable accuracy, by just measuring the unloaded resonance of the cantilever, assuming that one has accurate information on the length and width but is unsure of the thickness.

$$k = 2w \left(\pi l f_r\right)^3 \sqrt{\frac{\rho^3}{E}} \qquad (2.5)$$

where *l* is the length of the cantilever, *w* its width (not to be confused with ω in Eq. 2.4 above), ρ the density of the cantilever material, E the elastic modulus or (Young's modulus) of the cantilever material, and f_r is the measured resonant frequency. This is a useful equation because the length and width of micro-fabricated AFM cantilevers are well controlled by the high accuracy of the photolithography process (sub-micron), but thickness can vary significantly.

More recently a more accurate method, again requiring only that the unloaded resonance of the cantilever is measured, has been proposed (Sader *et al* 1995*)*. These authors present a detailed comparison of previous methods and, in addition, correct for the effects that air-damping and gold coating have on the measurement of the resonant frequency of AFM cantilevers. Perhaps most importantly they also demonstrate the importance of load position in relation to force constant. The force constant of an AFM cantilever depends quite strongly on where it is being loaded since, amongst other things, this determines the effective mass. For this reason some workers advocate a rather more direct measurement of the force constant of a cantilever by pushing on the actual tip itself (i.e. where the force acts upon the sample) using a wire of predetermined stiffness, and monitoring the relative deflection (Li *et al* 1993).

2.8.4. Calibration standards

Standards for calibration of the AFM in *x,y* and *z* can be obtained from several sources (NT-MDT Co., Russia; VLSI standards Inc., USA) and, by using a combination of these standards multiple point calibration can be carried out in each dimension, which is important because of piezoelectric non-linearity. Both companies manufacture calibration samples which have been specially designed to reveal tip shape, and to minimise probe broadening effects, so that calibration in the *x* and *y* dimensions are as accurate as possible.

Standards are generally made from silicon and so are incompressible and stable. However they can cause tip wear. Other useful materials which can be used for calibration include polystyrene latex, colloidal gold spheres (both available from any microscope accessory supplier) and tobacco mosaic virus (TMV). TMV is a cylindrical shaped virus particle with such a well defined diameter (18 nm) that it is used as a calibration specimen for the transmission electron microscope. It is also relatively hard, in biological terms, and so compression is not a problem as long as the imaging force is not too high. TMV will not damage the AFM tip like a silicon standard may and this makes it ideal for calibration of the z axis of the AFM scanner, particularly for biological applications.

2.8.5. Tips for scanning a calibration specimen

A few general points describing how to go about the calibration of an AFM are listed below.

•   Choose the appropriate calibration standards i.e. within the size range of your samples. It is pointless to calibrate the z dimension with a $1\,\mu m$ step if you are measuring samples of only 2 nm in height.

•   Choose the right gain setting for the feedback loop, particularly when calibrating z. Too much gain and feedback overshoot will distort the data. Too little and underestimation of heights will occur.

•   Scan at different speeds to determine over what range of scan speeds your calibration factor is reliable.

•   Scan several areas of different sizes (or heights) to give a multi-point calibration factor.

•   Perform forward (trace) and backward (retrace) scans to check for frictional effects, piezoelectric creep and scanner consistency.

Finally, although many modern commercial instruments have sophisticated (and in some cases automated) software algorithms for performing calibration, an appreciation of the fundamental problems associated with measurement by AFM is useful if things go wrong. After all the odd manual check of your instruments' calibration following the above procedure doesn't take very long and is better than relying upon blind faith! Taking all of the factors mentioned above into account it is obvious that measurement from AFM images is not a straightforward process and, if qualitative data is required, all of the potential systematic errors need to be considered and, if possible, quantified.

2.9. Integrated AFMs

Nowadays AFMs are often combined with other instruments and some examples are described below. This list is not intended to be exhaustive, since there are many highly specialised custom built AFMs, but there are a few instruments of particular interest to biologists which justify a brief description here.

2.9.1. Combined AFM-Light microscope (AFM-LM)

By far the most common integrated instrument is the combined AFM-light microscope. One of the main drawbacks of AFM is its slow image acquisition rate; a typical scan may take 1-2 minutes. This means that finding an area of interest on a specimen surface is not as easy as in other forms of microscopy, where one can track around quickly until the eye spots a recognisable feature. By combining the AFM with a good optical microscope this problem can be overcome. Most of the major manufacturers offer combined instruments although they are of course rather more expensive than stand-alone AFMs. They nearly all take the form of an inverted optical microscope with the AFM mounted on top of the sample stage, giving it unimpeded access to the sample. This configuration also allows the sample and AFM tip to be seen by the optical microscope at all times. The instruments generally scan by moving the AFM tip since the sample is mounted on the static stage of the light microscope. The samples can be presented on a glass slide or petri-dish for imaging in liquids and require no special treatment. Where optical beam detection methods are used for the AFM, filters can be placed in the optical path to prevent laser light entering the users' eyes through the eyepieces.

Combined AFM-LMs are excellent for large biological samples, such as whole cells or biological tissue, where there is an overlap in the magnification ranges required, and areas of interest can be located quickly and accurately with the light microscope before doing an AFM scan. Many of the techniques used in optical microscopy such as staining and fluorescence should be available on most combined commercial instruments to aid location and identification of structural features of interest but some of the contrast enhancing modes which rely upon overhead illumination may not be possible because of the location of the AFM head. This is a point worth checking before purchasing a combined instrument. In terms of absolute resolution the combined instruments tend to be inferior to stand-alone AFMs due to the compromises needed in the design. The AFM scanner must be moveable, so that any area of interest visible through the light microscope can be reached by the AFM, making the system less-rigid and so more prone to mechanical vibration. In practice this may mean, for example, that sub-molecular resolution cannot be obtained, but molecular resolution is within reach on a simple system such as DNA on mica. Nevertheless, such limitations are not a major problem because, for the larger and more complex samples suited to the combined AFM-LM, molecular resolution is in any case difficult to obtain for reasons of sample roughness (see section 3.5.5 for more details). Finally at least one company now builds a very compact AFM that can be fitted to any light microscope in place of one of the objective lenses via a standard screw mount (DME A/S, Denmark).

2.9.2. 'Submarine AFM' - the combined AFM-Langmuir Trough

Interfacial phenomena are an area of major interest in biology and one of the principle methods for their investigation is the Langmuir trough (see chapter 5). Perhaps the most important example of an interfacial system in biology is the cell membrane. In animal cells this is composed of a phospholipid bilayer and it controls many aspects of cellular function. The AFM is of course an interfacial technique itself, usually operating at a solid-liquid interface and so a combination of the two methods is a very attractive proposition. Recently just such an instrument has been built (Eng *et al* 1998). In a Langmuir trough the surface active molecules of interest are assembled at a planar air-water interface of reasonable area, in order to produce an interfacial film, and allow various physical parameters to be measured. In order to acquire an *in-situ* AFM image of the interfacial film the tip must approach from below, otherwise capillary forces would simply drag it through the air-water interface and into the bulk liquid. Hence the instrument was labelled a 'submarine AFM'. There is at present, however, a limit to what can be imaged in this manner; the interfacial film has to be relatively rigid, the required rigidity being governed by the softness (force constant) of the AFM cantilever. If the cantilever has too high a force constant the AFM tip will simply push through the interfacial film without deflecting. With currently available levers this necessitates high packing densities of the surface active molecules to produce an interfacial film in the 'solid-phase', a situation which would not be found in healthy natural cell membranes. Nevertheless the instrument is still being developed since it has demonstrated great potential.

2.9.3. Combined AFM-surface plasmon resonance (AFM-SPR)

Although AFM is capable of imaging dynamic events on a molecular scale, it is limited in its ability to provide quantitative data of such events due to factors such as scan size, and number of scans per unit time. This means that whilst the AFM is excellent for providing a description on a highly detailed scale it is less useful for measuring average properties of a system as a whole. Surface plasmon resonance (SPR), on the other hand, is a technique which is routinely used for quantifying the dynamics of biomolecular interactions, by monitoring changes in surface refractive index over relatively large areas of a sample. An excellent review of SPR can be found in the article by Silin and Plant cited at the end of this chapter (Silin and Plant, 1997).

Plasmons, or plasma waves, are charge density oscillations which propagate in a plasma; in the case of metals this is the 'gas' of free electrons. Plasmons which can occur at the interface of a metal and a dielectric material, usually gold and water in commercial instruments, are called surface plasmons. The general principle of operation is that the thicker an adsorbed layer of molecules on the gold surface the greater will be the resonance angle at which the plasmons are excited. In fact the technique is extremely sensitive to changes in optical thickness, and a resolution of better than 0.1 nm can be achieved. This translates to a detection limit of 0.5 ng cm^{-2} surface concentration of adsorbate. Although it has exceptional height resolution, its lateral resolution is only about 5 μm, and this is where AFM comes in.

Therefore the combination of AFM with SPR has provided a powerful and unique method for the quantitative study of dynamic surface events at the molecular level (Chen *et al* 1996). The capabilities of the AFM-SPR were demonstrated with studies of polymer degradation and protein adsorption. In both experiments the SPR provided qualitative kinetic data on surface dynamics and the AFM provided detailed spatial information on the nature of the events which were invisible to the SPR.

2.9.4. Cryo-AFM

Although the AFM derives much of its advantage over conventional electron microscopy by its ability to operate in ambient conditions, a low temperature AFM has been developed to enable higher resolution to be obtained on biological systems (Han *et al* 1995). The benefit of working at low temperature is that the molecules have a much higher mechanical strength, some 1000-10,000 times that of a hydrated protein at room temperature for example, and so tip induced distortion of the sample should be negligible. Basically the instrument consists of an AFM suspended in a cryostat containing liquid nitrogen. Operation of the AFM in the liquid nitrogen vapour allows temperatures in the range 77 to 220 K to be achieved. It was noted that below 100 K high resolution images of immunoglobulins, DNA and red-blood cell ghosts were possible.

Table 2.1. Summary of cantilever properties and uses.

Type of cantilever	k (N.m^{-1})	f_r (kHz)	Comments
contact (dc mode) usually V shaped	0.01-1.0	7-50	soft levers required for minimisation of force. V shaped to reduce lateral motion during imaging. Small levers required to prevent unwanted resonant oscillation and give maximum sensitivity.
non-contact (ac mode) V shaped and rectangular	0.5-5	50-120	ac mode means that the lever is oscillated near resonance so a high Q factor is important. Best achieved with a stiffer lever than for dc mode.
tapping (ac mode) rectangular	30-60	250-350	Very stiff lever required to give high Q factor and to overcome capillary adhesion between tip and surface if working in air. Note: for Tapping in liquid a soft V shaped lever is used.
force modulation (ac mode) rectangular	3-6	60-80	Used for mapping the compliance of a surface by applying a variable force during scanning and measuring the response using ac techniques. Therefore a high Q lever is important for sensitivity.

References

Albrecht, T.R, Akamine, S, Carver, T.E and Quate, C.F. (1990). Microfabrication of cantilever styli for the atomic force microscope. *J. Vac. Sci. Technol. A* **8**, 3386-3396.

Baselt, D.R, Clark, S.M, Youngquist, M.G, Spence, C.F. and Baldeschwieler, J.D. (1993). Digital signal control of scanned probe microscopes. *Rev. Sci. Instrum.* **64**, 1874-1882.

Binnig, G. and Smith, D. (1986). Single-tube three-dimensional scanner for scanning tunnelling microscopy. *Rev. Sci. Instrum.* **57**, 1688-1689.

Binnig, G, Quate, C.F. and Gerber, Ch. (1986). Atomic force microscope. *Phys. Rev. Letts.* **56**, 930-933.

Binnig, G, Gerber,C, Stoll E, Albrecht, T.R. and Quate, C.F. (1987). Atomic resolution with the atomic force microscope. *Europhys. Letts,* **3** 1281-1286.

Boisen, A, Hansen, O. and Bouwstra, S. (1996). AFM probes with directly fabricated probes. *J. Micromech. Microeng.* **6**, 58-62.

Chen, X, Davies, M.C, Roberts, C.J, Shakesheff, K.M, Tendler, S.J.B. and Williams, P.M. (1996). Dynamic surface events measured by simultaneous probe microscopy and surface plasmon detection. *Anal. Chem.* **68**, 1451-1455.

Cleveland, J.P, Manne, S, Bocek, D. and Hansma, P.K. (1993). A non-destructive method for determining the spring constant of cantilevers for scanning force microscopy. *Rev. Sci. Instrum.* **64**, 403-405.

Dai, H, Hafner, J.H, Rinzler, A.G, Colbert, D.T. and Smalley, R.E. (1996). Nanotubes as nanoprobes in scanning probe microscopy. *Science* **384**, 147-150.

Eng, L.M. Seuret, Ch, Looser, H. and Günter, P. (1996). Approaching the liquid/air interface with scanning force microscopy. *J. Vac. Sci. & Technol. B* **14**, 1386-1389.

Erlandsson, R, McClelland, G.M, Mate, C.M. and Chiang, S. (1988). Atomic force microscopy using optical interferometry. *J. Vac. Sci. & Technol. A* **6** 266-270.

Frisbie, C.D, Royzsnyai, L.W, Noy, A, Wrighton, M.S. and Lieber, C.M. (1994). Functional group imaging by chemical force microscopy. *Science* **265**, 2071-2074.

Göddenhenrich, T, Lemke, H, Hartmann, U. and Heiden, C. (1990). Force microscope with capacitive displacement detection. *J. Vac. Sci. &. Technol. A* **8** 383-387.

Han, W, Mou, J, Sheng, J, Yang, J, and Shao, Z. (1995). Cryo-Atomic Force Microscopy: a new approach for biological imaging at high resolution. *Biochemistry* **34**, 8215-8220.

Hecht, E. (1987). Interference. In *Optics,* pp. 333-388. Addison-Wesley Publishing Company, Reading, Massachusetts.

Martin, Y. and Wickramasinghe, H.K. (1987). Magnetic imaging by force microscopy. *Appl. Phys. Letts.* **50**, 1455-1457.

Martin, Y, Williams, C.C. and Wrickamasinghe, H.K. (1987). Atomic force microscope force mapping and profiling on a sub 100Å scale. *J. Appl. Phys.* **61**, 4723-4729.

Mckendry, R.A, Theoclitou, M, Rayment, T. and Abell, C. (1998). Chiral discrimination by chemical force microscopy. *Nature* **14**, 2846-2849.

Meyer, G. and Amer, N.M. (1988). Novel approach to atomic force microscopy. *Appl. Phys. Letts.* **53** 1045-1047.

Müller, D.J. and Engel, A. (1997). The height of biomolecules measured with the atomic force microscope depends on electrostatic interactions. *Biophys. J.* **73**, 1633-1644.

Pierce, M.L, Stuart, J.K, Pungor, A, Dryden, P. and Hlady, V. (1994). Specific and non-specific adhesion force measurements using AFM with a linear position sensitive detector. *Langmuir* **10** 3217.

Rugar, D, Mamin, H.J, Erlansson, R, Stern, J.E. and Terris, B.D. (1988). Force microscopy using a fiber-optic displacement sensor. *Rev. Sci. Instrum,.* **59** 2337-2340.

Rugar, D, Mamim, H.J. and Guethner, P. (1989). Improved fiber-optic interferometer for atomic force microscopy. *Appl. Phys. Letts.* **55** 2588-2590.

Sader, J.E, Larson, I, Mulvaney, P. and White L.R. (1995). Method for calibration of atomic force cantilevers. *Rev. Sci. Instrum.* **66**, 3789-3798.

Schönenberger, C. and Alvarado, S.F. (1989). A differential interferometer for atomic force microscopy. *Rev. Sci. Instrum.* **60** 3131-3134.

Silin, V. and Plant, A. (1997). Biotechnological applications of surface plasmon resonance. *Trends Biotechnol.* **15**, 353-359.

Tansock, J. and Williams, C.C. (1992). Force measurement with a piezo-electric cantilever in a scanning force microscope. *Ultramicroscopy* **42-44** 1464-1469.

Taylor, M.E. (1993). Dynamics of piezoelectric tube scanners for scanning probe microscopy. *Rev. Sci. Instrum.* **64**, 154-158.

Weisenhorn, A.L, Khorsandi, M, Kasas, S, Gotzos, V. and Butt, H-J. (1993). Deformation and height anomaly of soft surfaces with an AFM. *Nanotechnology* **4**, 106-113.

Wong, S.S, Joselevich, E, Woolley, A. T, Cheung, C.L. and Lieber, C.M. (1998). Covalently functionalised nanotubes as nanometre-sized probes in chemistry and biology. *Nature* **394**, 52-55.

Wong, T.M.H. and Welland, M.E. (1993). A digital control system for scanning tunnelling microscopy and atomic force microscopy. *Measurement Sci. Technol.* **4**, 270-280.

Useful information sources

SPMet - A European union network on the calibration and metrological use of scanning probe microscopes. Http://www.dfm.dtu.dk/SPMet

University of Nottingham SPM web page. This site provides a free service for tip deconvolution of AFM images and many other image processing/analysis techniques and can be found at:
http://pharm6.pharm.nottingham.ac.uk/new

Bioforce™ Laboratory, Inc, 2501 North Loop Drive, Ames, Iowa 50010, U.S.A.

NT-MDT Co., Zelenograd Research Institute of Physical Problems, 103460 Moscow, Russia. http://www.ntmdt.ru/

VLSI standards Inc., 3087 North First Street, San Jose, CA 95134-2006, USA
http://www.supersite.net/semiH2/vlsi

DME A/S, Herlev, Denmark, http://www.dme-spm.dk

CHAPTER 3

BASIC PRINCIPLES

3.1. Forces

How does the AFM actually record images and how can we ensure that they are of good quality? As briefly mentioned in chapters 1 & 2 the AFM images by 'feeling' the sample surface. The force between the tip and sample varies as the sample is scanned beneath the tip. Changes in force are sensed by the tip which is attached to a flexible cantilever. Depending on whether the cantilever is sensing repulsive or attractive forces, different imaging modes can be applied (section 3.2). The operation of the AFM depends on monitoring forces between tip and sample and in this section the different types of forces likely to be encountered in biological systems will be introduced. Finally, once an image can be obtained, there are sections on recognising artifacts, common problems, and image processing techniques.

As the name 'Atomic Force Microscope' suggests the important interactions between the tip and sample are due to one or more forces. What are these forces, and what are their origins?

3.1.1. The van der Waals force and force-distance curves

Using a classical model, the electrons within a substance are in continual motion, and travel extremely rapidly. Quantum physics treats them as waves rather than particles. Whilst a given substance may appear electrically neutral over conventional periods of time, over extremely short periods of time, say a snapshot, the distribution of electric charge due to the electrons is not perfectly symmetrical. This gives rise to subtle charge imbalances known as 'dipoles' or 'multipoles'. Each molecule therefore exhibits a slightly different distribution of charge within a given snapshot, which is also dependent on the number of electrons that the molecule possesses. The charge imbalance in one molecule can electrically induce a similar imbalance in a neighbour. The net result being that the slightly positive end of one molecule will be attracted to the negative end of another. Normally these effects are completely masked by the very much stronger electrostatic force. However, it is important to stress that van der Waals forces are present in all materials, even those that are electrically neutral.

It is possible to characterise both attractive and repulsive parts of the force-distance relationship between the tip and sample by modelling the interaction. This involves the variation of the potential energy of one particle, say at the apex of the AFM tip, due the interaction with a particle at the surface of the sample. As their separation (r) changes, so does the value of the potential energy, which can be described mathematically by the pair-potential energy function $E^{pair}(r)$. A special case of the well known 'Mie' pair-potential energy function is used to model this behaviour, called the 'Lennard-Jones' or '6-12' function:

$$E^{pair}(r) = 4\varepsilon\left[\left(\frac{\sigma}{r}\right)^{12} - \left(\frac{\sigma}{r}\right)^{6}\right] \tag{3.1}$$

where ε and σ are constants that depend on the material.

Incidentally, σ is approximately equal to the diameter of the atoms involved, and is sometimes called the 'hard sphere diameter'. Fig. 3.1 illustrates the variation of the pair-potential energy between two atoms. The $1/r^{12}$ term accounts for the steep increase in $E^{pair}(r)$ at small separations i.e. when $r < \sigma$ where the atoms strongly repel each other due to the Pauli exclusion principle. The $1/r^{6}$ term is responsible for the slower change in the attractive behaviour at relatively large separations, where the van der Waals force dominates.

Imagine what happens when you gradually approach this page of text with your hand and then push down onto its surface. You will not experience the initial attractive forces because they are only significant at very small separations – too small to be discernible in this case. But when your hand actually rests upon the page, and you try and push it down further, the repulsive forces abruptly stop you from making any further progress.

With some understanding of this force, let's replace the notion of using your hand as a primitive AFM tip with the real thing. Remember that the tip is suspended on the end of an extremely flexible cantilever. At relatively large separations, say a few hundred nm, any attractive forces are too small to exert a significant force between the atoms in the sample and those found in the very end of the pyramidal AFM tip. In addition, the springy nature of the cantilever ensures that no deflection is apparent (refer to Fig. 3.2, position 1). Yet as the separation is reduced the force between tip and sample rises rapidly. This happens even if the tip and sample are both electrically neutral. So although inter-ionic interactions need not occur the cantilever can still start to bend under the influence of attractive van der Waals forces; position 2.

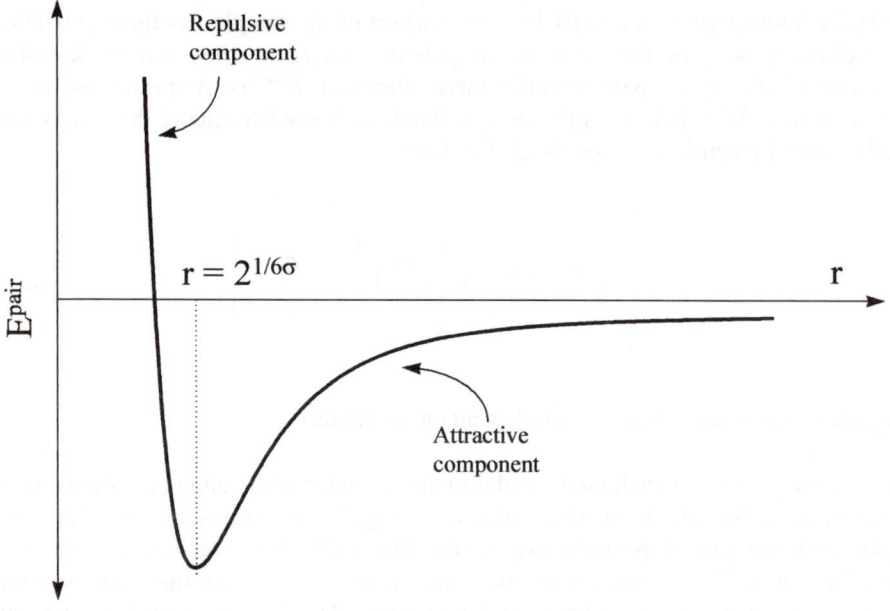

Figure 3.1. Schematic diagram showing the variation in pair-potential energy (E^{pair}) with separation (r) between two atoms as described by the Lennard-Jones function.

In fact, this happens over an extremely short distance from actual contact and the AFM tip appears to land on the surface of the substrate almost at once. This phenomenon is commonly known as 'jump to contact' or 'snap in'. At this point the tip lands on the surface of the sample, and the cantilever exhibits a significant bend in an attempt to pull the tip back off the sample; position 3. As the tip is pushed even closer towards the sample the bend on the cantilever straightens out again, as shown by position 4. This will always happen unless the sample is not very rigid and is easily deformed. When the cantilever is approximately straight the force between the sample and tip is nearly zero (position 5) and this is usually the ideal condition for general imaging because the force exerted on the sample is minimal. However if the tip is damaged this will not be true due to the presence of an 'adhesion force' as discussed later in section 3.1.3. It also follows that if the tip is forced further towards the sample the cantilever starts to bend in the opposite direction, as shown by position 6.

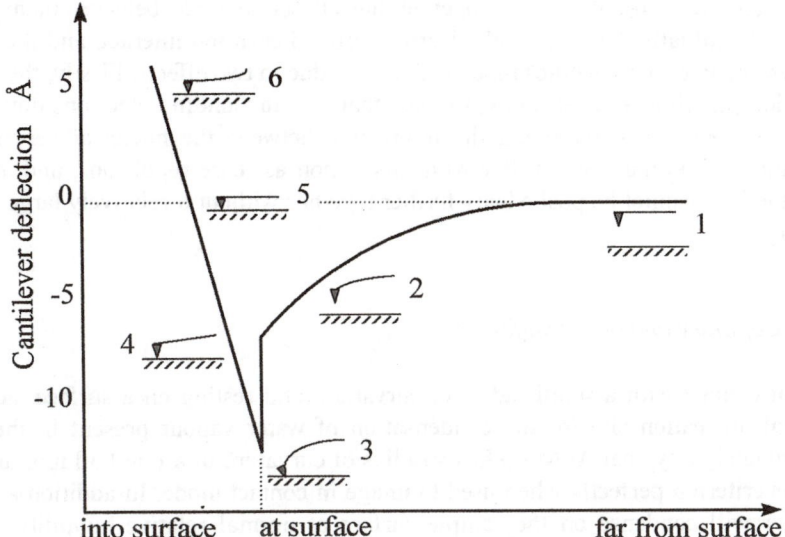

Figure 3.2. A idealised force-distance curve, illustrating how the bend of the cantilever changes at different distances from a rigid sample in a vacuum. Note that this diagram shows a magnified region, close to contact. Some features are exaggerated in order to clearly differentiate between the different stages of cantilever bending. A more typical experimental force-distance curve is illustrated later in Fig. 3.3 (right).

3.1.2. The Electrostatic force

The Electrostatic, or Coulombic, force present in ionic bonds has by far the largest physical influence of any of the intermolecular forces that we shall consider here. Visualise two oppositely charged ions, q_1 and q_2, in a vacuum. At a short separation (r) they are attracted to each other. The force between them follows the well known Coulomb force law, and is proportional to $1/r^2$.

$$F = \frac{1}{4\pi\varepsilon_o} \cdot \frac{q_1 q_2}{r^2} \tag{3.2}$$

The constant ε_o is referred to as the 'permittivity of free space'.

As the ions are brought closer together the attractive force between them rises sharply. Eventually the outer shell electrons around each ion interact, and the force between the two ions becomes repulsive. This is due to two effects. Firstly, the Pauli exclusion principle and, secondly, the fact that the surrounding electrons now do a relatively poor job of screening the interaction between the nuclei at very small separations. You may be familiar with this notion as 'core repulsion', and at this point the ions cannot be pushed any further together without a relatively large input of energy.

3.1.3. Capillary and adhesive forces

A point contact with a small radius of curvature, and resting on a surface, acts as an ideal nucleation site for the condensation of water vapour present in the air. Unfortunately a typical AFM tip has a radius of curvature of around 30 nm, and so fits this criterion perfectly when used to image in contact mode. In addition a layer of water will condense on the sample surface at normal relative humidity (RH). This means that when imaging in air the tip will be pulled down towards the sample by a strong liquid meniscus, giving rise to the so-called 'capillary force', which 'glues' the tip to the sample. This is a major problem because the capillary force is independent of the instrument settings, and therefore cannot be easily compensated for by the operator. The result is that the overall imaging force can now be large enough to destroy or move delicate samples over the substrate. Although it is possible to circumvent this effect by enclosing the AFM head in a sealed box fed with dry air to reduce the RH, this is hardly convenient as it hinders access to the instrument.

The presence of the capillary force also increases the importance of setting the correct amount of feedback gain when imaging in air. Since the tip is entrapped by a thin water layer at the surface of the sample, it has a greater tendency to oscillate uncontrollably with incorrect gain settings. This produces an effect known as 'gain ripple' in the image. In fact, on particularly humid days, you will probably find it almost impossible to completely remove all traces of gain ripple. This sets a practical scan size limit, beyond which scanning smaller areas is unlikely to produce more information, because the image is blurred by subtle gain ripple. As a rule of thumb, this scan size limit occurs at about 1 μm square when imaging in air with the tip in contact with the sample. Eliminating capillary forces has perhaps proved to be the most important step in successfully imaging many biological specimens.

Obviously AFM tips do not last forever. With time they become blunted, and contaminated with small amounts of the sample. Both these effects lead to a greater contact area between tip and sample, and ultimately to the presence of what is commonly known as an 'adhesive force'. This is a major problem when studying small objects such as discrete molecules since they are easily damaged by high forces. However, it is not so critical when studying large objects, because they can invariably withstand higher imaging forces. Fortunately, it is possible to check for the presence of adhesion by generating a force-distance curve, and examining it for any asymmetry in its inward and outward portions, see Fig. 3.3. below.

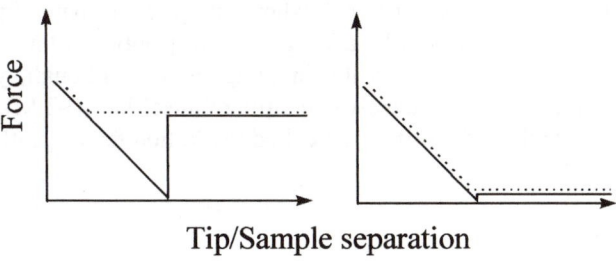

Tip/Sample separation

Figure 3.3. Two typical force-distance curves. The curve on the left shows the presence of a strong adhesion force whereas the curve on the right is reasonably ideal for most imaging situations. Dashed line: tip approach, solid line: tip retreat.

When adhesion forces become a significant problem the only real solution is to start afresh with a new tip. Unfortunately, the quality of new tips cannot always be guaranteed, and certain tips produce better images than others, even from the same wafer!

3.1.4. Double layer forces

This type of force arises when imaging under aqueous media, where mica is negatively charged and can attract oppositely charged ions from solution. This causes them to cluster at the solid-liquid interface leading to the formation of a positively charged layer, often known as an 'ionic atmosphere', since it can be visualised as being something like the earths' atmosphere. The potential decays exponentially away from the surface, and the "thickness" of this ionic atmosphere is known as the 'Debye length' $(1/\kappa)$. For low surface potentials (ψ_0), the potential at a distance (ψ_x) is related to the Debye length through the Debye-Hückel approximation:

$$\psi_x \approx \psi_0 e^{-\kappa x} \qquad (3.3)$$

For a complete explanation of the Debye length, double layers and many other surface and interfacial phenomena, see Israelachvili, 1995.

Generally, the greater the ionic strength of the imaging medium, the lower the electrostatic repulsion that an approaching AFM tip experiences (see section 6.4.2). Therefore, it is possible to electrically shield the AFM tip from the influence of the relatively large forces present when imaging in some liquids, such as buffers. This is usually achieved by adding a small quantity (a few mM) of a salt, containing divalent metal ions, to the imaging liquid. Although, you should be aware that the addition of an excessive quantity of metal ions (~2 M) is undesirable as it can exaggerate the effects of the so-called 'hydration force' (Butt, 1991).

3.2. Imaging modes

There are many imaging modes available when using AFM. These are often differentiated by the force interaction involved in each case. Each mode has its own specific advantages and disadvantages. Below is an outline some of the most important ones.

3.2.1. Contact dc mode

By far the easiest mode for the inexperienced user to get to grips with is that known as the contact, or dc mode. In this case the AFM tip is brought into direct contact with the surface of the sample, and the sample scanned beneath it (the tip-sample interaction is repulsive in nature). This process can be performed either in air or under a liquid, such as a biological buffer. The value of the pre-set imaging force is adjusted in the instrument software, which is equivalent to performing the entire scan with the cantilever bent by a small but fixed amount - hence this is known as 'constant force mode'. The larger the amount that the cantilever is bent, the higher the imaging force that the sample experiences. Of course, a similar level of bending on a cantilever with a large force constant will produce a higher resultant force than that from a cantilever with a low force constant. Additionally, by adjusting the force it is possible to vary the image contrast and reduce damage to the sample. Aspects of constant force mode are also discussed in the section 2.6.2.

One rather nice feature of this mode is that special tips are not required, in fact almost any reasonably flexible tip will do to get started. The most common types are perhaps those with a force constant of around 0.4 Nm^{-1}, and thankfully these are now reasonably cheap. All the more reason to start with this mode when you are just starting to use the AFM! By employing contact dc mode under liquid the capillary force can be eliminated, allowing much greater precision of applied forces. Solvents such as propanol and butanol are popular although greater understanding of ionic interactions has permitted the use of a wide range of biological buffers.

3.2.2. Non-contact ac modes

Another way of avoiding the problems caused by the capillary layer is to use the longer range attractive forces to monitor the tip-sample interaction. These attractive forces are weaker than the repulsive force detected in contact dc mode, and consequently different techniques are required to utilise them. There are two main types of ac mode: the first is often known as 'Tapping', whilst the second is usually called true non-contact mode. An important point is that these imaging modes can be significantly more complex for the novice user. Therefore, it is generally harder to get good images than with traditional contact mode. This is partly due to the fact that more information is available, and partly due to the extra parameters it is necessary to consider and adjust.

Tapping in air

In this mode a relatively stiff rectangular or 'beam type' cantilever is used when the instrument is operated in air. The purpose of using this mode is to prevent the AFM tip from being trapped by the 'capillary force' caused by the extremely thin film of water surrounding samples in air. Here the cantilever is deliberately excited by an electrical oscillator to amplitudes of up to approximately 100 nm, so that it effectively bounces up and down (or taps) as it travels over the sample. Often the AFM tip is simply secured with a blob of glue to a small block of piezoelectric material and excited directly with the signal from a precision signal generator. Tapping mode also reduces the lateral (side to side) force on the sample because the tip spends less time on the surface of the sample, which is useful because it means that delicate samples such as networks of molecules can be imaged without severe distortion.

Tapping under liquid

In this case there is no 'capillary force' to cause imaging difficulties because the sample is immersed under a liquid, so a super stiff cantilever is not required. Another difference is that the cantilever can be driven into oscillation indirectly, which is fortunate since electrical devices, such as piezoelectric materials, will obviously not function correctly under many liquids.

Q: *"If there is no capillary force present when imaging under liquids, why would we want to use Tapping mode when traditional contact mode is simpler?"*

A: *Even under low force instrument conditions, contact mode exerts a significant shear or lateral force on the sample. Although for many samples this is not a problem, if we subsequently want to look at a sample that is weakly attached to the substrate, dynamic, or even living, high lateral forces can scrape it off the substrate - remember that the sample may have a poor affinity for the substrate.*

In liquid Tapping the cantilever can be excited by applying a small sinusoidal electrical signal onto the z-channel input of the high voltage amplifier. This causes the main piezoelectric tube to vibrate up and down in the vertical (z) direction, whilst still performing its normal task of responding to signals from the control loop. Therefore the sample, and the liquid surrounding it, begin to vibrate. This vibration is also communicated to the cantilever, which is immersed in the liquid, by viscous coupling. Hence we have another form of Tapping mode. Alternatively a small piezoelectric oscillator can be attached to the outside of the liquid cell and used to excite the cantilever. The advantage of using such a small oscillator is that the main piezoelectric tube is limited in its vibrational characteristics because of its relatively large size. At anything above 10-15 kHz the amplitude of vibration of the piezoelectric tube becomes highly non-linear. Since the resonant frequency of even the most flexible of levers under liquid is > 15 kHz, improved performance can be obtained by using the smaller piezoelectric oscillator.

    Cantilevers coated with a magnetic material can be excited using a nearby electromagnet, this is known as 'magnetic ac mode' or MACmode[TM]. The electromagnet is usually a few loops of copper wire wrapped around the liquid cell and driven by a signal generator. The principal advantage of using this method is that the cantilever is vibrated directly, so that both the liquid and sample remain undisturbed.

True, non-contact ac mode

In this imaging mode the oscillating cantilever never actually touches the surface of the sample, but hovers a few nm above it, with an oscillation amplitude of only about 5 nm. The relatively long range van der Waals attractive force between the sample and AFM tip produces a dampening effect on the oscillating cantilever. This makes the cantilever behave as though it were slightly stiffer, and consequently its natural resonant frequency is increased slightly. However, the frequency of the electrical excitation remains unchanged so that the dampened cantilever is now oscillating a little bit further from its new resonant frequency. This is important because efficient energy transfer only occurs when the frequency of the electrical oscillator exactly matches the resonant frequency of the cantilever. Clearly, whilst the cantilever is subjected to the influence of the van der Waals force there will be a deviation away from the optimum frequency. This has the result of reducing the cantilevers amplitude of oscillation. So when the separation between the AFM tip and the sample is reduced, say when the tip encounters a protruding feature on the sample, the oscillation amplitude of the cantilever falls. This effect can by used by the instrument software at each image point to detect and construct the topography of the sample.

Q: *"So I now know what the difference is between the various ac modes, but what are the advantages and disadvantages of using true non-contact ac mode?"*

A: *"Since in true non-contact ac mode there is obviously no contact with the sample, the forces exerted on it are extremely low. This results in low deformation and shear. Perhaps, more importantly, the image contrast can be remarkably high producing striking images. These are the advantages but there are, as always, disadvantages. First of all, it is usually necessary to choose a fairly flexible cantilever with a low resonant frequency, say around 30 kHz. In fact, a contact dc mode cantilever will often do. This is because the relatively weak van der Waals force would have little dampening effect on a stiff cantilever. If the AFM is not configured optimally right from the outset there is a good chance that the tip will hit the sample at some point during the scan. When imaging in air this is disastrous because the capillary force will easily trap such a flexible cantilever, leaving it 'glued' to the sample, and abruptly halting its oscillation. If this happens repeatedly the end of the pyramidal tip can be damaged or contaminated by fragments of the sample. In other words, the tip can easily be blunted by inadvertent high speed contact with the sample. A blunt tip is useless in any AFM mode because it invariably produces poor resolution images".*

3.2.3. Error signal or deflection mode

In order for the tip to track accurately across the sample, whilst still exerting a low force, it is necessary for the gain of the system control loop (i.e. the control or feedback loop) to be set as close as possible to its optimum value. However, in error signal mode the gain is deliberately set to a relatively low value in order to make the control loop response extremely sluggish. In this case the image is certainly <u>not</u> recorded at a constant force. In fact the AFM tip literally crashes into the features on the sample, rather than lifting gently over them. So what is recorded by the AFM in this case is a force map (or 'force image') of the sample, with large features represented by regions of high force, since they cause a greater degree of cantilever bend. The origin of the name 'error signal' or 'deflection' mode comes from the fact that the deflection of the cantilever is used as the input signal into the control loop, so that any deviation, or 'error', away from the cantilever's approximately straight position is continuously corrected.

 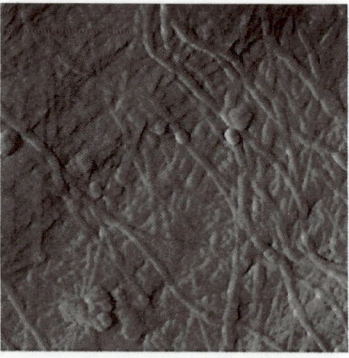

Figure 3.4. A comparison of topography (left) and error signal images (right) taken over the same area of nettle cell wall material. The higher frequency information available in the error signal image yields considerably more detail although height information is lost. Scan sizes: 2.5 x 2.5 μm.

Q: *"I understand how the error signal mode is set up, but it seems to represent the exact opposite of what we are trying to do in order to get a decent image. Does it really have any practical value?"*

A: *"Oh yes. The error signal mode is an extremely useful technique, particularly when used to image rough samples, such as in Fig 3.4. There are many samples that are simply too rough for the AFM to ever get a good quality image without continuously adjusting the gain throughout the scan. In this case, the force exerted on the sample is going to be fairly high anyway because the gain is never at its optimum value. In standard imaging modes the sample (or tip) is moved vertically by the piezoelectric tube under instruction from the control loop. Although this is the traditional way of dynamically controlling the imaging force, it has its flaws. In particular, this electromechanical technique has to move a relatively heavy mass. If the sample is rough, yet fairly rigid, the actual imaging force is not so critical. By setting the gain of the control loop to a very low value we are effectively restricting it from moving the piezoelectric tube and sample in the z direction. Now the main thing that moves is the cantilever as it bends up and down in response to features on the surface of the sample. Since the cantilever has a much lower mass than the tube/sample combination, the frequency response is higher when using the error signal mode. This means that there will be more fine detail in the error signal image than in the equivalent topography image. However it's main application is for imaging rough but fairly rigid samples such as cells and bacteria."*

3.3. Image types

This section discusses different ways of displaying the data and is somewhat dependent on what sample properties you are interested in measuring. Image types, as opposed to imaging modes, can often be combined during a single scan. For example, it is relatively easy to obtain frictional data on a sample, whilst at the same time recording its topography.

3.3.1. Topographical

This is by far the most common way of recording images using the AFM. The information used by the instrument software to create an image is simply the vertical movements of the piezoelectric tube. Ideally the whole scan should be performed with a constant applied force i.e. a constant bend on the cantilever, and the way of compensating for any cantilever bend is to move the sample or tip up and down via the piezoelectric tube. From this type of image it is possible to measure the heights of objects in the image, say by using 'line profile' software. Whilst this may sound obvious, it is actually worth stressing because other image types use differing contrast mechanisms, so that the height information is not recorded.

3.3.2. Frictional force

This type of operation is also known as lateral force imaging. Remember from chapter 2 that the face of a typical photodiode detector is split into four areas, or quadrants. In most imaging situations the difference signal between the top two and the bottom two quadrants is all that is needed to determine the extent of vertical cantilever deflection. In frictional force imaging, however, the side to side difference signal is used to determine the twisting behaviour of the cantilever under lateral forces. That is, the difference between the laser intensity received by the two left hand side quadrants is compared to that received by the two on the right hand side. When scanning an area of a sample with a significant frictional component, the AFM tip is restricted in its motion by the lateral force, so twisting of the cantilever occurs and the friction is detected.

3.3.3. Phase

Phase, or more specifically phase lag, is a quantity that can be recorded in the ac Tapping mode. In this mode the control loop uses the drop in amplitude of the oscillating cantilever to determine the vertical movement of the piezoelectric tube.

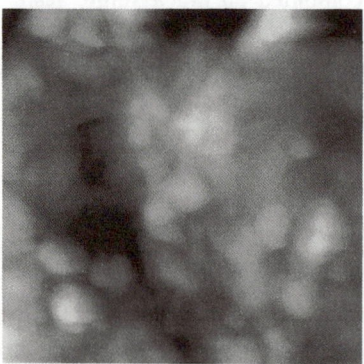 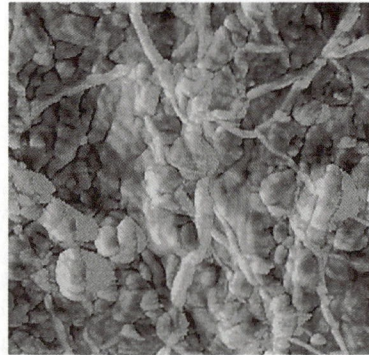

Figure 3.5. An example of the power of phase imaging illustrated using a two-part epoxy resin sample that was deliberately poorly mixed. The conventional topography image (left) shows little information because the sample is essentially flat. However the simultaneously recorded phase image (right), is considerably more detailed and highlights areas with different elastic properties. Scan sizes: 4 x 4 μm.

In addition, when the tip actually strikes the sample, its phase of oscillation is disturbed and it is no longer precisely in step with the phase of the electrical oscillator that is driving it. This is principally because each time the tip strikes the sample, it transfers a small amount of energy to it. Just how much energy depends on the elastic nature of the sample. Fig. 3.4 shows a synthetic sample prepared from epoxy adhesive, note how much more detail there is in the phase image when compared to the somewhat unimpressive topographical image. When you first start using the phase technique you might like to try this sample for yourself, since it is easy to prepare, and has the bonus of being virtually indestructible! Mix together a small amount of two-part epoxy adhesive - but ensure that you do it poorly! By doing this the hardener and the adhesive do not form a homogeneous mix, so that when the mixture sets some areas are more elastic than others. However, before this occurs, sandwich the mixture between two sheets of PTFE to produce a surface that is flat enough to image. After about 20 minutes, peel back the PTFE to reveal the rigid epoxy slab.

A reasonable level of contrast in a phase image is primarily dependent on there being at least two components in the sample with sufficiently different elastic properties. The ideal material would perhaps be something like small rigid metal particles embedded into a flexible polymer matrix. Unsurprisingly, biological systems fall a long way short from this ideal, so the contrast can, in many cases, prove to be unacceptable. This is nothing more than a confirmation of the notion: "different samples almost certainly require different imaging techniques".

3.4. Substrates

In order to image a material by AFM it needs to be deposited and secured onto a rigid substrate. The most common types of substrates used in biological AFM experiments are discussed below.

3.4.1. Mica

This is the most popular AFM substrate, partly due to its wide availability and low cost. It consists of a series of thin, flat crystalline plates that can easily be split apart ('cleaved') by inserting a pin at its edge, or even using adhesive tape. The result is a truly fresh surface that has not been exposed to the atmosphere since it was originally formed millions of years ago. In addition, mica sheets are atomically flat over large areas (typically a few microns), which is essential if molecular resolution is to be achieved. There are many different types of mica, which can be distinguished from each other by noting which metal ions they contain. Perhaps the most common type used as an AFM substrate is known as 'muscovite' mica [(K $Al_2(OH)_2$ $AlSi_3O_{10}$)]. One point that you should remember is that mica is negatively charged in aqueous liquids.

3.4.2. Glass

Glass, usually in the form of polished coverslips, is an ideal substrate for imaging larger samples, such as cells, where molecular resolution is often unnecessary. Glass coverslips can be surprisingly flat considering their low cost, and their roughness can be as low as a few nm over areas of several μm. It is, nevertheless, wise to rinse each coverslip in isopropanol, or acid, before use in order to remove any contaminants. Cells and bacteria can be cultured directly onto the coverslips before imaging, although any remaining media will need to carefully washed away. This promotes a reasonably strong attachment to the glass, and helps ensure that the sample is not desorbed when imaging under aqueous media. Glass is also negatively charged under aqueous liquids.

3.4.3. Graphite

This material has been used since the early days of STM, where the substrate had to be conducting. This is clearly unnecessary for AFM unless you need to simultaneously acquire STM data. Graphite can be cleaved using adhesive tape, although some practice is necessary if you are to avoid removing thick layers. One point to consider is that graphite is very hydrophobic, so that samples deposited from aqueous solution both 'wet and spread' poorly. However, graphite can still be useful in AFM, particularly for performing control experiments where it is suspected that the sample conformation is affected by interaction with the mica surface.

3.5. Common Problems

3.5.1. Thermal drift

This is generally only apparent when imaging under a liquid. If there is a significant difference in temperature between the liquid and the liquid cell, then thermal currents can evolve. A typical cantilever is composed of a layer of reflective gold applied onto silicon nitride. If the gold layer happens to be unusually thick, this composite can behave like a bimetallic strip and bend with temperature. This happens because each layer has a different coefficient of expansion, and therefore expands by different amounts at any given temperature. In addition, internal stresses created during manufacture can produce substantially larger thermal bending of cantilevers (Radmacher *et al* 1994). Thermal drift is best avoided by leaving both the instrument and the sample to equilibrate for a short period of time after filling the liquid cell. Since the presence of thermal drift partly depends on the thickness of the gold layer, you may find that it affects a whole wafer of tips, but then does not occur at all with tips from a different wafer.

3.5.2. Multiple tip effects

Even more frustrating is a very common effect known as a 'double tip'. As the name implies, the sharp point at the very apex of the tip can sometimes be accompanied by others, usually as the result of damage or contamination. In theory there can be any number of extra tips but, in the majority of cases, no more than a total of two are actually observed. This produces two copies of the image, offset by a distance equal to the gap between the tips (Fig. 3.6), typically a few tens of nm. Whilst this sounds like a serious source of artifacts, it is reassuringly easy to determine when a double tip is present because the image adopts an obvious symmetry; Fig. 3.7. illustrates two examples. If the double tip is caused by contamination it can sometimes be rectified by scanning over a large area at high speed. This can remove the contaminants, improving the subsequent images.

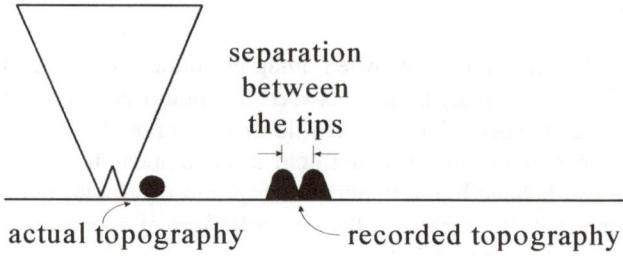

Figure 3.6. A 'multiple tip' usually leads to more than one copy of the real topography being recorded.

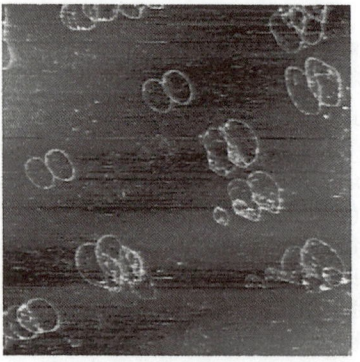 

Figure 3.7. Examples of multiple tip images, which are surprisingly easy to detect. Left: circular configuration of a polysaccharide arabinoxylan extracted from wheat, scan size: 2 x 2 μm. Right: Valonia cellulose, scan size: 2 x 2 μm.

3.5.3. Tip convolution and probe broadening

When part of a sample has a very low height compared to the radius of curvature of the tip, for example an individual DNA molecule, it seems reasonable to conclude that it only interacts with the extreme apex of the AFM tip. However, larger samples, such as a cell, ought to interact with the side walls of the tip in addition to the actual apex of the tip, as discussed in section 2.8.2. As a result the final image will be a convolution between the real topography of the sample and that of the tip. The most obvious effect is that samples with vertical sides are depicted as having sloping sides at an angle equal to the slope of the sides of the pyramidal tip. Unfortunately, even small molecules like DNA are affected by this problem, simply because the radius of curvature of the tip (i.e. its sharpness), and its apex angle are larger than the molecular diameter. This explains why the width of molecules measured from AFM images is often much larger than their actual width, sometimes by up to a factor of 10. This effect is commonly known as 'probe broadening'. Perhaps the easiest solution to these molecular width anomalies is to assume the molecules are cylindrical, and then obtain a better estimate of their true width by measuring their height. Software is available that provides both automatic image deconvolution and an indication of which parts of an image are composed of real data. Image deconvolution is discussed in section 2.8.2.

3.5.4. Sample roughness

A typical pyramidal tip is only around 3 µm tall, which places a strict limitation on the roughness of any sample that the AFM can successfully image (Fig. 3.8). If the sample is very rough and possesses features taller than 3 µm, then the flat underside of the cantilever will touch the sample instead of the actual tip. This makes an unwelcome contribution to the imaging process, and is effectively like using the instrument with the ultimate blunt tip. This is unfortunate because many interesting samples are simply too rough to image by AFM. One solution to this problem is to partially embed the sample into a matrix in order to effectively reduce its height (Thompson *et al* 1994). Examples of this approach are discussed in sections 4.4.4 & 4.7.11. Alternatively, it is possible to purchase taller high aspect ratio tips (such as those outlined in section 2.3.2) although they are currently somewhat more expensive than ordinary tips. In addition, their use assumes that the tip height really is the limiting factor, but this is not necessarily true. Most piezoelectric tubes have a relatively modest vertical range of only a few µms, so this factor can also determine the maximum allowable sample roughness of a sample that can successfully be imaged.

Figure 3.8. This SEM image of a 200 μm long cantilever highlights the importance of ensuring that the sample is not excessively rough. It also illustrates how incredibly small the tip really is.

It is possible, in some circumstances, to use 'sectioning' techniques developed for histology. By encasing the sample in a resin, then slicing it into fine layers with a microtome, it may be possible to image it using AFM, see section 7.1.1.

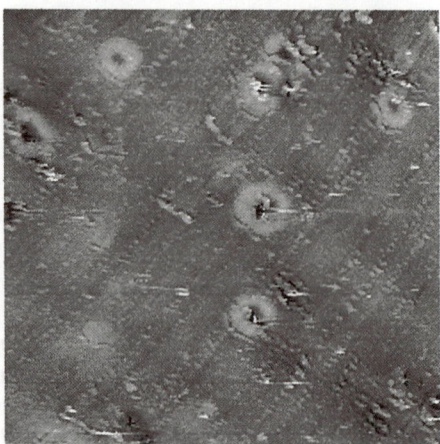

Figure 3.9. Starch granules embedded in an epoxy matrix and then thinly sliced using a microtome. The sample is now considerably smoother and can be successfully imaged, although there are obviously severe disadvantages in choosing this method. The angled striations are caused by small flaws in the edge of the microtome blade. Scan size: 10 x 10 μm.

The principal disadvantage with this method is that the AFM is easily powerful enough to resolve any flaws in the microtome blade which may cause image artifacts: Fig. 3.9. However, when sectioned in this way the sample can acquire a degree of anisotropy which can usefully provide extra contrast, see section 7.10.

3.5.5. Sample mobility

Samples are generally immobile when imaged in air, since they are dried onto the substrate. However this situation changes when imaging under liquids, especially aqueous buffers. Most biological materials have a high affinity for water and swell or become mobile in its presence. This is obviously a major problem when imaging in aqueous media because the sample can desorb off the substrate, and therefore become impossible to image. The main weapon to combat this problem is to use charge screening techniques as discussed under *Double layer forces* (section 3.1.4). Successful imaging in natural media is a challenging problem and a variety of solutions have been devised depending on the sample characteristics.

3.5.6. Imaging under liquid

Working in liquids is generally more difficult than working in air, but the points listed below should help you avoid some of the common pitfalls that await the inexperienced!

Good thermal equilibrium is crucial when working in liquids as convection currents can severely upset the stability of the AFM tip. If possible, store the liquids in the air-conditioned, temperature controlled, laboratory where the AFM is sited so that their temperature is not greatly different from the instrument itself. If using a buffer which needs to be stored in a refrigerator, take it out and place it in the laboratory for a couple of hours before you begin the experiment. Be careful when filling the cell, as leakage can be very expensive, as well as dangerous. Many instruments have the high voltage electrodes (typically up to 500 V) that are connected to the piezoelectric tube, sited underneath the liquid cell. Open 'bucket' type cells need particular care, although even so-called 'sealed' cells can leak if they are filled too quickly. The technique of sandwiching a drop, without the O-ring in place, can also be susceptible to leakage if the surface tension of the liquid is reduced by the addition of surface-active compounds, such as detergents, lipids or proteins.

Liquid cells can be difficult to 'wet' due to the high surface tension of aqueous liquids. This is a particular problem for bucket type cells where the liquid sits as a discrete hemispherical drop. However, this can usually be circumvented by filling the cell, and then gently agitating it before securing it into the instrument. Obviously it is absolutely essential that all the components are scrupulously clean. With sealed cells, air bubbles can sometimes become trapped between the tip and glass sight-plate during filling. This can manifest itself in one of two ways. Firstly, it becomes impossible to find the reflected laser spot when the liquid has been added, even after relentless adjustment of the photo-diode. Secondly, the reflected laser spot can be found but appears to 'flicker', producing a rapidly varying output signal from the photo-diode, making the instrument impossible to operate.

3.6. Getting started

If you have just acquired an AFM and are unsure where to begin, then this section is principally aimed at you. It is not intended to be a *'tour de force'* of operating AFM. Rather, its main purpose is to quickly get you up to speed by giving you a few tips and covering the most common pitfalls, without you having to read through a vast amount of literature. Obviously all instruments are different in their exact layout, and just as importantly, their software is likely to vary considerably. Therefore, it will be impossible to discuss specific features that are unique to individual instruments. However, assuming you have got to grips with the basics i.e. putting in the tip, aligning the laser, and getting onto the surface you are well placed to start on a real biological sample.

3.6.1. DNA

DNA is an excellent sample to begin learning about the AFM for a variety of reasons:

1. It is extremely well characterised, so you should have some idea about what you expect it to look like, and if its observed size and shape actually seems reasonable.

2. It is widely available. Therefore you should have no difficulty in obtaining a relatively pure sample.

3. It is a fairly delicate sample, so you can quickly see the results of correctly setting the instrument parameters, such as the force and gain, in order to

obtain a good quality image. Then you can learn about what happens when you vary one of the imaging parameters to obscure values.

4. Although delicate, DNA is surprisingly robust when compared to many other discrete molecules, which is due to the fact that it forms a double helix. This means that it should not prove impossibly difficult to image, which could easily be the case if you were to choose a sample at random.

You will need:

- A purified 3-5 µg mL^{-1} suspension of DNA, and definitely not more concentrated than 10 µg mL^{-1}

- A short cantilever of about 100 µm in length, with a force constant of approximately 0.4 Nm^{-1}

- A few mL of propanol or butanol, preferably redistilled. This is to be used as the imaging medium, and is considerably easier than starting with an aqueous buffer since the DNA does not redissolve off the mica surface when imaged under alcohol. In fact it is widely used as a precipitant in DNA preparations.

- A small piece of mica, say 10 mm square, that has been freshly cleaved, by inserting a pin or something similar, at one edge.

- It is also worthwhile getting copies of the following papers as both give useful information for imaging DNA: Hansma *et al* 1992, and Li *et al* 1992.

If you are fortunate enough to have an AFM system with a range of scan heads that cover different scan sizes, then you will generally be better off choosing one of the smaller ones. This is because there is no advantage in performing scans over about 5 µm since any individual molecules will then be too small to be seen. Additionally a smaller scan head will typically have a better signal to noise ratio than a larger one. Using one of the freshly exposed mica surfaces, deposit a small drop of the DNA suspension onto its centre. You should use a clean micropipette and apply no more than 5 µL. Make sure that the mica really is freshly cleaved, so <u>do not</u> use it if it is more than a few minutes old, otherwise organic molecules from the atmosphere could interfere with the adsorption of the DNA. Next, using tweezers, pick up the mica and gently swirl the drop around by hand for about 20 seconds to produce a reasonably even coverage. This is to prevent the drop from drying down into one tiny, but extremely concentrated spot, leaving no molecules anywhere else. Finally leave to air dry for about 10-15 minutes at room temperature.

After securing the sample in the liquid cell, you will probably need to refer to the instrument documentation in order to determine how much propanol, or butanol, to use as the imaging medium. About 200 µL is probably about average, but check anyway as you could damage the piezoelectric tube from leakage caused by overfilling as discussed earlier. After filling the liquid cell do not be surprised if you loose track of the laser spot. As air and alcohol have different refractive indices, the laser path can change quite dramatically after introducing the liquid. You will therefore need to readjust the position of the photodiode, so that it once again captures the reflected laser spot.

You are now ready to approach the surface of the sample. Most systems tend to have some combination of automatic and manual sample approach. So this step should be fairly straightforward, assuming the default instrument parameters are reasonable. If they are not you may observe one of the following:

- Severe oscillation of the tip as it sits on the surface. This is generally caused by having the gain parameter(s) set too high, so that the control loop is excessively sensitive and constantly overshoots in its attempts to stabilise the tip position.

- The indicator showing the laser intensity received at the photodiode becomes very dim or changes to a low value. Congratulations! You have probably just "crashed the tip" (self explanatory). No matter, we have all done it. The number one cause of this is the operator forgetting to switch on the power supply to the high voltage amplifier. This means the piezoelectric tube has no power source in order to control the tip-sample separation. Alternatively, the control loop gain was ridiculously low causing the vertical movement of the piezoelectric tube to be extremely sluggish. Or perhaps you were simply approaching the sample far too rapidly - in which case even a properly set up control loop could not react fast enough to prevent the tip from colliding with the sample.

- The instrument hunts around but never quite seems to settle the tip onto the sample. This is probably due to the value of the force set point in the software being too low. Try increasing it a little.

- You successfully get onto the surface but sometime later, perhaps part way through a scan, the tip pops off the surface again. Again try increasing the force a little. If this happens repeatedly, then you are probably experiencing thermal drift. In this case you will have to leave the system to equilibrate for at least 20 min.

After successfully getting the AFM tip onto the surface of the sample, modest scan sizes should be selected, say 1 or 2 μm. This will enable you to ascertain the degree of molecular coverage over the mica substrate. You should aim to adjust both the scan line density and speed so that a whole image composed of 256 lines is acquired in about 1.5 to 2 minutes, thus ensuring that tip is not moving so fast that it simply rips through the sample. Slower scan speeds are obviously needed when acquiring images with a line density of more than 256.

If you are fortunate you will find that the DNA molecules are reasonably evenly spaced across the mica substrate. However, you may find that there is a degree of aggregation leading to some areas of bare mica, with other areas containing relatively large clumps of molecules. If the second description fits your initial observation, you are probably better off preparing a new deposit, taking care to gently swirl the mica in the air when held with tweezers as described earlier. However, if you repeatedly observe aggregates after preparing new deposits, it is likely that the stock solution has aggregated. This can happen when the sample is frozen or stored at low temperatures. Gentle heating in a water bath at around 40-50 °C can often provide sufficient energy to break up aggregation.

3.6.2. Troublesome large samples

Normally with a conventional microscope, the larger the sample, the easier it is to image. Surprisingly, AFM turns this rule on its head, principally because of the problems associated with tip convolution and surface roughness - such as ensuring that there is adequate clearance between the underside of the cantilever and the sample. For extremely large samples, there is no justification in using AFM when traditional forms of microscopy would be easier to apply, and almost certainly yield better results. However it is sometimes desirable to obtain high resolution data over a small region of a relatively large sample, for example, receptors on a cell surface - yes, in AFM terms a cell really is a fairly large sample! In this case AFM can be useful provided that the tip is not continuously forced to climb over outrageously large features. The packing together of particulate samples into a monolayer ensures that the tip simply skips across the very top of the sample without having to travel back down to the bare mica substrate between features. Hence the apparent roughness is reduced. Materials that fall into this group of larger samples include cells, bacteria, starch granules and latex spheres, amongst others. As they are all discrete, separate particles, up to a few μm in size, the same general rules apply to imaging them. The various schemes used to minimise roughness for cellular materials are discussed in detail in section 7.1.1.

Firstly, and most importantly, concentrate on preparing your sample so that it forms a confluent monolayer in order to reduce the sample roughness. If the sample forms stacks of particles it will be impossible to get good quality, meaningful images. In the case of bacteria, as with cells, the form of the monolayer deposit can usually be controlled by manipulating the culture conditions (Gunning *et al* 1996). For particulates such as starch granules, apply a thin dusting of the sample onto double sided adhesive tape attached to the sample holder, followed by a gentle blast with a jet of air to remove any excess. However, do not attempt to image the larger varieties with diameters > 5 µm, as AFM is simply not the appropriate technique unless you resort to embedding. In the case of latex spheres, these can usually be encouraged to form monolayers by the addition of a small amount of detergent to the stock suspension. Of course it is still possible to image isolated samples of the type discussed here but this depends on them being securely attached to the substrate.

The importance of ensuring that samples that are composed of large particulates are as close to a continuous monolayer as possible cannot be stressed too heavily. Image distortion problems that occur by not taking this advice seriously can be summarised as follows:

- Mistracking and sporadic jumping of the AFM tip - indicated by streaky images.

- Artifacts resulting from severe image convolution. For example curved surfaces can appear severely flattened.

- Subsequent images taken over the same area yield different results - Isolated particles, not locked into a continuous layer, can be displaced by the tip - microscopic football!

Here are some useful guidelines to consider when imaging this class of material:

- Begin imaging with conservative scan sizes, say around 2 µm. Then if you encounter an unexpectedly large surface feature it won't bring about the end of the experiment through tip damage.

- Most large samples are surprisingly resilient, so achieving ultra low forces is no longer such a priority.

- If you can afford them, use high aspect ratio tips.

- Scan more slowly than usual. The control loop has more work to do in negotiating large features, and reducing the scan speed acts rather like increasing the gain of the control loop, but without causing 'gain ripple' shown in Fig. 3.10. Note that when sizeable objects are present in the image gain ripple tends to occur at their leading edge.

- Try using error signal mode.

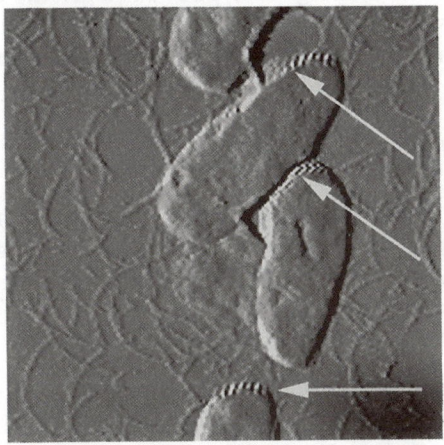

Figure 3.10. An error signal image taken on a portion of a bacterial biofilm displaying rippling artifacts (arrowed) at the leading edge of the bacteria due to excessive control loop gain. Note that the scan direction (left to right) is indicated by the presence of shadows on the far edge of the bacteria. Scan size: 6 x 6 μm.

3.7. Image optimisation

3.7.1. Grey levels and colour tables

This is simply a way of depicting image information (be it in terms of height, force, or phase etc) as regions of different brightness. The usual convention employed for a topographical (height) image is to set the lowest point as black, the highest as white, with everything else in between as a linearly changing shade of grey. It is fairly standard practice to utilise 256 different grey levels (0-255) in an image, although the definition of whether level 0 represents black or white does vary between manufacturers. This is not a problem as it is a trivial matter to invert the image to obtain correct contrast when swapping between different software packages.

It is also possible to artificially colour the image, either in the instrument software itself, or through an image enhancement package such as PhotoShop® (Adobe Systems Inc). This can be achieved by applying a monochrome colour table onto the image, for example green linearly graduated through to black. However, colour tables containing more than one colour can be difficult to implement unless you are fairly experienced in using this type of package. The advantage of applying colour to an image is that the eye is far more sensitive to changes in wavelength (colour), than it is to the intensity variation in greyscale images.

3.7.2. Brightness and contrast

This is probably the easiest way to optimise an image, either in the instrument software or by using a separate image enhancement package. The effects are relatively subtle but worthwhile, particularly prior to printing. Adjustment of either parameter beyond +/- 15% is to be avoided.

3.7.3. High and low pass filtering

This is fairly self-explanatory. The application of a high pass filter removes the lower frequency information from the image. This is particularly useful if you want to flatten the image background, because it appears like a rolling landscape, without loosing any of the fine detail. Low pass filtering, on the other hand, removes everything but the rolling landscape.

3.7.4. Normalisation and plane fitting

When an image is recorded by AFM it is acquired gradually, line by line. Each individual line can contain up to 256 grey levels, although subsequent lines may not necessarily attribute the same grey level to the same height, force or phase value. Therefore, it is necessary to ensure that a given grey level represents the same height throughout that particular image - this process is called 'normalisation'.

Plane fitting or 'Plane subtraction' involves using a least squares technique over the whole image to calculate an imaginary plane. This is performed because the sample surface is almost certainly not perfectly parallel to the X-Y plane of the AFM scanner. By subtracting this imaginary plane from the original image we are effectively removing any underlying tilt. This technique is usually followed by a normalisation procedure as described above.

3.7.5. Despike

Small bright white spots in the image, which can be single pixels, are very often produced by tip jumps. A tip jump occurs when the AFM tip is forced violently upwards and away from the sample, due to momentarily getting stuck on a surface feature. This adversely affects the resultant contrast because the tip jump is easily the tallest part of the image, so that all the interesting parts of the image are compressed into just a few of the darker grey levels. The despike routine searches for small areas that have wildly varying gradients - a spike, which can then removed by replacing it with the median value of the pixels that make up its immediate neighbours. The image is then subjected to a normalisation procedure.

3.7.6. Fourier filtering

This technique has unfortunately earned itself a bad reputation in many branches of image processing. This is unjustified since it is an extremely powerful tool when used correctly. Problems usually only begin to appear when the technique has been misapplied, or used in an extreme manner. When it is used incorrectly it is quite literally possible to generate something out of nothing. Fourier filtering is generally only ever useful for optimising images containing periodic information, such as atomic lattices or membrane proteins.

Introductions to Fourier filtering tend to be highly mathematical since the technique itself is necessarily mathematical and, unfortunately, complex. This approach is rather indigestible, particularly if you have not come across it before, and is certainly outside the aim of this book. Fortunately, it is possible to get a good grasp of the technique by considering it purely from the optical perspective, which is the approach adopted here. To get a sense of how this technique operates, consider an experiment where a narrow beam of monochromatic light from a laser falls onto a narrow slit. As the beam exits on the other side of the slit it spreads out, or 'diffracts'. This effect can be captured by projecting it onto a screen where it is displayed as a number of bright fringes, known as a 'diffraction pattern'. By varying the width of the slit, the number of slits present, or their orientation, it is possible to significantly alter the diffraction pattern. The interesting point is that there is an inverse spatial relationship between the separation of the fringes in the diffraction pattern and the width of the slit that caused them. In more general terms this can be expressed as: "Fine detail leads to widely spaced fringes - or fringes at the outer edge of the diffraction pattern".

The Fast Fourier Transform (FFT) routine, generally present in AFM software, provides an automatic mathematical method of producing an equivalent diffraction pattern for any given digital image, although it is now usually referred to as a 'power spectrum' or 'frequency domain'. Two examples are shown above their respective masks in Fig. 3.11. Highly periodic information in the image causes bright spots to appear in the power spectrum. Once again, any spots at the

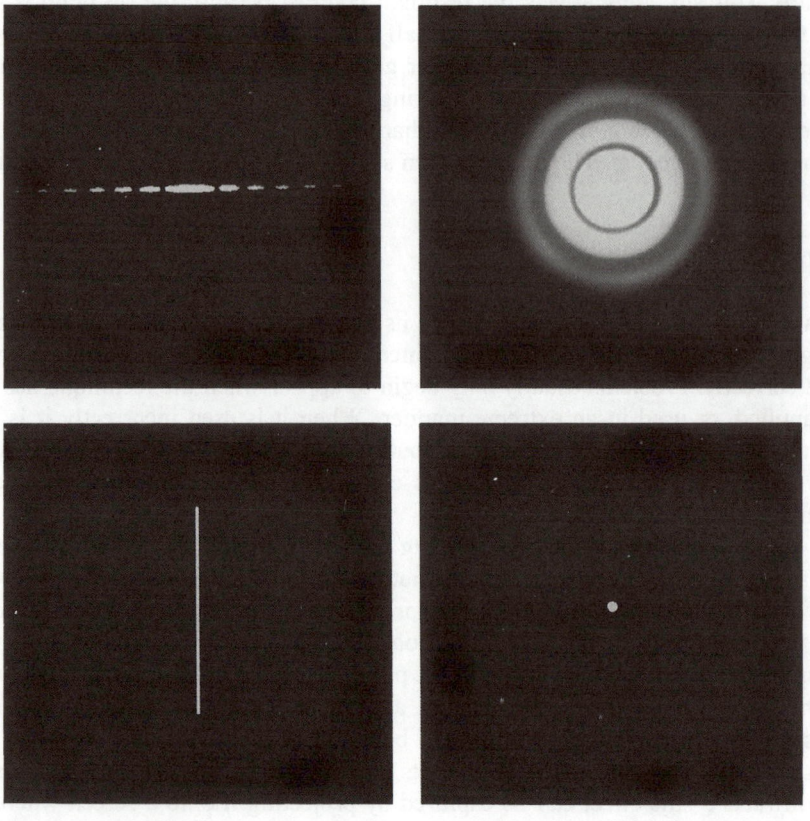

Figure 3.11. Two examples of power spectra (top), and the masks from which they were produced (bottom).

edge of the power spectrum are caused by fine detail (high frequency) present in the original image, whereas spots closer to the centre of the power spectrum are from coarser detail (low frequency) in the image. It is now possible, by using the FFT software, to edit the power spectrum, which provides an opportunity to remove any non-periodic random noise before reapplying the Fourier transform. The reapplication of the Fourier transform is known as a 'reverse' or 'back' transform. Hopefully, the result of the reverse Fourier transform should be a somewhat clearer image. This is where the danger lies; if the power spectrum is very noisy it could simply be that the original image is noisy, or that the image contains no significant periodic information. If the power spectrum is now edited injudiciously to enhance any areas that appear even remotely periodic, then the reverse transform will always produce an image with strong periodic structure, whether it was actually there or not!

3.7.7. Correlation averaging

This image processing technique, which has its roots in electron microscopy, is frequently applied to periodic structures such as 2D-protein crystals or even isolated macromolecules. Initially a unit feature is identified which is repeated throughout the image. A number of these unit features are then chosen and aligned, for example by rotation. By summing over several unit features an averaged representation can be produced which has an improved signal to noise ratio, and therefore an improved contrast. This technique has been used in studies on membrane proteins as discussed in section 6.3.5.

3.7.8. Stereographs

A stereograph, or a 'stereo-pair', is simply two images of a subject taken and viewed from slightly different viewpoints. This is in order to simulate the subtle variation in perspective observed by the left and the right eye. When the images are placed side by side and then viewed through a stereoscope, a simple frame containing a collector lens for each eye, the brain superimposes them and a single image is produced containing crude 3D information. This technique was originally popular in Victorian times but has subsequently been used for SEM, TEM and, to a lesser extent, AFM images (Shao and Somlyo, 1995). In the case of AFM data, the original image is electronically processed to produce two subtly different images which appear to have been acquired at different viewpoints.

3.7.9. Do your homework!

The simple fact that AFM literature contains images means that you should take special care of how they are finally reproduced. It is absolutely vital that you take the trouble to maximise the impact of your images, as it can dictate whether your research gets read or passed over. Unfortunately this does not end with using image optimisation techniques such as those described above. Even if you start out with excellent, high contrast, images there is no guarantee that they will appear in print exactly as they were submitted. Surprisingly most publishers actually concentrate more on the text (spelling, grammar, style and so on), rather than the images. This is partly because it is the printers who have the final say in what appears on paper. Most readers see library photocopies rather than original copies or reprints, so if the originals are poor - the photocopies will be terrible, sometimes just a black box. Dedicated microscopy journals nearly always do the work justice, but to reach the widest audience you may do better by targeting a journal according to the sample, or problem under investigation.

Colour images are generally reproduced well, although not all journals offer this option, and those that do usually charge. In order to ensure good quality image reproduction; **demand to see image proofs**, don't accept fax or photocopies. Additionally it is always worthwhile studying back issues of journals, paying attention not only to the quality of the images but also to the actual paper used.

References

Butt, H-J. (1991). Measuring electrostatic, van der Waals, and hydration forces in electrolyte solutions with an atomic force microscope. *Biophys. J.* **60**, 1438-1444.

Gunning, P.A, Kirby A.R, Parker. M.L, Gunning, A.P. and Morris, V.J. (1996). Comparative imaging of Pseudomonas putida bacterial biofilms by scanning electron microscopy and both DC contact and AC non-contact atomic force microscopy. *J. App. Bact.* **81**, 276-282.

Hansma, H.G, Vesenka, J, Siegerist, C, Kelderman, G, Morrett, H, Sinsheimer, R.L, Elings, V, Bustamante, C. and Hansma, P.K. (1992). Reproducible imaging and dissection of plasmid DNA under liquid with the atomic force microscope. *Science* **256**, 1180-1184.

Israelachvili, J.N. (1985). *Intermolecular and Surface Forces*. Academic Press.

Li, M-Q, Hansma, H.G, Vesenka, J, Kelderman, G. and Hansma, P.K. (1992). Atomic force microscopy of uncoated plasmid DNA: nanometer resolution with only nanogram amounts of sample. *J. Biomolecular Structure & Dynamics* **10**, 607-617.

Radmacher, M, Cleveland, J.P. and Hansma, P.K. (1995). Improvement of thermally induced bending of cantilevers used for atomic force microscopy. *Scanning* **17**, 117-121.

Thompson, N.H, Miles M.J, Ring S.G, Shewry P.R. and Tatham A.S. (1994). Real-time imaging of enzymatic degradation of starch granules by atomic force microscopy. *J. Vac. Sci. & Technol. B* **12**, 1565-1568.

Shao, Z. and Somlyo, A.P. (1995). Stereo representation of atomic force micrographs: optimizing the view. *J. Microscopy* **108**, 186-188.

CHAPTER 4

MACROMOLECULES

4.1. Imaging methods

To study isolated macromolecules by AFM it is necessary to confine them to a suitable surface or interface. Ideally, in order to gain the full advantages of the AFM, one would wish to investigate these molecules under natural conditions: usually in aqueous or buffered environments. Finally, the imaging process itself should not damage or displace the molecules. Achieving all these conditions simultaneously is very difficult and certain compromises need to be made. The methodology used will depend on the type of molecule under study and the sort of information required on molecular structure. It is convenient to discuss first methods of imaging in general terms.

4.1.1. Tip adhesion, molecular damage and displacement

In the earliest studies on macromolecules, originally through the use of scanning tunnelling microscopy, the major problems were sample damage, artifacts and irreproducibility. The origins of these problems are common to both STM and AFM. Most of these problems arose because samples were imaged in air, after adsorption to a suitable substrate. This is because, except at low relative humidity, a thin layer of water will be present on the sample surface and, depending on its radius of curvature, on the probe tip. When these two surfaces are brought sufficiently close together then these liquid layers will coalesce. The result is an adhesive force which effectively glues the tip to the surface (section 3.1.3). When the tip and surface are scanned relative to each other it is difficult to lift the probe over the deposited molecules. The tip either tears through the molecules, or pushes them across the surface. At high adhesive forces the molecules can be pushed across the surface and it was this which led to irreproducibility in imaging: molecules tended to be seen only when they became trapped at defects on the sample surface. Groups of molecules tended to be seen aligned at steps or other features. Not only did sample displacement account for irreproducibility, but it also probably explained the high level of artifacts reported in early studies. If molecules are only seen when they are trapped at defects then the chance of observing a defect will be similar to that of seeing a molecule. Unfortunately some defects can give images which can be mistaken for macromolecules: the most publicised examples are the STM images of grain boundaries, which can be mistaken for helical molecules such as DNA (Clemmer and Beebe, 1991). The earliest methods of imaging, many of which were developed empirically, were designed to reduce molecular mobility and prevent sample damage and displacement.

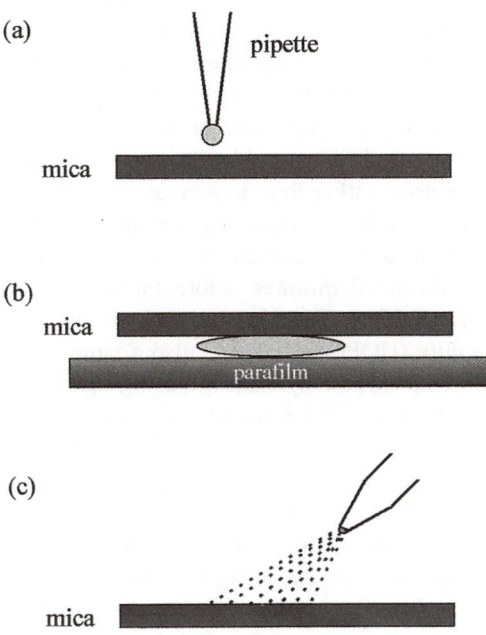

Figure 4.1. Schematic diagram showing different methods for depositing macromolecules onto substrates. (a) drop deposition, (b) sandwich method and (c) spray deposition.

4.1.2. Depositing macromolecules onto substrates

Biopolymers are normally prepared as aqueous solutions and then deposited or spread onto a suitable substrate. The substrate needs to be flat with respect to molecular dimensions, clean, cheap and easy to prepare. The most commonly used substrate is mica. Mica is a non-conducting layered material. It is cheap and can easily be cleaved, usually with a pin or sometimes cellotape, to produce clean, atomically flat surfaces up to even millimetres in size. The commonest form of mica is Muscovite [$KAl_2(OH)_2AlSi_3O_{10}$]. The minimum step size which can be observed on the surface is the thickness of an individual layer (1 nm) and the hexagonal lattice constant within the layers, which can be used for calibration, is 0.52 nm. The commonest methods for introducing the molecules onto the substrate are illustrated in Fig. 4.1. A pipette can be used to place a drop of solution onto the surface (Fig. 4.1a), a drop of solution can be formed on a hydrophobic surface, such as parafilm, and the mica touched onto the surface of the drop (Fig. 4.1b), or the solution can be sprayed as an aerosol onto the substrate (Fig. 4.1c). With drop deposition the spreading on the substrate can be enhanced by addition of small quantities of surfactant to the aqueous solution. Samples can be sprayed directly from aqueous solution, or in the presence of glycerol. The glycerol is not essential, and is probably only present as a preservative or cryoprotectant for the sample, rather than to increase sample viscosity. In any case any glycerol present needs to be removed (e.g. by drying in vacuum) in order to prevent it coating the molecules, or contaminating the tip. The samples are then normally air dried for about 10 minutes before further preparation or imaging. Mica is an hydrophilic surface. An alternative hydrophobic surface is highly oriented pyrolytic graphite (HOPG). HOPG is also a simple layered structure and flat, clean surfaces can be produced by cleavage using cellotape. Graphite surfaces tend to contain more steps than that found on mica, and common defects on HOPG are grain boundaries. The layer spacing for graphite is 0.355 nm and the minimum step size is 0.669 nm. The true hexagonal lattice spacing for carbon atoms in the layers is 0.142 nm. However, alternate layers are staggered and the spacing of equivalent carbon atoms in each layer is 0.246 nm. Larger macromolecular complexes can be imaged after deposition onto rougher substrates: for example chromosomes are usually spread on glass and collagen fibres can be imaged on substrates as rough as filter papers. The roughness of the substrate must be small compared to the 'height' of the sample.

4.1.3. Metal coated samples

This approach borrows methodology used in transmission electron microscopy (TEM) and adapts it for use in probe microscopy. Rather than generating and imaging metal replicas, metal coated samples can be examined directly by AFM. The molecules are deposited onto a substrate, usually mica, air dried, inserted into a vacuum coating unit, evacuated and metal coated. The metal coated samples can be imaged directly, or stored for imaging at a later time. Provided the initial deposition conditions are correct then the surface concentration will be sufficiently low to permit visualisation of individual macromolecules (Fig. 4.2). The metal coating freezes molecular motion and protects the underlying molecules from damage or displacement. Thus images can be obtained under constant force (dc) conditions in air. Information can be obtained on molecular size and shape for whole populations of molecules. Even if the metal coating is fairly uniform it will still be difficult to obtain information on the heights, or diameters, of the molecules. The major disadvantages of this approach are possible denaturation of the molecules during sample preparation, and the loss of resolution due to the finite size of the metal grains coating the sample. Molecular distortion during preparation is a problem well studied by electron microscopists, and the elegant methodology developed in this field can be adapted for AFM studies. Thus samples can be freeze dried, or thin specimens cooled very rapidly and the water subsequently removed by sublimation below the glass transition temperature. These methods can be used to preserve the native structure of the molecules. By optimising coating conditions, and suitable choice of the coating material, it is possible to minimise the size of the metallic grains. However, this factor will always limit the potential resolution achievable by AFM.

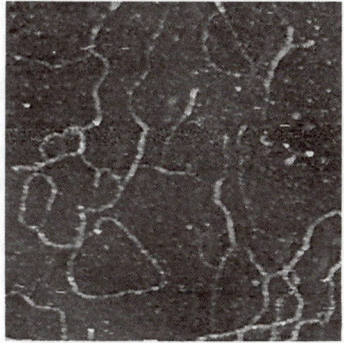

Figure 4.2. AFM image of a metal coated fibrous polysaccharide (xanthan) imaged under butanol. Scan size 1.6 x 1.6 μm.

4.1.4. Imaging in air

To obtain images of molecules in air it is necessary to eliminate the adhesive forces. One approach is to remove the surface layers of moisture. This can be done by subjecting the sample to a low vacuum or by storing the sample in a desicator under controlled low relative humidity (typically RH < 40%). The samples should then be imaged in the AFM at low RH: the AFM head is enclosed and flushed with dry nitrogen. Under these conditions it is possible to achieve imaging forces in the dc constant force mode which are high enough to generate contrast in the image, but not too high to cause sample damage or displacement (typically between 1-10 nN).

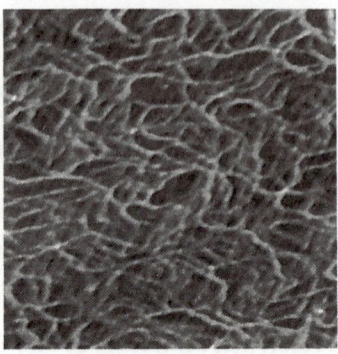

Figure 4.3. Tapping mode image of an entangled array of the polysaccharide xanthan imaged in air. Scan size 1.2 x 1.2 μm.

An alternative approach is to image in the Tapping mode. Imaging can be improved by working under an atmosphere of dry air, helium or nitrogen. The cantilevers are normally operated at a frequency just below (say 0.1-2 kHz below) the resonant frequency of the tip-cantilever assembly. Figure 4.3 shows a Tapping mode image in air for a polysaccharide sample. It has been noticed in our laboratory that, under similar deposition conditions, the concentration of polymers observed on the surface by Tapping is often higher than that seen when imaging in the dc constant force mode. It is possible that deposition leads to several layers: a most tightly bound primary layer covered with less tightly bound secondary layers, which are then displaced by the less controlled dc contact mode imaging conditions. Reports in the literature suggest that lower concentrations are often used for studies made by Tapping mode. Imaging under Tapping mode also tends to reveal more 'dirty' substrates. Again debris on the surface may more easily be displaced or picked up by the probe in the dc mode and thus is not seen in the image.

4.1.5. Imaging under non aqueous liquids

Another approach for eliminating adhesive forces is to image under liquids. The chosen liquid displaces the water layer allowing the imaging force to be controlled during imaging. It is common to image under non-solvents such as alcohols. A variety of alcohols can be used including butanol, propanol and propan-2-ol. These appear to function by displacing water, thus inhibiting molecular motion on the substrate surface, and by preventing desorption from the surface. Imaging can be carried out using either dc constant force conditions or Tapping mode. Under dc conditions it is still necessary to precisely control the imaging force. Figure 4.4

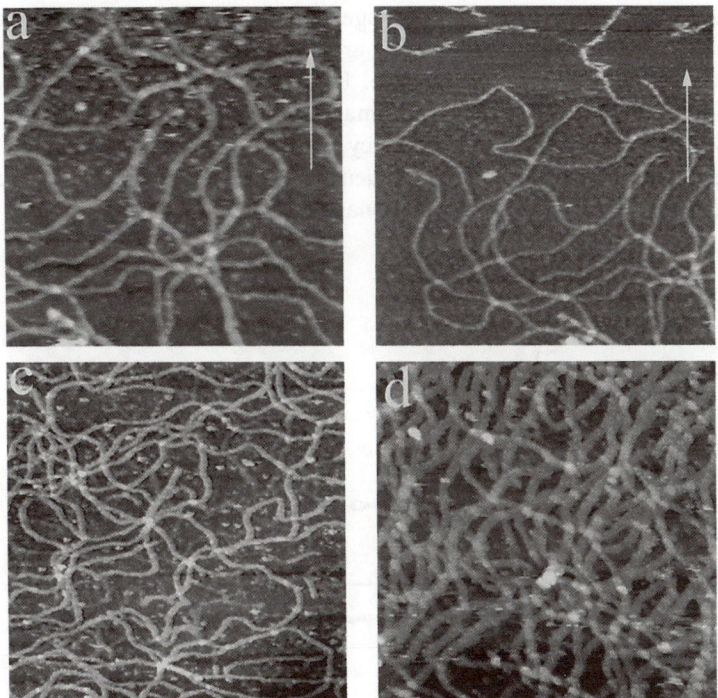

Figure 4.4. Contact mode images of xanthan polysaccharide imaged under alcohol. (a) Force decreases from the bottom (3 nN) to the top of the image. Force is about 1 nN at arrow and image contrast fades. Scan size 600 x 600 nm. (b) Force increases from bottom (3 nN) to top of image. Force at arrow is about 10 nN and damage and displacement occurs. Scan size 700 x 700 nm. (c) Optimised contrast. Scan size 1.2 x 1.2 μm. (d) Image degradation due to adhesive force. Scan size 700 x 700 nm.

shows the effect of varying the imaging force on the quality of the image. In Fig. 4.4a when the imaging force becomes too low (< 1 nN) the molecules merge into the background. Fig. 4.4b shows the effect of allowing the imaging force to increase during scanning. At sufficiently high imaging forces (> 10 nN) evidence appears for the damage and displacement of the molecules. Under optimum conditions reproducible images with good contrast are obtained (Fig. 4.4c). Figure

4.5a shows the force-distance curve corresponding to the image shown in Fig 4.4c. The force distance curve is perfectly reversible with no evidence of adhesion. Even under these conditions the quality of the image will begin to decay with time; typically over periods of hours, although the timescale involved is very sample dependent and can be as short as minutes. Figure 4.4d shows such an image in which the quality has begun to decay, and Fig. 4.5b shows the corresponding force-distance curve. The decay in image quality is related to the appearance of an increasing adhesive force, presumably arising as the tip accumulates molecules or other debris from the sample surface. Images can also be obtained by Tapping under liquid conditions. Setting the imaging conditions is slightly more difficult because the single resonance frequency of the tip-cantilever assembly is usually replaced by a range of resonance frequencies. The chosen conditions are normally about 0.3-2 kHz below the largest resonance frequency.

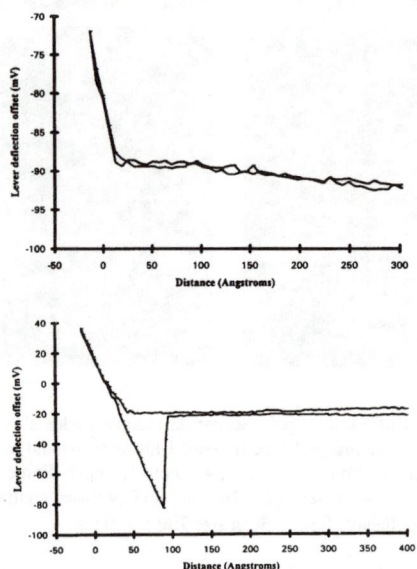

Figure 4.5. Force-distance curves for the images shown in Fig. 4.4. Top corresponds to Fig. 4.4c and bottom corresponds to Fig. 4.4d.

4.1.6. Binding molecules to the substrate

Instead of simple physisorption of molecules onto substrates like mica it is possible to specifically modify the substrate in order to improve molecular attachment. This is particularly important if subsequent imaging is to be in aqueous or buffered media, and as a method of avoiding molecular distortion or denaturation on drying. The main approach involves the use of self-assembled monolayers on the substrate. The background literature in this field is vast and only the general features appropriate to scanning probe microscopy will be considered here. The earliest attempts to modify substrates were made for STM studies which require a conducting substrate, and concerned modification of epitaxially grown gold layers on mica.

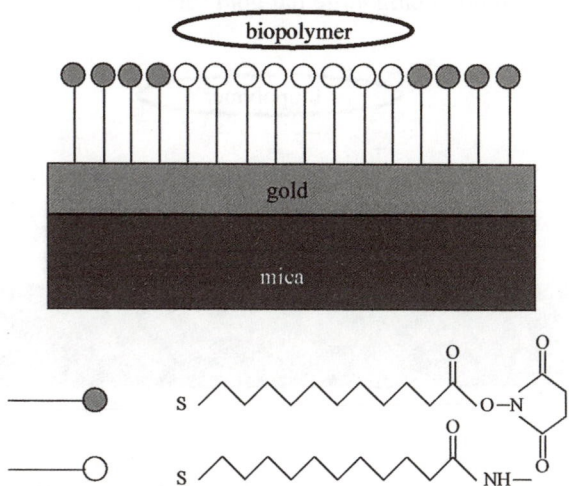

Figure 4.6. Schematic diagram showing the use of derivatised thiols for binding biopolymers to gold coated mica. The process is illustrated for DSU.

The growth of flat islands of gold on mica is challenging, but standard methods are published in the literature (Chidsey *et al* 1988; Clemmer and Beebe, 1992). An approximately 200 nm thick gold layer is thermally evaporated onto freshly cleaved preheated (300 ^{0}C, overnight) mica. Small plates or discs of coated mica (about 1 mm in size) are cut or punched. The gold coated mica fragments are derivatised by incubating immediately with a solution of an alkane-thiol derivative. The derivatised mica is removed, washed, dried under nitrogen and then used immediately for immobilising the biopolymer. The biopolymer solution, in an appropriate buffer, is placed as a drop on a piece of parafilm and the activated coated mica placed on top of the drop. The assembly is covered to prevent evaporation and allowed to incubate for a suitable period of time. The biopolymer coated sample is washed with buffer and placed directly in the AFM liquid cell without drying, and then imaged using dc contact, ac or Tapping modes. Different types of thiol derivatives can be used to bind different types of biopolymers. Thus, for example, proteins, phospholipids or molecules containing amino sugars can be covalently attached (Fig 4.6) using dithiobis (succinimidylundecanoate) (DSU) (Wagner *et al* 1994; 1996). Anionic biopolymers can be bound by use of 'positively' charged 2 dimethylaminethanethiol (Allison *et al* 1992a,b). The limitation of this method for AFM is the difficulty in producing large flat gold surfaces.

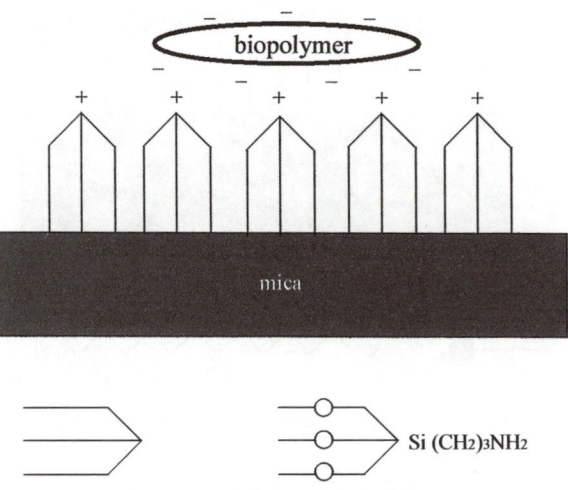

Figure 4.7. Schematic diagram illustrating the use of silane derivatives to bind biopolymers to mica. The process is illustrated with APTES.

An alternative is to derivatise mica surfaces directly. The main constituent of mica is SiO_4 tetrahedra and thus this can be achieved by covalent attachment of silanes onto the mica surface. Covalent attachment of 3-aminopropyltriethoxysilane (APTES) can be used to generate a positively charged surface which will bind anionic biopolymers (Lyubchenko *et al* 1992a,b). The mica is freshly cleaved and placed at the top of a desicator containing a solution of the silane. The desicator is briefly evacuated, purged with dry argon, and the reaction allowed to proceed for about 2 hours. The silane solution is removed, the desicator again briefly evacuated and then purged with dry argon to allow storage of the activated mica until use. The biopolymers are attached (Fig. 4.7) by either immersing the substrate in the biopolymer solution or by placing a drop of the solution on top of the substrate. The surface is then rinsed and dried either under vacuum or dry gas. By omitting the drying step it is possible to prepare samples for imaging which have been maintained in a liquid environment. By using other silane derivatives it is possible to selectively bind particular biopolymers.

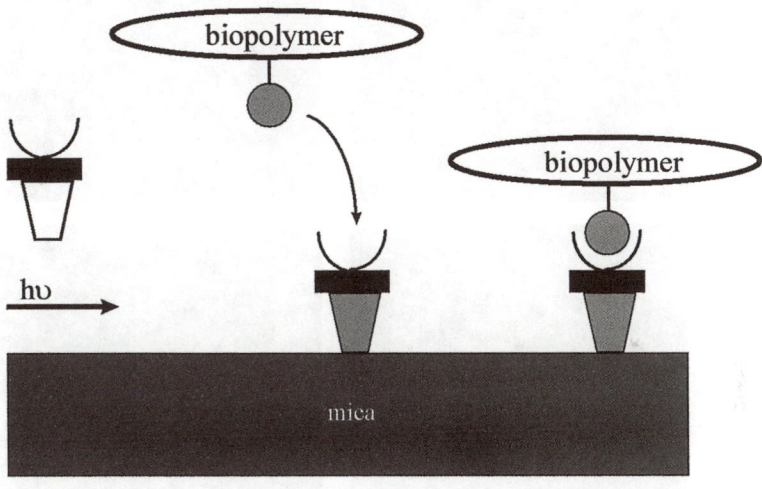

Figure 4.8. Schematic diagram illustrating the use of photoactive bifunctional reagents for binding biopolymers to mica. The reagent is first bound to the mica and then captures the biopolymer by binding a target site on the biopolymer.

Photochemical reactions can be employed to immobilise biopolymers (Luginbuehl and Sigrist, 1998) onto bare substrates such as mica, or to bind macromolecules to the thermochemically derivatised mica described above. Direct attachment to bare substrates can be carried out using heterobifunctional photoreagents such as trifluoromethyl aryldiazirines (Fig. 4.8). The target groups for attachment of biopolymers are primary amines or thiols. Freshly cleaved mica is rinsed with ethanol and dried under nitrogen. A solution of the photoactive agent is dropped onto the substrate surface and the solvent evaporated under vacuum at room temperature. The substrate is agitated and irradiated with a suitable light source: at 350 nm carbines are generated from the trifluoromethyl diazirenes which covalently link to the surface.

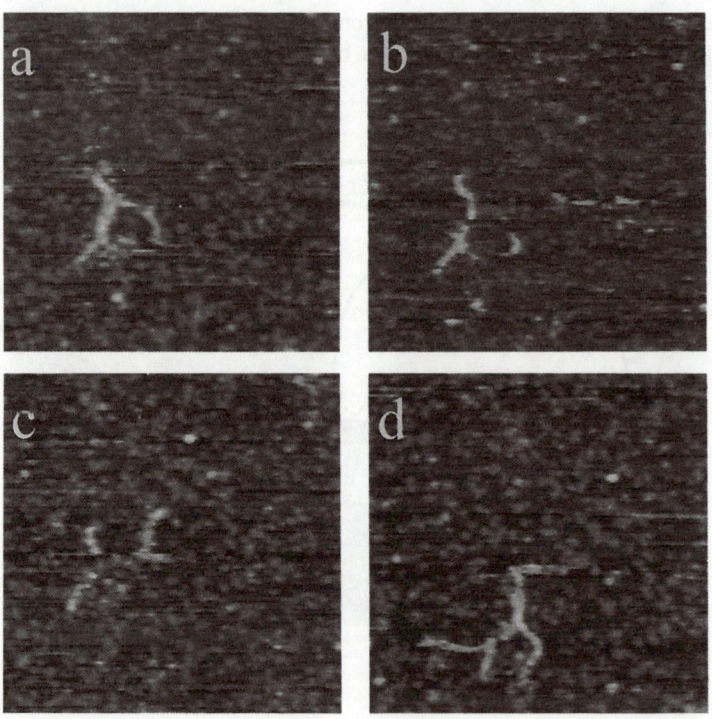

Figure 4.9. Tapping mode AFM image of a water soluble wheat pentosan polysaccharide imaged under 10mM HEPES buffer plus 2 mM ZnCl₂ (pH 7). Scan size 600 x 600 nm. Sequence of images showing 'motion' of the molecules.

Excess crosslinker is removed by rinsing and sonication, and the substrates dried under mild vacuum. Biopolymers are attached by incubating a solution of the biopolymer with the substrate, washing and sonicating, and then imaging under buffer. An alternative approach is to use these photocrosslinkers to attach biopolymers to derivatised substrates containing covalently attached alkylthiols or alkylamines. The experimental conditions need to be optimised, particularly in order to avoid aggregation of the crosslinkers due to photopolymerisation reactions. Assessment of non-specific binding to these substrates can be made by using separate controls, or photo masking techniques.

4.1.7. Imaging under water or buffers

Imaging molecules and molecular processes under water or buffers is perhaps one of the most exciting aspects of the use of AFM. Although it was at first felt that the molecules would need to be covalently attached to the substrates to prevent their desorption, damage or displacement there is now growing evidence that these restrictions need not be as severe as first thought. For fibrous molecules such as nucleic acids or polysaccharides it is difficult to achieve low enough applied imaging forces in order to image in the dc mode. Presumably the weakly attached physically adsorbed molecules are displaced from the surface by the lateral or frictional forces imposed by the probe during scanning. The use of the Tapping mode certainly makes imaging much easier. Imaging can also be improved by enhancing adsorption or inhibiting desorption of the molecules from the surface. For example the samples can be air dried onto mica, vacuum dried or even washed with alcohols prior to imaging under buffer. Fig. 4.9 shows a sequence of images of a water soluble polysaccharide which has been deposited onto mica and then imaged by Tapping under buffer. Not only can the molecule be clearly resolved but successive images reveal an apparent motion of the molecule on the mica surface. As will be explained in sections 4.2.1 & 4.4.1 this type of motion is the result of the molecules trying to desorb from the surface. However, the ability to obtain such images of molecules in motion has allowed the production of 'molecular movies' of biological processes. In the case of certain physically adsorbed proteins the binding to substrates such as mica is sufficiently strong, and the proteins structures are sufficiently resilient to distortion, to allow imaging of individual proteins under water or buffer even in the dc contact mode, provided minimal normal forces are applied (section 4.5.1). It remains to be seen whether the use of Tapping mode will improve the imaging of these types of protein systems. As will be shown in sections 4.5.1 and 6.2.2 methods are slowly being developed to allow complex membrane proteins to be examined under physiological conditions in environments close to their natural state. These advances are slowly realising one of the original aims of applying probe microscopy to biological systems.

4.2. Nucleic acids: DNA

DNA is the biopolymer most studied by scanning probe microscopy methods. In fact, the first published STM image of a biopolymer was of a DNA molecule deposited onto a coated Si wafer (Binnig and Rohrer, 1984). DNA was one of the first, if not the first biopolymer to be imaged by AFM. It is not difficult to understand why DNA has received such attention. DNA is a well characterised biopolymer of exceptional biological importance. Its characteristic size and shape make it a good model system for evaluating the quality and reliability of SPM images. The availability of linear DNA, circular plasmids, well defined restriction fragments, together with the possibility for synthesising specific sequences of defined structure and length, make these molecules ideal candidates for study. The early studies of DNA, primarily by STM, laid the foundations for routine imaging by AFM; now the method of choice for studies on DNA. Indeed the studies on DNA have probably played a pivotal role in the development of general methodology for imaging biopolymers. The driving force for studies on DNA lie in the potential high resolution of the AFM under natural or physiological conditions. The possibility for using AFM methods for automatic sequencing or manipulation of DNA has provided impetus for studies on nucleic acids. Other goals include the prospects of studying how small changes in sequence, or the binding of small molecules (drugs, carcinogens), alter the structure of the molecule, and the very real possibility of characterising, under physiological conditions, the many biologically important protein-DNA interactions.

4.2.1. Imaging DNA

It is now possible to obtain reliable and reproducible AFM images of DNA: the method used depending on the type of information required. Basically there are two types of experiment which are of importance. Firstly, static studies where the molecules are immobilised, and the interest is in determining size, shape or details of molecular structure, or molecular interactions. Secondly, dynamic studies in buffered or physiological media, where the molecules are imaged in motion.

The most common substrate for AFM imaging of DNA is mica or derivatised mica. The molecules are generally dropped onto mica from solution and then air dried prior to imaging. The major problem in obtaining reliable images of immobilised DNA molecules appears to be tip damage or displacement due to 'adhesive forces' (sections 3.1.3 & 4.1.5). Although 8 - 10 nm resolution images have been obtained in air (Bustamante *et al* 1992; Vesenka *et al* 1992a) the most reliable and reproducible images are acquired by imaging under alcohols (Hansma *et al* 1992a,b; Lyubchenko *et al* 1993), where the adhesive forces are eliminated, and the applied force can be controlled during imaging. The development of the Tapping mode provides an alternative method of eliminating adhesive forces and has allowed imaging of DNA in air, gaseous media such as dry helium, water and buffered media. Since DNA is soluble in aqueous media the trick in imaging involves enhancing the binding of the DNA to the mica without inhibiting molecular motion. This can be achieved in several ways. Air drying the sample, vacuum drying or washing the deposit with alcohols before imaging in the aqueous medium appears to improve the surface coverage. Addition of certain divalent cations to the imaging medium enhances binding without the need for a drying stage. It is not clear whether these cations act by directly binding the molecules to the charged mica substrate, or act by altering the surface charge, or electrical double layer, so as to promote adsorption and/or inhibit desorption. In this context it should be noted that the effects are sensitive to cation type and size, and not just ionic strength. The use of Tapping mode in water and buffered medium has allowed imaging of DNA molecules in motion and their interactions with various enzymes.

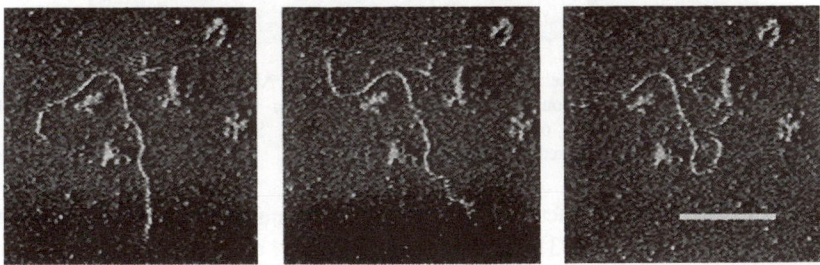

Figure 4.10. Phase images showing motion of a ϕX-174 DNA molecule. Scan size 2 x 2 µm. Reproduced with permission of the authors and Oxford University Press from Argamen *et al* 1997.

Phase imaging has also proved useful for studying DNA under static and dynamic (Fig. 4.10) conditions. For impure or 'dirty' samples, where the DNA is present within a thick contamination layer on the mica, the phase images highlight the stiffer, less compressible nucleic acids, which are not resolved in the topographic images (Hansma *et al* 1997). Phase imaging allows molecular resolution at lower forces and also at higher scan rates; permitting faster processes to be studied (Argaman *et al* 1997).

4.2.2. DNA conformation, size and shape

Are the images of DNA adsorbed to mica representative of the 'natural' state of the molecule, or does the adsorption perturb or denature the molecules? In general the resolution achieved is not sufficient to visualise the helical structure directly, although, quite often, exceptional images show structural features compatible with

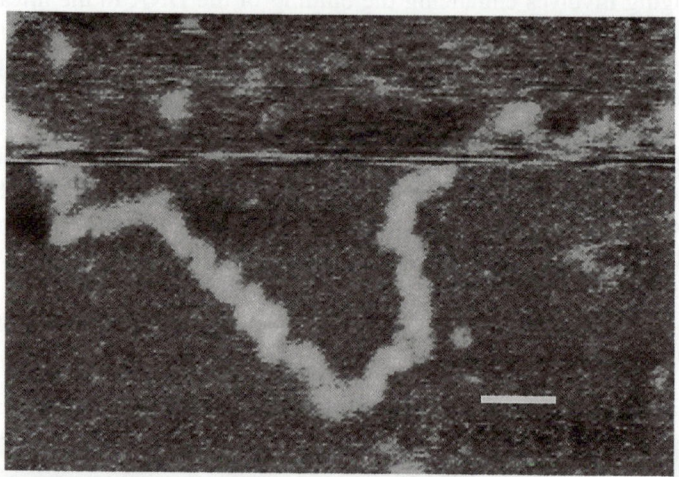

Figure 4.11. Right handed DNA helix imaged using Tapping mode AFM. The double-stranded DNA has been adsorbed to mica under aqueous buffer and then imaged under propanol. The spacing of the helix turns is comparable to the spacing of the major groove of B-DNA. Scale bar 10 nm. Reproduced with permission of H.G. Hansma and based on Hansma, 1996.

the turns of the helical structure (Fig. 4.11). Thus images obtained under butanol (Hansma and Hansma, 1993) or propanol (Hansma *et al* 1995) are consistent with the adoption of the helical structure. Measurements of the length of cyclic plasmids deposited from aqueous solution and then imaged under propanol are consistent with the adoption of the B-DNA helical structure (Hansma *et al* 1993a). The highest spatial resolution images of DNA (Fig. 4.12) have been obtained for samples deposited onto a cationic bilayer as close packed arrays and then imaged in dc contact mode under aqueous buffer (Mou *et al* 1995a). The images revealed a right handed helix of pitch 3.4± 0.4 nm, in good agreement with the known pitch of the double helix for B-DNA. The use of cryo-AFM (Zhan, Y *et al* 1996; Han *et al* 1995) offers the prospect of improved resolution but, as yet, it has not proved possible to consistently resolve the helical structure (Han *et al* 1995). The diameter of the DNA helix is known and measurements of molecular widths or heights should help to confirm the presence of the helical structure. Measurements of widths are complicated by probe broadening effects . When probe broadening effects are estimated then the sizes observed are generally consistent with the expected helical structure, and the measured values in close packed arrays, where probe broadening effects should be minimised, approach the values expected for the helical conformation.

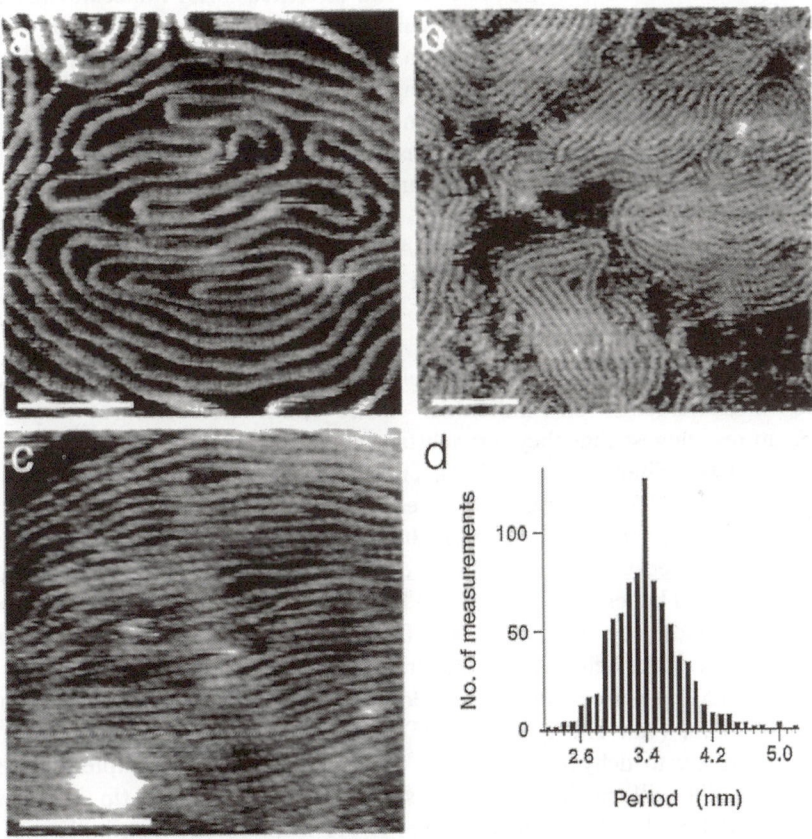

Figure 4.12. High resolution images of three DNA samples (a) pBR 322 (4.36 kb), (b) pBR 325 (6 kb) and (c) HaeIII restriction fragments of φX174. Data reproduced from Mou *et al* 1995 with permission. This reference contains full details of the preparation and imaging methodology. All three specimens showed periodic modulations (measurements recorded in the histogram (d)) of mean value 3.4 ± 0.4 nm which is consistenet with the pitch of the DNA double helix.

In general measurements of heights are smaller than expected. The values vary with sample preparation conditions and with the applied force. Measured heights tend to be smaller than expected for the dimensions of the helix. The explanation for this discrepancy is not fully understood at present. Given the evidence above for retention of a 'native' helical conformation the most likely explanation is that the helices are compressed during scanning. Certainly changes in height consistent with changes in structure or conformation have been reported: AFM studies (Thundat *et al* 1994) of relaxed and stretched DNA molecules revealed lower heights for the stretched molecules and comparative studies on single, double and triple helical DNAs (Hansma *et al* 1996) showed smallest heights for the single stranded form and largest heights for the triple helical form. Changes in sequence also lead to changes in measured height: the height of G-DNA, containing guanine tetrad motifs (G quartets), was close to the expected value obtained from X-ray and nmr data, suggesting that these molecules are less compressible than B-DNA (Marsh *et al* 1995). Indeed height measurements can be used to probe unusual features of DNA molecules. For example, measurements of length and height have been used in the characterisation of enzyme-resistant DNA fragments, and have been taken to suggest that such structures are unusual multi-stranded forms of DNA (Li, J *et al* 1997). Studies using cryo-AFM will eliminate thermal motion and may yield more reliable molecular dimensions. Force measurements on individual DNA macromolecules at cryogenic temperatures suggest that they are substantially stiffer than at room temperature (Han *et al* 1995; Zhang, Y *et al* 1996). Thus height measurements should be more realistic in cryo-AFM. At present the measured heights appear to be almost 1 nm smaller than room temperature values, although there is good evidence that this is, at least in part, because the molecules are buried in a thin layer of frozen solution on the mica, and better measurements of heights will await development of freeze-etching methods for exposing the DNA molecules. Shear force SNOM measurements of the height of double stranded dsDNA yield higher values ($\approx$ 1.4 nm) than AFM measurements and are closer to the expected diameter ($\approx$ 2 nm) of the DNA helix (van Hulst *et al* 1997). In this case the reduced heights are attributed solely to dehydration effects. The measurement of sample heights in aqueous media will be dependent on sample-tip and substrate-tip interactions (Muller and Engel, 1997). For example, if the local charge distribution on the molecule is different to that of the substrate then the force-distance curves will be different for the sample and surface. Thus the estimated 'height', measured under 'constant force' conditions, will be not be simply determined by the diameter of the molecule. Electrostatic interactions can be screened (Butt, 1992a,b) and under these conditions better estimates of height should be obtained. However such effects have not been systematically studied for DNA.

Images of DNA can provide information on molecular size and stiffness of the molecules. Sizing of DNA is an important and widely used tool in molecular biology; lengths of DNA molecules are used in restriction mapping, +/- screening, fingerprinting and genotyping. It has been shown that contour lengths determined by AFM correspond to lengths expected from the sequence and number of base pairs (Hansma and Hoh, 1994; Bustamante and Rivetti, 1996; Shao *et al* 1996; Thundat *et al* 1993). However, the lengths of the molecules bound to mica are found to be sensitive to the type of cation species used for binding the DNA to the mica and optimum and standardised methods need to be used for sizing. Methods are being developed for automating the solid state DNA sizing (SSDS) process (Fang *et al* 1998). This basically involves three steps: sample preparation, sample imaging and image analysis. The most difficult step to automate is the sample preparation, which requires automatic pipetting onto mica. Automated methods of imaging are well established for uses of AFM in the semiconductor industry and image analysis software for recognition and sizing has been developed (Fang *et al* 1998). SSDS has performed well in comparison with conventional gel electrophoresis for +/- screening of a set of P1 artificial chromosomes. The SSDS method is potentially much faster than conventional methods, uses less sample, and would permit a reduction in the number of polymerase chain reaction (PCR) cycles needed in the amplification step. A possible problem in developing fully automated systems for biological systems is the contamination of the tip, which introduces adhesive forces and limits the working lifetime of the tip. Methods for exchanging or cleaning tips may need to be devised. Automated imaging has potential for mapping sequence specific decoration of DNA molecules: examples may include chemical modification, hydridisation with short oligomers, or binding of sequence specific proteins, such as restriction enzymes, for health care or forensic screening applications.

The shape adopted by the molecules on the substrate is determined by sequence specific factors, which dictate the stereochemistry of the molecule. For the simplest type of polymer the monomeric units can freely rotate about the inter - monomer linkages and the molecules adopt a random coil configuration. The overall size of the polymer depends on the number and size of the monomer units. Stereochemical restrictions on free rotation about the linkages, such as the adoption of a helical conformation, increase the overall size of the molecule. The simplest treatment of such effects for extended polymer chains defines molecular stiffness (or extension) in terms of a parameter called the persistence length L_P. The persistence length is a measure of the separation of points within a chain which are randomly distributed in space and is usually measured by light scattering (Burchard, 1994), or by electrooptic methods such as transient electric birefringence or dichroism (Hagerman, 1988). For polyelectrolytes there are essentially two contributions: charge-charge repulsions, which enhance stiffness at low ionic strength, and can be eliminated by screening to reveal a second 'intrinsic' contribution. Persistence lengths can be determined for DNA by AFM (Bustamante and Rivetti, 1996; Rivetti *et al* 1996; Hansma *et al* 1996; Hansma *et al* 1997). The procedure involves measuring the angle (θ) between tangents to the molecule at points separated by a contour length (L). The reciprocal of the slope of a plot of the arithmetic mean of the square of the angle $<\theta^2(L)>$ versus L gives the persistence length L_P. These experimental plots are not linear and the value of L_P depends on the size range studied, possibility accounting for the variability in measured L_P from sample to sample. The L_P values for a given sample have been found to be substrate dependent, suggesting that surface interactions affect the shape and extension of the molecules. Such surface interactions mean that the L_P values found by AFM may not be equivalent to those determined for DNA in solution. However, provided the surface interaction makes only a small contribution to the molecular extension then measurements of shape by AFM can still be used to investigate structural changes of the DNA molecules. In fact differences in L_P have been observed for different DNAs: single stranded DNA has been seen to be more flexible than normal double stranded DNA itself more flexible than protonated double stranded poly $(A)^+$ (Hansma *et al* 1997). Measurement of overall size provides an alternative assessment of molecular stiffness. This type of measurement has been used to investigate the effects of molecular interactions such as ligand binding to DNA. For example, by using AFM, it has been possible to observe the straightening of abnormally bent kinetoplast DNA (the mitochondrial DNA from trypanosomes and related parasitic protozoa) due to the addition of the ligands distamycin and the microgonotropen (MGT-6b), and the bending of normal DNA on addition of MGT-6b (Hansma *et al* 1994). An alternative approach to the measurement of the stiffness of DNA molecules, and fibrous molecules in general, is to use the AFM to measure force-distance curves, and to analyse such data to determine the stiffness of the molecules. In this type of experiment the molecules are deposited onto the substrate, parts of the molecule are picked up by the tip and the force-extension curve measured. For DNA the force-extension curve can be described by a worm-like chain model (Bustamante *et al* 1994b) and the interpretation of such data to determine persistence lengths is described by Bouchiat and coworkers (Bouchiat *et al* 1999). This approach offers the prospect of analysing molecules over a wide range of ionic strengths not easily accessible in a purely microscopic study.

The advantage of using microscopy to study molecular shape is that it allows localised changes within individual molecules, or variations of structure for molecules within a population, to be visualised and quantified. The likely advantages of the use of AFM over the use of EM is that the sample preparation is simpler and the samples can be examined under more natural conditions. Processes which involve changes in the shape of the DNA molecules can be studied. Examples of this type of study are surface induced condensation of DNA into toroids and rods (Fang and Hoh, 1998) and supercoiling (Tanigawa and Okada, 1998; Lyubchenko and Shlyakhtenko, 1997; Samori *et al* 1993; Henderson, 1992) of DNA. Attempts have been made to look for distortions of DNA due to the effects of anti-cancer drugs, such as cisplatin (Rampino, 1992), although the interpretation of the results is not clear cut. However, more recent studies using STM in blind trials do appear to have been successful (Jeffrey *et al* 1993), confirming the potential of probe microscopy for this type of study.

4.2.3. DNA-protein interactions

A major area of interest is the study of DNA-protein binding (Lyubchenka *et al* 1995; Hansma, 1996; Kasas *et al* 1997a,b; Hansma *et al* 1995) and the consequent effects on DNA condensation (packaging) and processing. Because of the extensive literature on AFM studies on DNA, and DNA-protein interactions, there are a number of good review articles covering this research area. The articles by Lyubchenka and coworkers (Lyubchenka *et al* 1995) and Bustamante and coworkers (Bustamante *et al* 1993) are particularly useful, and provide detailed descriptions of methods for imaging DNA, and DNA-protein complexes. Some examples of the types of DNA protein complexes studied by AFM are given below.

RNA polymerase complexes

This is a major area of interest and the purpose of these studies is to investigate the transcription of DNA. RNA polymerase is a high molecular weight protein (465 kD) which should be relatively easy to observe attached to DNA, and early studies (Bustamante *et al* 1993; Zenhausern *et al* 1992b) showed that the enzyme can be visualised bound to DNA. Comparative EM and AFM images (Zenhausern *et al* 1992) revealed two RNA polymerase enzymes specifically bound to sites on circular plasmid DNA. Early studies of the binding of RNA polymerase to DNA as a transcription complex were made in air, after drying the sample onto mica (Rees *et al* 1993; Bustamante *et al* 1993). Bending of the DNA was observed in both open promotor complexes and in elongation complexes containing a nascent 15-nucleotide transcript. The more marked bending of the elongation complexes has been attributed to conformational changes of the DNA, believed to accompany maturation of an open promoter region into an elongation region. The ability to image DNA and DNA-protein complexes under water or buffers has led to spectacular advances in probing transcription and transcription complexes. Key advances were the use of selected cations for binding DNA onto mica, and the use of Tapping mode for imaging the molecules (Thomson *et al* 1996a; Hansma and Laney, 1996). Under buffered conditions it is possible to observe DNA molecules in motion (Fig. 4.10) (Hansma *et al* 1995; Shao *et al* 1995). Time lapse films show changes in shape of linear or circular DNA. In some images parts of the molecules are missing, suggesting that these regions have detached from the surface, and that their motion in solution makes them impossible to image. This would seem to suggest that the preparation conditions force the DNA onto the surface and that the motion is a reflection of the desorption-adsorption of parts of the molecule, as the entire molecules are trying to escape from the surface into the buffered medium. This belief is reinforced by the observation that some DNA molecules eventually completely disappear and new molecules appear on the surface (Hansma *et al* 1995). The ability to image under natural conditions has allowed real-time imaging of the assembly of RNA polymerase-DNA complexes (Guthold *et al* 1994) and visualisation of RNA polymerases transcribing linear dsDNA (Fig. 4.13) and single stranded ssDNA circular templates (Kasas *et al* 1997a,b; Hansma, 1996). The rate of the transcription process was slowed by using a low nucleotide triphosphate (NTP) precursor concentration. It has been possible to visualise the translocation of DNA templates by the RNA polymerase, to image stalled complexes by controlling the supply of specific NTPs and, after initiating its binding to the mica, to visualise the RNA produced in the process. The demonstration that changes in protein shape during enzyme activity can be detected by AFM as fluctuations in height (Radmacher *et al* 1994), coupled with the development of 'tracking' methods for observing such changes on individual proteins (Thomson *et al* 1996b), mean that it is possible to detect changes in the shape of the polymerase during the transcription process.

Chromatin

An understanding of the molecular structure of chromatin is of importance in explaining mechanisms of gene replication and expression in the cell. AFM is a tool well suited to the investigation of DNA condensation induced by complex formation with nucleoproteins. In fact AFM studies of reconstituted (Allen *et al* 1993a; Vesenka *et al* 1992b) and native (Zlatanova *et al* 1994; Martin *et al* 1995; Leuba *et al* 1994; Fritzsche *et al* 1994; 1997; Allen *et al*1993b) chromatin have played an important role in resolving controversies on chromatin structure. Chromatin is an important structural unit of chromosomes. The basic units of chromatin are nucleosomes; 10 nm particles composed of 146 bp of DNA molecule wrapped around the histone octamer. Recent studies (Woodock *et al* 1993) had raised doubts about the proposed regular 'solenoid-like' (Finch and Clug, 1976) or 'twisted-ribbon like' (Woodock *et al* 1984; Bordas *et al* 1986) geometry for the chromatin fibres. The AFM data suggests that the fibres consist of an irregular 3D array of nucleosomes. In studies of marsupial spermatozoa by AFM differences in the organisation of chromatin packaged by histones or protamines has been detected (Soon *et al* 1997): the nucleoprotamine particles appearing as tighter bundles than the nucleohistones particles. Isotonic cell lysis has been used to prepare and image largely intact chromatin from chicken erythrocytes (Fritzsche and Henderson, 1996a,b; Fritzsche *et al*1994) and human B lymphocytes (Fritzsche *et al* 1995a,b; 1997). Such studies partially preserve the spatial relationships between neighbouring chromatin sites permitting investigation of the proposed compartmentalisation of chromatin in interphase nuclei.

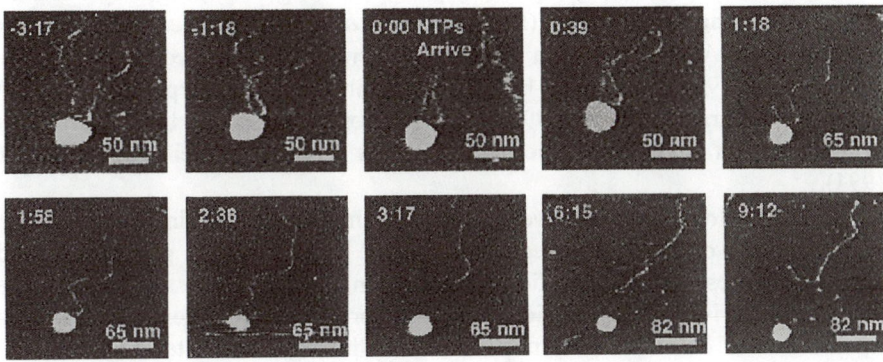

Figure 4.13. Time lapse series of AFM images showing the transcription of a 1047 double stranded DNA template by an RNAP molecule. The first two images show that before addition of ribonucleoside 5'-triphosphates (NTPs) the DNA is mobile on the mica substrate. The following 6 images after NTP addition, from time 0.00 onward are sequential and show 'reading' of the DNA molecule until release (02.38). Data reproduced with permission from Kasas *et al* 1997b. Copyright (1997) American Chemical Society.

Other DNA-protein complexes

Examples of AFM studies of such complexes include investigations of the binding of single stranded DNA binding (SSB) protein (HG Hansma *et al* 1993b) and RecA protein (Lyubchenko *et al* 1995; Hansma *et al* 1995) to DNA. SSB binds only to single-stranded DNA (ssDNA) and is important in DNA replication and recombination. The AFM images show a uniform coating of the protein along the DNA chain. RecA protein is important in the molecular mechanism of conjugal recombination in *E. coli*. The protein binds to both ssDNA and double stranded DNA (dsDNA). AFM images of RecA complexes with dsDNA circular plasmids suggest complete coverage of the DNA with the RecA protein. Studies on ssDNA by AFM indicate a compact folded structure. The binding of the RecA protein has been observed to extend the ssDNA complex, which can then bind to dsDNA forming extended filaments.

There are a number of specific protein-DNA interactions in which conformational constraints on binding can lead to bending, unwinding or overwinding of the DNA. Such interactions are of considerable interest, and there is a reservoir of background crystallographic, spectroscopic and biochemical data on these complexes. Not surprisingly a number of these complexes have been probed by AFM. Examples include studies on the specific and non-specific binding of Cro protein, a key regulatory protein in the growth of λ-phage, and the resultant bending of the DNA (Bustamante *et al* 1994a; Eire *et al* 1994). The bending of DNA on interaction with the bacteriophage φ 29 connector protein (Valle *et al* 1996) and with binding of antibodies to incorporated Z DNA sequences (Pietrasanta *et al* 1996) has been observed, as has the stiffening or straightening of DNA due to the binding of Fur repressor protein (Le Cam *et al* 1994).

DNA looping is believed to play an important part in the molecular interactions involved in the regulation of the expression of prokaryotic and eukaryotic genes, of site specific recombination, and of DNA replication. The phenomenon arises when individual proteins, or protein complexes bind to two separated sites on the DNA molecule, thus linking these sites together and forming loops in the DNA molecule. Here AFM complements or extends EM work. There are at least two AFM studies showing looping induced by eukaryotic transcription factors (Wyman *et al* 1995; Becker *et al* 1995). In the case of heat-shock transcription factor 2 (HSF) looping was observed due to cross-linking of HSF binding sites on the DNA. The use of AFM provided more information than previous EM studies because measurements of height permitted the stoichiometry of the binding complexes to be determined: linkage is considered to occur due to the interaction of HSF trimers bound to specific sites on the DNA (Wyman *et al* 1995). Multi-enzyme complexes have also been observed in AFM studies (Becker *et al* 1995) of DNA looping using the transcription complex (PIC). PIC binds to the promoter region and looping only occurs in the presence of Jun protein, which binds to a distant AP-1 site. It was shown that deletion of the AP-1 site prevents looping and protein binding to the DNA, suggesting that binding of Jun protein stabilises the PIC complex.

AFM has been used to visualise enzymatic interactions involved in replication and degradation of DNA. Images have been obtained showing the binding of a DNA polymerase to ssDNA (Yang *et al* 1992) and DNA replication has been observed (Hansma, 1996; Argaman *et al* 1997). Phase imaging has allowed the time dependent replication of DNA to be visualised directly in a buffered environment (Argaman *et al* 1997). The ssDNA appears as globular structures attributed to base pairing within parts of the molecule. With time elongated dsDNA molecules appear in the images. Another exciting example of real time imaging is the visualisation of enzymatic degradation of DNA with the nuclease DNase I (Bezanilla *et al* 1994; Hansma, 1996; Ikai, 1996).

The tertiary structure of DNA may be of importance in protein binding. AFM has provided evidence for the importance of such effects in the regulation of heat-shock protein production (Ohta *et al* 1996).The *Staphyloccus aureus* heat shock operon (HSP70) contains an inverted-repeat sequence, between the promoter and the first structural protein (ORF37) gene, designated CIRCE (controlling inverted repeat of chaperone expression) which is believed to form a stem-loop structure. AFM images of the promoter region have revealed a protrusion, positioned correctly for the CIRCE region, and interpreted as a novel pairing of stem-loops. The ORF37 protein has been shown to bind to this structure, providing a molecular basis for a 'feed-back' process of regulation of heat shock gene expression (Ohta *et al* 1996).

4.2.4. Location and mapping of specific sites

The tagging of specific sites on DNA offers a route to locating and mapping these regions. This is just the type of microscopic study which should be well suited to the use of AFM. Combination of such labelling with a rapid and automated procedure for collecting and analysing images, of the type described earlier for sizing DNA (section 4.2.2), would provide a powerful physical tool for mapping DNA. A variety of tags can be envisaged, some of which have been studied experimentally. Labelling may involve chemical modification, hybridisation of short oligomers to create triple helical regions (Fang *et al* 1998) or the binding of site specific proteins. Certainly the binding of sequence specific proteins (Schleif and Hirsh, 1980) and the identification of probes hybridised to ssDNA (Wu and Davidson, 1975) have been used as aids in the mapping of DNA by TEM. As discussed earlier, proteins can be imaged bound to DNA and thus can be used as markers. In the case of hydrolytic enzymes it may be possible to modify them at their active sites to allow binding but not cleavage. An example of such a study is the mapping of individual cosmid DNAs (Fig. 4.14) by direct AFM imaging (Allison *et al* 1997). It has been shown that wild type EcoRI endonuclease can be visualised bound to DNA and the precision of the mapping using mutant enzymes, able to bind specifically but not cleave, has been assessed by mapping well characterised λ DNA, and then applied for mapping a cosmid clone.

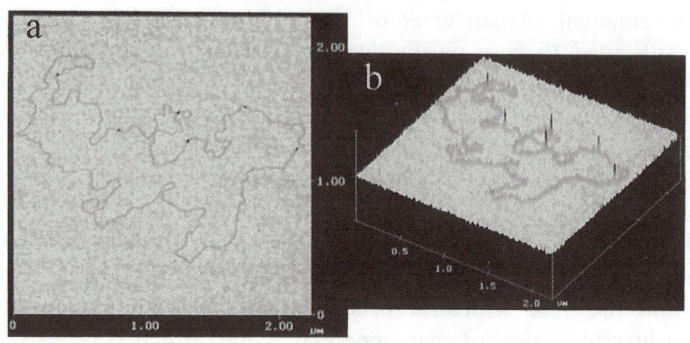

Figure 4.14. Mapping of a 35 kb cosmid clone with a mutated *Eco*RI endonuclease. (a) Image showing the 6 *Eco*RI binding sites. (b) Computer-generated image emphasising the binding sites. Data provided by the authors and publisher and based on Allison *et al* 1997.

The use of biotin labelled nucleotides has been used as a basis for binding sequence specific tags to DNA (Murray *et al* 1993; Shaiu *et al* 1993a,b). Streptavadin gold conjugates have been used as labels to target biotinylated nucleotides incorporated into linear or cyclic DNA (Shaiu et al 1993a,b). A chimeric protein fusion between streptavadin and two IgG binding domains of *Staphylococcal* protein A was used as a label to identify biotin labels at the ends of DNA fragments (Murray *et al* 1993).

Many proteins involved in transcriptional regulation of gene expression have been cloned and their functionalities are well characterised. Such proteins can be used as probes for specific sequences of nucleotides. An example is the investigation by AFM of the binding of protein Ap2 to DNA (Nettikadan *et al* 1996). The distance of these binding sites from the ends of the molecule, as measured from the AFM images, were found to be consistent with the known nucleotide sequence data. A low level of non-specific binding was also noted in these studies.

Antibodies also provide high specificity. An anti-Z DNA IgG antibody has been used to label the left handed Z DNA conformation of a d(CG)$_{11}$ insert in a negatively supercoiled DNA plasmid (pAN022) for imaging by AFM. The antibody binding caused bending of the DNA, and the position of the bound antibodies was consistent with known nucleotide sequence, and the adoption of the B DNA helix (Pietrasanta *et al* 1996). Location and mapping on chromosomes is discussed later in section 4.2.5.

The pros and cons of using SPM methods for sequencing DNA are discussed elsewhere (Hansma and Hansma, 1993; Morris, 1994). The prospects for automating the AFM procedures seems good but, at present, the resolution achievable with AFM would need to be improved by at least an order of magnitude in order to achieve the level required for sequencing. Although AFM has good prospects as a tool for high resolution mapping it seems unlikely that it will challenge standard biochemical methods of sequencing in the near future. AFM probes have, however, been used to dissect DNA molecules into smaller fragments (Henderson, 1992; Hamsma *et al* 1992b). As will be discussed later (section 4.2.5) there are suggestions that, following such dissection, DNA fragments become attached to the probe and can be isolated or transferred by such methods. Related studies of this kind on chromosomes are discussed in the next section (section 4.2.5).

Even if AFM is not used for DNA sequencing then it may have applications related to gene therapy. It has been shown that AFM has potential for assaying DNA condensation, which is important for different modes of gene therapy. Studies have been reported on glycoprotein-polylysine complexes of DNA designed to target the asialoglycoprotein receptor of liver cells (Hansma *et al* 1998). A variety of shapes were seen to be adopted by the DNA, including short rods and toroids. These studies show that AFM can be used to assess optimum condensation, potentially related to the efficiency of uptake by the receptor. It has been suggested that AFM could also be used to assay DNA condensation with liposomes (Hansma *et al* 1998) although no such studies have been reported at the present time.

4.2.5. Chromosomes

Cytogenetics, the study of chromosomes, is basically a visual science. Established microscopic methods, such as light and electron microscopy, have been widely used to study chromosomes. The use of hypotonic methods results in good spreading of chromosome preparations allowing the determination of the number and morphology. Introduction of banding and hybridisation techniques has permitted the identification and mapping of chromosomes. Goals in cytogenetics are improved specification and localisation of probes, and more rapid analysis. AFM has potential in both these areas and has joined the range of microscopic techniques used to study chromosomes. The chromosomes are normally spread on glass and can be imaged in air or under liquid.

Polytene chromosomes are an amplified form of interphase chromosomes present in the nuclei of specific dipteran cells. They are substantially larger than normal mammalian metaphase chromosomes and contain densely condensed banded regions, and loosely condensed interbanded and puff regions. These features can be imaged in the AFM, the bands appearing as high regions, and the interbands and puffs as low or flat regions (Li *et al* 1996; Jondle *et al* 1995; Mosher *et al* 1995; Vesenka *et al* 1995; Puppels *et al* 1992). The resolution obtained depends on the type of tip used, but features as small as 1 nm have been reported in some images (Jondle *et al* 1995). Colloidal gold as a calibration standard has been used for reconstructing images of chromatin (Fritzsche *et al* 1996). By increasing the imaging force it has been possible to dissect polytene chromosomes. The image quality was found to decrease after dissection, and this was attributed to the adsorption of chromatin fragments onto the tip. It is reported that DNA can be recovered from the tip and amplified by PCR methods (Jondle *et al* 1995). More detailed and controlled studies of this type have been made on metaphase chromosomes.

Metaphase chromosomes imaged by AFM have revealed structures similar to those reported by light and electron microscopy. Cytogenetic abnormalities, expressed as changes in the length of the chromosome (McMaster *et al* 1994), or bi-armaed chromosomes (Oberleithner *et al* 1996), which are believed to play crucial roles in tumor development, have been identified by AFM. AFM has been used to image human (Heckl, 1992) and Chinese hamster (DeGrooth and Putman, 1992) chromosomes. Metaphase chromosomes consist of a 30 nm fibre folded to form a tandem array of radial loops, which are packaged into a fibre of overall diameter between 200-250 nm. High resolution AFM images of the surface of the chromosomes revealed structural features in the size range of 30-100 nm which may correspond to the loops of the 30 nm fibre (DeGrooth and Putman, 1992). Other authors (Winfield *et al* 1995; McMaster *et al* 1996a,b) have reported features as small as 10-20 nm which could correspond to individual nucleosomes. Banding is used to classify, or karyotype, chromosomes by microscopy. AFM images of plant chromosomes revealed that the C and N bands are observed as regions of high relief on a slightly collapsed chromosome structure (McMaster *et al* 1996a,b; Winfield *et al* 1995). It has been possible to identify features equivalent to G banding patterns in untreated human metaphase chromosomes (Musio *et al* 1994) and to use this for classifying the chromosomes. This suggests that, at least in this instance, the high resolution of the AFM has allowed an intrinsic banding pattern to be visualised which, for viewing by other microscopic methods needs to be enhanced by accumulation of stain. Combined

AFM and SNOM studies have been made on banded human chromosomes demonstrating the higher resolution achievable using near field, rather than far field optical microscopy for chromosome classification (Wiegrabe *et al* 1997). Although chemical banding is the basis of karyotyping it has been shown by EM that the relative volume of metaphase chromosomes is chromosome specific, and thus can be used for classification (Heslop-Harrison *et al* 1989). The volume of plant (McMaster *et al* 1996a) and human chromosomes (Fritzsche and Henderson, 1996c) have been measured using AFM. The height of air dried chromosomes increases with rehydration in aqueous buffer (DeGroth and Putman, 1992; Fritzsche *et al* 1994; 1996c; Rasch *et al* 1993), and this swelling is accompanied by changes in their viscoelastic properties, as measured by AFM (Fritzsche and Henderson, 1996b). For plant chromosomes it has been shown that the volumes determined on dried specimens can be used for classification (McMaster *et al* 1996a). Hybridisation techniques have been developed for mapping chromosomes. AFM has been used for localisation of an *in situ* hybridisation probe on a human chromosome (Putman *et al* 1993b; Rasch *et al* 1993). Biotinylated DNA probes were used for mapping, and the specific sites visualised by detecting the changes in topography induced by a peroxidase-diaminebenzidine reaction. In studies on cereal chromosomes the D genome specific probe pAs 1 was detected using AFM to image changes in height due to biotin-avidin-fluorescein isothiocyanate complexes formed as a consequence of using a standard fluorescent *in situ* hydridisation (FISH) procedure (McMaster *et al* 1996). SNOM images of FISH studies of human chromosomes have demonstrated detection of the maxima in fluorescence at nanometre resolution raising the possibility of detecting individual probe molecules (van Hulst *et al* 1997).

The ability to directly dissect and collect DNA from cytogenetically recognisable regions on chromosomes offers the potential for producing *in situ* hydridisation probes, the generation of band specific libraries for chromosomes, and mapping of chromosomes using such probes. It has been demonstrated that AFM can be used as a microdissection tool for extracting DNA from selected sites on human chromosomes (Thalhammer *et al* 1997). Microdissection of chromosome 2 was achieved by raising the imaging force for a single line scan (Fig 4.15). The resultant DNA attached to the probe was isolated, amplified by a modified PCR procedure and used, via a FISH study, to demonstrate specificity for the cut region of chromosome 2. The use of AFM offers the prospect of isolating smaller probes than those obtainable with current methods such as dissection with glass needles or laser cutting, allowing a more detailed mapping of chromosomes.

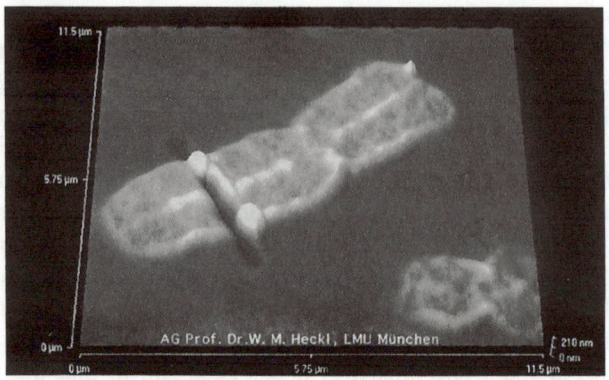

Figure 4.15. Microdissection of a human chromosome 2 using the AFM. Image provided by Prof. W.M. Heckl and reprinted with permission. Methodology is as described by Thalhammer *et al* 1997.

4.3. Nucleic acids: RNA

RNA is the Cinderella of the nucleic acids: far less data is available possibly due to the much greater difficulties involved in the handling of RNA, and in adsorbing RNA to mica. RNA is an attractive candidate for study by AFM because of the complex three-dimensional structures adopted by RNAs and RNA-protein complexes. Images of RNA released by osmotically shocked viruses, such as influenza virus (Shao and Zhang, 1996), have been obtained by cryo-AFM. Early studies describe methodology for binding RNA to chemically modified mica (Lyubchenko *et al* 1993a) and report studies on reovirus dsRNA (Lyubchenko *et al* 1993ab). The molecules were found to show convoluted shapes and, in some instances, compact structures. Data has been reported which is interpreted as showing ternary structure for ribosomal RNA (Wu and Liu, 1997). AFM images of ssRNA deposited from water onto mica showed a globular structure. Treatment with formamide and deposition onto silanised mica captured the linear denatured structure (Fritz *et al* 1997).   Data has also been reported for polyA and polyG ssRNA (Smith *et al* 1997; Hansma *et al* 1996) using the Tapping mode for air dried samples on mica. The polyA samples show a heterogeneous population of sizes with 'blob-like' structures located at the ends of the RNA strands. The heights of the 'blobs' are larger than the remaining ssRNA and may represent regions of base-pairing or helical RNA. PolyG samples were very different in appearance, showing amorphous 'gel-like' aggregates. Similar amorphous aggregates are reported for mRNA and this has prevented studies of the binding of polyA binding protein (PABP) to the polyA tails of mRNA (Smith *et al* 1997). PABP-RNA binding has been observed for polyA; PABP was observed to bind randomly, and the 'blobby' features seen on the isolated RNA disappeared, suggesting an alteration of the RNA structure on protein binding (Smith *et al* 1997). Formation of protein dsRNA complexes are important molecular events in host defence against viral infections. The vaccina virus inhibitor (protein p25) of protein kinase (PKR) is characterised by unusual binding specificity to dsRNA. AFM studies of p25-dsRNA complexes have revealed binding of the protein to the RNA and evidence for p25 induced aggregation of the RNA. The authors (Lyubchenko *et al* 1995) speculate that such condensation may prevent PKR accessibility to dsRNA thus inhibiting the dsRNA dependent induction of interferon synthesis. Ternary complexes of both linear and cyclic DNA with RNA polymerase and the nascent RNA have been imaged (Kasas *et al* 1997a,b; Hansma *et al* 1993a; Rees *et al* 1993; Zenhaussern *et al* 1992b) and the transcription process observed in real time (Fig. 4.13) (Kasas *et al* 1997b). Protein-RNA complexes have also been visualised by AFM for transcriptional complexes (Fritzsche *et al* 1997) prepared by hypotonic spreading from *Pleurodeles* oocytes, and in ribosomal-RNA polysomes hypotonically spread from *Chironomus pallidivittatus* salivary gland cells (Fritzsche *et al* 1997).

AFM has been used to image intercellular RNA after cytochemical detection (Kalle *et al* 1996). Rat 9G and HeLa cells were hybridised with haptenised probes for ribosomal and messenger RNA. These were detected with peroxidase labelled antibody and visualised with diaminebenzidine (DAB). The DAB precipitate can be detected by AFM, although the resolution is just marginally better than that obtained by optical microscopy, even on dried cells. Crystallisation of tRNA has been followed by AFM (Ng *et al* 1997) and this is discussed further in section 6.1.3.

4.4. Polysaccharides

Polysaccharides are ideal candidates for study by AFM. There are good reasons for studying polysaccharides. They are vital structural components of plant cell walls, starch and other cell wall polysaccharides are important food components, and polysaccharide extracts from bacterial, land plant and algal sources are widely used industrially, with cellulose being perhaps the most important example. Many polysaccharides are of importance in understanding the development and treatment of disease: particular examples are the role of hyaluronic acid in the stiffening of joints with age and as a consequence of diseases such as rheumatoid arthritis, the role of bacterial alginate in cystic fibrosis, and the continuing use and development of bacterial polysaccharides as vaccines. Polysaccharides secreted by bacteria are of wide interest because they are important virulence factors for plant and human pathogens, they promote bacterial adhesion to surfaces and, in addition to medical uses as vaccines, these polysaccharides are used commercially as industrial polymers. Like nucleic acids polysaccharides are fibrous polymers but, unlike nucleic acids, they are often complex irregular structures. The structures may be branched or multiply branched. There are examples of natural block copolymers, and minor substituents in polysaccharide chains can radically alter their shape, conformation, or mode of interaction with other biopolymers. Most of these heterogeneous features are difficult to study by current biophysical methods and probe microscopy should play an important role in this area.

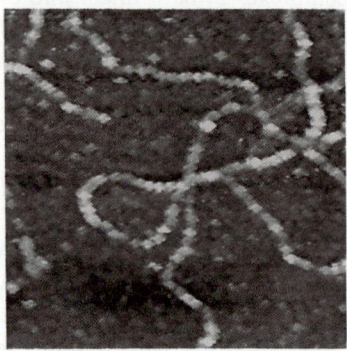

Figure 4.16. AFM image of a metal coated xanthan polysaccharide. Scan size 700 x 700 nm. The image shows the restricted resolution caused by the finite size of the metal grains.

4.4.1. Imaging polysaccharides

The earliest probe microscopy studies on polysaccharides used STM. These data were generally fairly poor, and failed to produce new information on polysaccharide structure (Morris, 1994). The challenges to imaging polysaccharides by AFM are similar to those faced in imaging nucleic acids. The methodology developed to image polysaccharides parallels the methods developed for nucleic acids. The major problem which had to be solved in order to obtain reproducible, reliable images was the need to eliminate adhesive forces, and the resultant damage and displacement of the molecules. The difficulties arose because the molecules were either sprayed or drop deposited onto mica, air dried and then imaged in air. The simplest solution, first developed for STM studies

(Wilkins *et al* 1993) and then applied to AFM studies (Gunning *et al* 1995; Kirby *et al* 1996a), was to deposit the molecules onto mica either by spraying (Wilkins *et al* 1993) or drop deposition (Gunning *et al* 1995; Kirby *et al* 1996a) and then to metal coat the sample. The metal coating prevents damage or displacement of the molecules. The samples can be stored and imaged at leisure, and provide information on molecular size and shape (Fig. 4.2). The major disadvantages of this approach are the need to evacuate the samples and the limitation on resolution imposed by the grain size of the metallic coating (Fig. 4.16). The next simplest solution is to image under a liquid thus eliminating the adhesive forces. As with nucleic acids it is convenient to image under liquids such as alcohols which inhibit desorption of the molecules from the substrate (Kirby *et al* 1996a). Provided the applied normal force is controlled, in order to optimise image contrast and to prevent molecular damage or displacement (section 4.1.5), then high quality images (figure 4.4c) can be obtained reliably and reproducibly. Image quality slowly decays with time due the gradual appearance of an adhesive force as the probe tip accumulates debris from the sample surface (section 4.1.5, figure 4.5b). An alternative method for imaging polysaccharides is to use non-contact ac methods. Samples sprayed or drop deposited onto mica, and then air dried, can be imaged in air by Tapping or non-contact mode atomic force microscopy (NCAFM) (Gunning *et al* 1996a; Cowman *et al* 1998a; McIntire and Brant, 1997a). By using Tapping it is also possible to image polysaccharides in an aqueous or buffered environments. Fig. 4.9 shows an image of a fibrous water soluble pentosan deposited onto mica and imaged by Tapping under buffer. Close inspection of the image shows that parts of the polysaccharide chain appear to be missing (Figs. 4.9b,c). This is because the polysaccharide is actually trying to desorb from the surface and those parts (loops), which are sampling the bulk medium, are moving too fast to be imaged. The visible parts (trains) of the molecule are those sections which are still in contact with the mica. The result of this process is an apparent motion of the molecule on the mica surface. This is shown in Fig. 4.9 which shows several 'stills' from a sequence of images taken at different times. With time different sections of the molecule desorb, and then re-absorb in a different position, leading to changes in shape of the molecule. Eventually the molecule finally wins the struggle, and is able to completely desorb from the surface. This sequence of images can be linked together in order to produce a molecular movie of this molecular motion. The ability to image polysaccharide molecules under natural conditions offers the prospect for studying molecular interactions, or processes such as the enzymatic breakdown of complex carbohydrate structures.

4.4.2. Size, shape, structure and conformation

Two main methods are used for imaging polysaccharides. In the first case the molecules are dropped onto mica, air dried and then imaged in the dc contact mode under alcohols (Kirby *et al* 1996a). In the second case the molecules are sprayed onto mica, air dried and then imaged in the ac non contact mode (McIntire and Brant, 1997a). In the instances where both techniques have been used to study the same systems there has been good agreement (Kirby *et al* 1996a; Gunning et al 1996b, 1997; McIntire and Brant, 1997a,b).

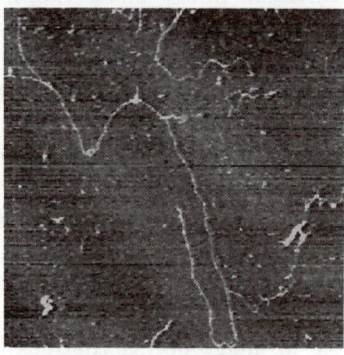

Figure 4.17. AFM image of a high methoxyl pectin molecule showing evidence of branching. Scan size 2 x 2 μm. The molecule was imaged on mice under butanol.

The AFM provides information on the shape and size of the molecules. Random coil polysaccharides such as dextrans are highly mobile and the AFM images (Tasker *et al* 1996; Frazier *et al* 1997a,b) show globular structures. It is possible to image dextrans passively adsorbed to mica although the published images are of thiolated derivatives bound as monolayers. The size of the 'blobs' varies as expected with molecular weight (Tasker *et al* 1996) and swelling/deswelling was observed with hydration/dehydration of the monolayers (Frazier *et al* 1997a). AFM studies have been used to complement 'in situ' monitoring of dextran degradation with dextranase (Frazier *et al* 1997b).

As the persistence length of the molecules increases the molecules become more extended, less mobile and easier to image. The images reveal more detail about the molecular structure. For pectic polysaccharides, generally considered to be linear polymers, it is possible to show that a percentage of the polymers are actually branched structures (Round *et al* 1997). The level of branching detected is too low to be detectable by conventional chemical or enzymatic methods of structural analysis, so the AFM images are providing new information on molecular structure. Figure 4.17 is an image of an exceptionally high molecular weight, high methoxyl pectin showing the presence of branches.

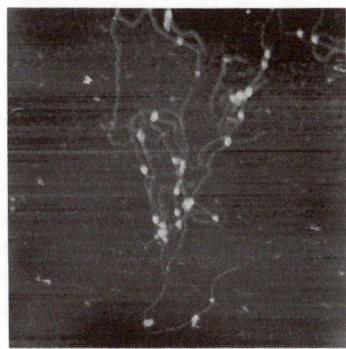 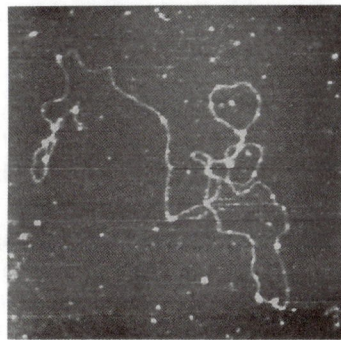

Figure 4.18. AFM image of entangled hyaluronan molecules. (Left) Deposited onto mica and imaged under butanol. Scan size 2 x 2 μm. Image obtained in authors laboratory. (Right) Tapping mode in air image of an hyaluronan chain showing simple cross points. Scan size 1.2 x 1.2 μm. Data reproduced from Cowman *et al* 1998a with permission of the authors and the Biophysical Society.

The images obtained can be analysed in detail to give information such as distributions of contour length, branch length, number of branches per molecule, or branch separation, and these data can be used to investigate the effects of chemical or enzymatic modifications of the pectins. Pectins are important constituents of the plant cell wall and play important roles in elongation and growth. A detailed understanding of pectin structure and interaction is of importance in the construction of molecular models of cell walls and for explaining their biological function. There are a few studies on carrageenans, an analogous structural polysaccharide found in algal cell walls (McIntire and Brant, 1997b; Kirby *et al* 1995; Morris *et al* 1997; Gunning *et al* 1998). Some of these studies show individual molecules (McIntire and Brant, 1997b) but the others (Kirby *et al* 1996a; Morris *et al* 1997; Gunning *et al* 1998) reveal aggregated structures which are dealt with in section 4.4.3. The published images of individual kappa carrageenan molecules (McIntire and Brant, 1997b) are highly extended, even though in the sol state prior to deposition they should be in the denatured 'coil' state suggesting, either that spray deposition extends the molecules, or that they reform the helical state on deposition. Images of the extracellular matrix polysaccharide hyaluronan can be obtained after deposition onto mica either by dc contact mode imaging under butanol (Fig. 4.18) or by Tapping mode in air (Fig. 4.18) (Cowman *et al* 1998a,b). At moderately high stock concentrations the molecules form network structures indicative of the tendency of the molecules to associate in solution (Gunning *et al* 1996c). At sufficiently high dilutions individual stiff elongated structures can be seen (Cowman *et al* 1996a,b). It is also possible to image a variety of more complex structures corresponding to various levels of self-association (Cowman *et al* 1996a).

The polysaccharides which have been studied most are the very stiff helical bacterial polysaccharides: xanthan (Meyer *et al* 1992; Gunning *et al* 1995; 1996a; Kirby *et al* 1995a; 1996a; McIntire and Brant, 1997b; Capron *et al* 1998), acetan (Gunning *et al* 1995; Kirby *et al* 1995b; 1996a), the acetan variant CR1/4 (Ridout *et al* 1998), gellan (Gunning *et al* 1996b; 1997; McIntire and Brant, 1997a,b) and scleroglucan (schizophyllan) (McIntire *et al* 1995; McIntire and Brant, 1997a,b; 1998; Brant and McIntire, 1996). Apart from the early studies of Meyer and coworkers (Meyer *et al* 1992), which showed periodic structures with no clearly identifiable individual molecules, the recent studies on xanthan, acetan, CR1/4 and scleroglucan show distributions of highly extended stiff fibrous biopolymers. As well as just imaging the molecules it is possible to quantify the data by generating contour length distributions for the polysaccharides (McIntire and Brant, 1997b; 1998; Ridout *et al* 1998) providing information on molecular size and polydispersity. If the conformation of the molecule is known, then its mass per unit length is known, and the contour length distributions can be converted into molecular weight distributions. Conversely, if the conformation is not known, then the AFM data can be combined with light scattering data to determine the mass per unit length, and hence the molecular conformation. This approach was used to demonstrate a double helical conformation for the polysaccharide CR1/4 (Ridout *et al* 1998). Fairly accurate data can be obtained by imaging several hundred molecules and, although the measurement of contour lengths is tedious, the whole process is probably as quick as the use of gel permeation chromatography (gpc). The use of gpc is often unreliable for many of these polysaccharides which are prone to aggregation. As has been mentioned for hyaluronan (Cowman *et al* 1996a) it is possible to monitor and analyse unusual structures formed by intra-molecular association (McIntire *et al* 1995; Brant and McIntire, 1996; McIntire and Brant, 1997a,b; 1998). In particular AFM has been used to monitor, record and analyse the linear triple helix to circular triple helix transition for scleroglucan (McIntire *et al* 1995; Brant and McIntire, 1996; McIntire and Brant, 1997a,b; 1998). Fig. 4.19 shows a mixture of both linear and cyclic scleroglucan together with some hairpin structures. These types of cyclic structures are seen with polysaccharides which can adopt multiple helical structures. If the helical structures are first denatured, and then allowed to reform, then either intra- or inter-molecular association may occur. The cyclic structures result from intra-molecular association (Brant and McIntire, 1996; McIntire and Brant, 1998). Inter-molecular association will be discussed later in section 4.4.3.

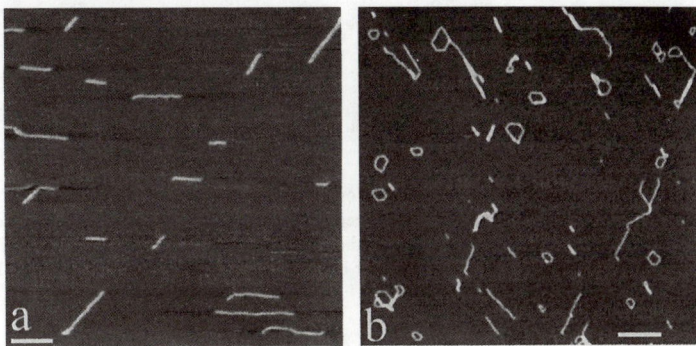

Figure 4.19. AFM images showing the transition between (a) extended helical and (b) cyclic forms of the polysaccharide scleroglucan. (a) Scale bar 200 nm. Data shown was provided by McIntire and Brant and published with permission.

The elongated structures of these bacterial polysaccharides suggest that they retain their helical structure when deposited onto mica. At the present time no detailed analysis has been made to determine whether meaningful values for persistence length can be extracted from such images. As for DNA measurements of molecular width are generally too large, due probe broadening effects, and can not be used to determine the conformation of the polysaccharide. If reasonable estimates are made for the size and shape of the probe tip, then the measured widths for polysaccharides such as xanthan are consistent with the dimensions expected for the helical form. If arrays of polysaccharides are imaged then the periodicity of the 'lattice' is measured. In this case the probe broadening effect disappears and the true width of the molecules is measured (Kirby *et al* 1995b). The widths measured for the acetan polysaccharide were consistent with the expected diameter of the helix. In addition it was also possible to observe (Fig. 4.21a) a periodicity along the polysaccharide molecules consistent with the pitch of the known helical structure (Kirby *et al* 1995b). Similar results can be obtained for aggregates of other bacterial polysaccharides such as xanthan (Fig. 4.21b)

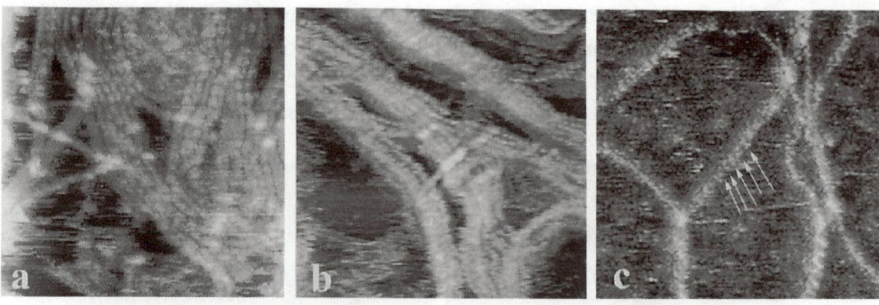

Figure 4.21. AFM images of polysaccharides showing evidence of the helical structure. (a) Acetan microgel scan size 140 x 140 nm, air dried and imaged under butanol. (b) Xanthen aggregate scan size 200 x 200 nm, air dried and imaged under 1,2 propanediol. (c) Individual xanthan molecules scan size 135 x 135 nm, air dried, evacuated and then imaged under butanol. Turns of the helix are indicated by the white arrows.

It is occasionally possible to observe the helical structure on individual polysaccharides (Fig. 4.21c). This tends to happen when 'double tipping' occurs. Presumably the sharper points on the end of the tip responsible for 'double tipping' allow higher resolution images of the polysaccharide chain. In the absence of direct visual evidence for a helical structure then measurements of heights are generally more realistic tests of the presence of the polysaccharide helix than measured widths. Typically height values of the order of about 66% of the expected helical diameter are usually obtained (Kirby *et al* 1996a; McIntire and Brant, 1997b). The reason for this effect is still not known, but it is undoubtedly the same as that seen with nucleic acids (section 4.2.2). The most likely origins of this effect are the compression of the molecule during scanning, or the breakdown of the assumption that imaging at constant force directly yields heights of molecules. If the charge density on the molecules is different to that of the substrate then the displacement of the probe to maintain constant cantilever deflection (constant force) may not be directly relatable to molecular thickness. Recent publications (Muller and Engel, 1997; Muller *et al* 1999) address these issues in more detail.

Measurement of force-distance curves for individual molecules provides a measure of molecular elasticity, and theories for worm-like coils (Bouchiat *et al* 1999) are available for calculating persistence lengths. Such data has been reported for polysaccharides such as native and derivatised dextrans (Rief *et al* 1997; Li, H *et al* 1998; Frank and Belfort, 1997), cellulose derivatives (Li, H *et al* 1998), extracellular polysaccharide from *Pseudomonas atlantica* (Frank and Belfort, 1997) and xanthan (Li, H *et al* 1998). Analysis of the data on dextrans suggested Kuhn lengths ($L_K = 2L_P$) of 0.6 nm (Rief *et al* 1997). In the case of xanthan, measurements were made on native and denatured xanthan. Denatured xanthan behaved similarly to the structurally related carboxymethyl cellulose, whereas native xanthan showed quite different behaviour, which was attributed to helix formation (Li, H *et al* 1998). In recent investigations of the force-distance curves of a number of polysaccharides (amylose, pullulan, dextran, pectin and methylcellulose) the role of the pyranose ring in determining the elasticity of the polymer chain has been considered (Marszalek *et al* 1998). For dextran, pullulan and amylose it was shown that the enthalpic component of the elasticity was eliminated by periodate cleavage of the glucose rings, suggesting that force-induced distortion of the sugar ring and a chair-boat conformational transition determine this contribution. In the case of polysaccharides such as pectin or amylose which adopt ordered helical secondary structures periodate oxidation will inhibit helix formation and this may contribute to the loss of the enthalpic contribution. Whereas polysaccharides such as dextran, pullulan, amylose, and pectin all show enthalpic contributions to their elasticity carboxymethyl cellulose and methylcellulose behave entropically (Li, H *et al* 1998; Marszalek *et al* 1998). In the cellulose derivatives the pyranose rings are already fully extended and cannot contribute further to extension of the chain.

4.4.3. Aggregates, networks and gels

Interactions between polysaccharides play an important role in determining the functional behaviour of the polysaccharides in their native biological state or when used as industrial additives. AFM provides a useful means of studying such interactions. Thus the complex structures formed by intra-molecular association of hyaluronan (Cowman *et al* 1996a) are probably alternatives to the more complex inter-molecular association formed at higher concentrations (Gunning *et al* 1996c). Hyluronan is responsible for the useful viscoelasticity and lubrication of joints. In diseases such as arthritis the molecular structure is broken down. One form of proposed treatment is the injection of hylan, a cross-linked hyaluronan 'microgel'. This increases the viscoelasticity of the joint and, by their very nature, the cross-linked structures are more resistant to breakdown. Fig. 4.22 shows AFM images believed to be a hylan microgel (Gunning *et al* 1996c). The image shows a complex point cross-linked network structure with parts of individual hyaluronan molecules extending from the aggregated microgel. Here the AFM is providing new information on a complex heterogeneous aggregate which could not easily be obtained by other means. Furthermore, AFM could be used to monitor the breakdown of such structures with time during trials of their use as a treatment for arthritis. Similar types of structures are formed by bacterial polysaccharides such as xanthan (Morris, 1998a; Morris *et al* 1999) and are responsible for determining the solubility and thixotropy important for their use as thickening and suspending agents. A number of AFM studies on xanthan have reported evidence for molecular aggregation (McIntire and Brant, 1997b; Capron *et al* 1998) and it is generally quite difficult to prepare true solutions of polysaccharides such as xanthan: usually aggregates have to be removed by centrifugation or filtration.

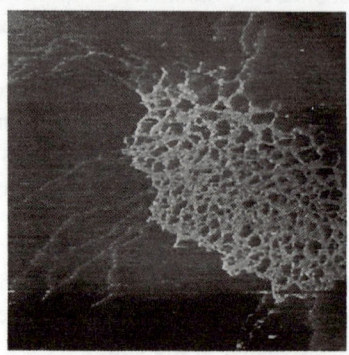

Figure 4.22. AFM images of a Hylan aggregate. The image shows individual hyaluronan molecules extending out from the cross-linked structure. Sample was air dried onto mica and then imaged in the dc contact mode under butanol. Scan size 1.8 x 1.8 μm.

Figures 4.5c and 4.23 shows AFM images of xanthan prepared under different conditions. In Fig. 4.5c the xanthan has been dried down onto mica as an entangled solution of individual molecules. The ends of the molecules can be seen and the bright spots in the image show the doubling of height when one molecule lies on top of another. Figure 4.23 shows an image of a xanthan microgel. These structures are formed when the polysaccharide is concentrated, precipitated or dried under conditions where the helical structure is partially denatured. The process of concentration promotes further helix formation and inter-molecular rather than intra-molecular linkages are probably formed. The result is the microgel particles shown in Fig. 4.23. The solubility of xanthan is determined by the ease of swelling of these particles. Aqueous xanthan preparations are actually dispersions of these swollen particles and it is this which dictates the rheology of the sample. Despite elaborate attempts to clarify acetan samples it was the presence of small amounts of microgel which gave rise to the molecular arrays which allowed imaging of the acetan helix (Kirby *et al* 1995b). The microgels are a continuous branched aggregates within which it is difficult to locate ends of individual molecules (Fig. 4.21a). The observation of the helical structure in the acetan microgels suggests that helix formation is the molecular basis of the intermolecular association.

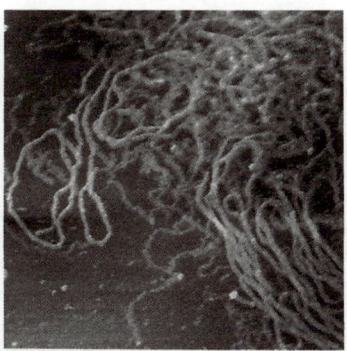

Figure 4.23. AFM image of a xanthan microgel particle, air dried onto mica and imaged in the dc contact mode under butanol. Scan size 1.4 x 1.4 µm. Figure 4.21b shows that the aggregate is based on interacting helical molecules.

Similar forms of aggregation are believed to be responsible for the gelation of polysaccharides. Gels are often considered as useful models for the natural biological roles of these polysaccharides. The slimy extracellular polysaccharides secreted by bacteria play roles in preventing dehydration and in assisting adhesion to surfaces. In plants polysaccharides play an important role in determining the structural integrity of the cell wall. It is possible to follow this type of polysaccharide-polysaccharide association, to test mechanisms of gelation and to investigate the long range structures of gels. This type of study is best illustrated through studies on the gelation of gellan gum (Gunning *et al* 1996b; 1997). The strong tendency for gellan molecules to associate means that it is in fact quite difficult to image individual molecules: AFM studies of gellan sols generally show aggregated structures (Gunning *et al* 1996b; 1997; McIntire and Brant, 1997a,b; Morris, 1998b). This can be turned to advantage and used to investigate gelation mechanisms. By studying such aggregates (Fig. 4.24) it is possible to investigate the effects of the 'coil-helix' transition and selective cation binding on molecular association (Gunning *et al* 1996b; 1997; Morris, 1998b). In the absence of gel-promoting counterions helix formation alone results in intermolecular association and aggregates (filaments) of constant height and width (Fig. 4.24a). When gel-promoting cations are present side-by-side aggregation of these filaments occurs resulting in branched fibrous structures of variable width and height (Fig. 4.24b). Deposition onto mica from more concentrated sols results in the formation of aqueous films. The network structures formed in these films can be visualised

(Gunning *et al* 1995; 1996b; 1997; Morris, 1998b,c) (Fig. 4.24c) and can be regarded as models for structures formed in 3D gels.

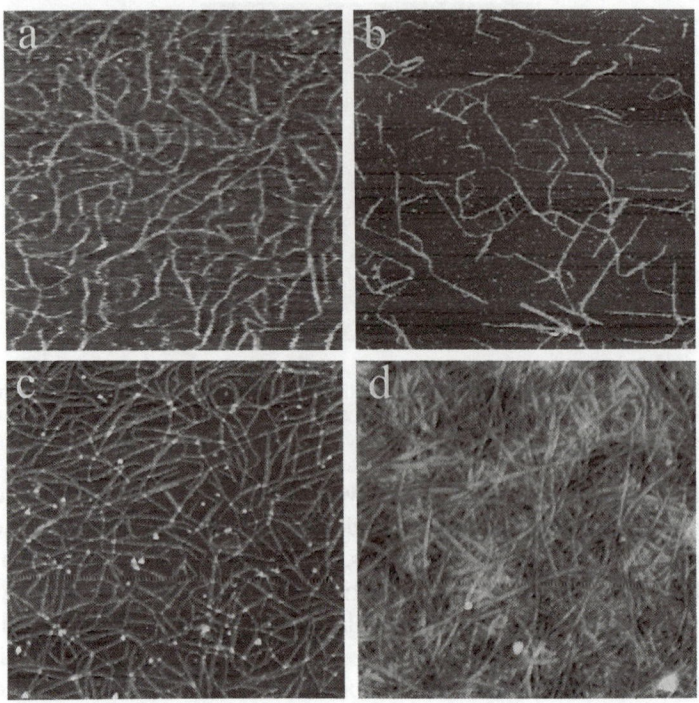

Figure 4.24. AFM images of gellan gel precursors, gellan network and gel. (a) Aggregates of TMA gellan molecules. Scan size 1 x 1 μm. (b) Gel precursors formed from potassium gellan. Scan size 800 x 800 nm. (c) Aqueous gellan network. Scan size 800 x 800 nm. (d) Aqueous acid set gellan gel. Scan size 2 x 2 μm. The samples a-c were deposited from aqueous sols onto mica, air dried and then imaged under butanol. For sample d the gel was set on mica and imaged wet under butanol. Data reproduced from Gunning *et al* 1996b; 1997 with permission.

The networks are continuous branched fibrous structures consistent with the modes of association deduced from the studies on the gel precursors. The only ends of molecules which can be seen are short embryonic stubs which have not grown into larger branches. Imaging 3D gels is more difficult because the gel surfaces appear to deform during scanning blurring the images. Gellan gels can be prepared directly on the mica and the gel surface imaged under butanol. Scanning the gels rapidly makes it is possible to minimise distortion and reveal some details of the molecular structure of the gel (Morris, 1998b), although the images are poor showing periodic striations; a typical artifact of scanning too quickly. By setting gellan gels at acid pH it is possible to produce very stiff gels (1.2% gel, shear modulus $\approx 10^4$ Pa). For these stiff gels the distortion is minimal and the molecular network within the hydrated gel can be seen (Gunning *et al* 1996b; 1997; Morris, 1998b) (Fig. 4.24d). The fibrous network observed in the hydrated gel is similar to that seen in the hydrated film. The use of AFM has confirmed mechanisms of gellan gelation deduced from physical chemical studies, and provided new insights in the long range structure within the gels and the molecular origins of the elasticity of the gels (Gunning *et al* 1996b; 1997; Morris, 1998b). Gellan can be considered as a model system for studying thermoreversible polysaccharide gels and the behaviour of the gellan system appears to be typical of other polysaccharide systems. Thus reported studies on iota and kappa carrageenans also reveal aggregated gel-precursors and aqueous films containing fibrous networks (Kirby *et al* 1996a; Morris *et al* 1997; Morris *et al* 1998b,c). Semi-refined carrageenans are extracted from algal cell walls using a milder extraction process, which doesn't completely remove all of the cellulose component of the cell wall. The AFM images of semi-refined carrageenan (Gunning *et al* 1998) show interpenetrating networks in which the carrageenan and cellulose components can be easily distinguished (Fig. 4.25).

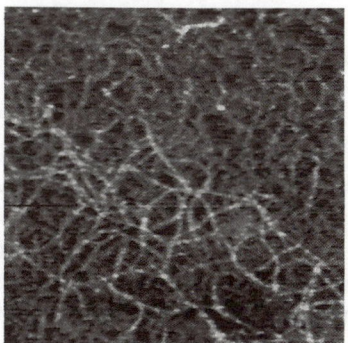

Figure 4.25. AFM image showing a mixed network of carrageenan and cellulose molecules. The cellulose molecules are stiffer and thicker than the carrageenan molecules, and appear brighter in the image. Scan size 700 x 700 nm. Reproduced with permission from Gunning *et al* 1998.

The stiffer, thicker cellulosic fragments of the disrupted cell wall appear brighter than the thinner carrageenan molecules. AFM has been used to image unperturbed gels of the related algal polysaccharide agarose under TBE (tris-borate-EDTA). The studies (Pernodet *et al* 1997; Maaloum *et al* 1998) revealed fibrous bundles of agarose but were mainly focussed on a detailed study of the pore distribution of the gels. The deduced variation of pore size with agarose concentration was found to be consistent with that deduced from electrophoretic mobility measurements using reptation models (Maaloum *et al* 1998). A detailed study of the surface roughness of synthetic polymer gels has been reported by Suzuki and coworkers (Suzuki *et al* 1996; 1997; 1998). These studies include observations of changes in surface structure attributable to temperature induced phase transitions of the gels. An alternative method for gelling polysaccharides is chemical cross linking. Power and coworkers have used AFM to study gels formed by cross linking hydroxypropylguar with borate (Power *et al* 1998a,b). Gels were air dried before imaging and then imaged using dc contact or Tapping mode under butanol or in air, and Tapping mode images in air were found to give the best images (Power *et al* 1998b). The authors report changes in gross morphology for gels prepared under quiescent conditions or under different shearing regimes. Given that the gels are supposed to be formed by point cross linking of a flexible polysaccharide, then the fibrous structures observed are surprising, even for the quiescent gels. The fibres between branch points appear to be too extended and too broad to be sections of individual polysaccharides. However, it is possible that in these xerogels the polysaccharide chains do become highly extended and rigid, and that the widths are artificially enhanced due to probe broadening effects

4.4.4. Cellulose, plant cell walls and starch

Cellulose

Cellulose is the main structural component of plant cell walls. It is also one of the few polysaccharides to be biosynthesised in a crystalline form. Within the cell wall it occurs as microfibrillar structures of varying degrees of crystallinity. There are a number of AFM studies on isolated cellulose fibres (Hanley *et al* 1992; 1997; Baker *et al* 1997; 1998; van der Wel *et al* 1996; Kuutti *et al* 1994) some of which report high resolution images and analysis of the surface structure (Hanley *et al* 1992; 1997; Baker *et al* 1997; 1998; Kuutti *et al* 1994). The sizes of the fibres are consistent with those observed by electron microscopy, provided probe broadening effects are taken into consideration. Comparative EM and AFM images of microfibrils from *Micrasterias denticulata* provided evidence for a right handed helical twist of the microfibrils (Hanley *et al* 1997). The highest resolution images have been obtained for *Valonia ventricosa* cellulose crystals where it has been possible to resolve individual glucose molecules and features attributable to the hydroxmethyl groups (Baker *et al* 1997). Use of AFM to investigate the crystal structure of cellulose is discussed in more detail in section 6.1.1.

Certain bacterial species also produce cellulose and studies of cellulose production by *Acetobacter xylinum* have played an important role in understanding the genetics and biosynthesis of cellulose. The bacterial cellulose filaments appear to be more ribbon-like (Gunning *et al* 1998) than those extracted from plants (Fig. 4.26). Both the plant and the bacterial cellulose exist in the cellulose I form but so far no high resolution AFM images have been obtained for bacterial cellulose.

Plant cell walls

Cellulose is the main structural component of plant cell walls and it is also possible to image the cellulose microfibril network in isolated cell wall fragments (Kirby *et al* 1996b: Round *et al* 1996; Morris *et al* 1997; van der Wel *et al* 1996). Extracts of root hair cell walls from *Zea mays* and *Rhaphanus sativus* have been visualised after extraction of the cell wall matrix material with H_2O_2/HAc (van der Wel *et al* 1996). The samples were imaged in air after air drying onto glass slides although poly-L-lysine coated glass slides were also used as substrates for imaging these cell walls under water (van der Wel *et al* 1996). The size and orientation of the microfibrils was consistent with that observed for platinum/carbon coated specimens imaged by EM. Similar information has been obtained on cell wall microfibrils from the sporangia of *Linderina pennispora* (McKeown *et al* 1996). For plant tissues such as carrot, potato, apple or Chinese water chestnut it is possible to isolate cell wall fragments by ball milling, wash them free of cytoplasmic components and then deposit them onto mica (Kirby *et al* 1996b; Round *et al* 1996; Morris *et al* 1997). A problem using the AFM alone is locating fragments which are not buckled, but lay flat on the mica surface and use of a combination AFM/ optical microscope would be an advantage in this type of study. It was also difficult to image these fragments under water because they tend to float off the mica (Kirby *et al* 1996b). For flat fragments it is possible to blot off excess liquid and then image by dc contact mode in air before the fragments dehydrate. The method of preparation presents the face of the cell wall previously adjacent to the plasma membrane (Fig. 4.26). The images reveal layered arrays of cellulose microfibrils whose orientation varies as one looks down through the cell wall towards the middle lamella. The surface of the fragment is rough and the images show light (high) and dark (low) regions apparently devoid of molecular structure (Fig. 4.26a). The microfibrillar structure is more clearly visible in the brighter (higher) regions: in these regions the eye can more easily perceive differences between grey levels whereas the whole image can contain more shades of grey than the eye can distinguish. The error signal mode image of the same area of surface reveals more detail (Fig. 4.26b).

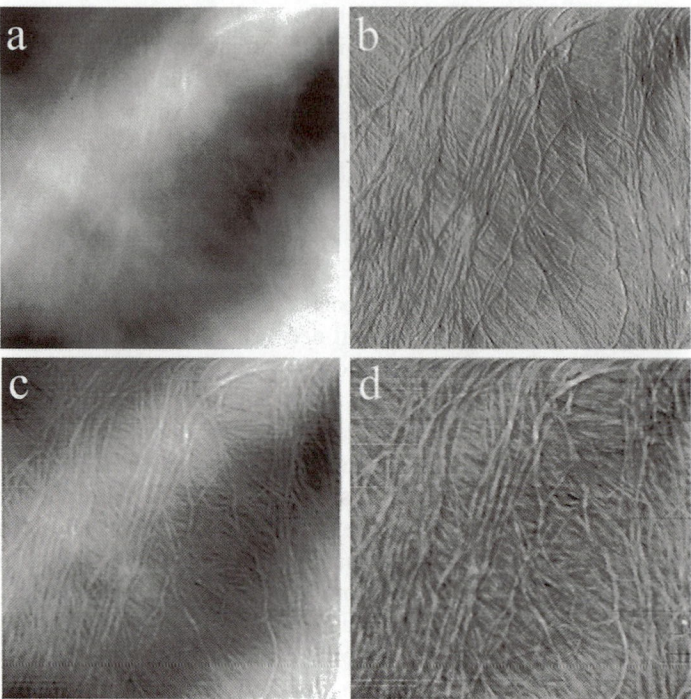

Figure 4.26. Composite set of images showing the effect of different processing steps on AFM images of hydrated Chinese water chestnut plant cell walls. Scan size 2 x 2 µm. (a) Topographic image: the bright and dark bands reflect the roughness of the cell wall surface. (b) Error signal mode image of the region shown in 'a'. This mode of imaging emphasises the 'high frequency' molecular structure within the image. (c) High pass filtered image corresponding to the region shown in 'a'. (d) Background subtracted image of the region shown in 'a'. This form of processing selects the molecular structure and projects it onto a flat plane. The image resembles that shown in 'b' but in this case the heights of the molecules can be measured. The figure is based on data presented in Kirby *et al* 1996b and Round *et al* 1996 and is reproduced with permission.

This form of imaging effectively represses the low frequency background curvature of the surface emphasising the high frequency molecular information. Although the error signal mode image displays molecular detail it is not strictly a 'real' image. The heights represent momentary changes in force and are effectively a differential function of the topographic image. To generate an image showing fine detail in a measurable form it is necessary to process the topographic image (Round *et al* 1996; Morris *et al* 1997). High pass filtering improves the image (Fig. 4.26c) but a better approach is to subtract the low frequency background curvature from the topographic image effectively projecting the molecular detail onto a flat plane (Round *et al* 1996; Morris *et al* 1997) (Fig. 4.26d). The background function can be generated by locally smoothing the topographic image to remove the fine (high frequency) detail. The level of detail seen by AFM in the cell wall preparations is comparable to that seen by EM studies. Woody tissues are lignified and phase images of wood pulp fibres reveal bright patches, attributed to residual lignin, which are invisible in the normal topographical images (Hansma *et al* 1997). The lignin is considered to be more hydrophobic than the cellulose thus giving rise to the difference in contrast. The AFM does not see the more mobile components of the cell wall such as pectin, or hemicellulosic cross-linking of the cellulose fibrils. However, the advantage of imaging under aqueous conditions offers the prospect of probing enzymatic degradation of cell walls, an area of importance for understanding utilisation of plant waste material, gut fermentation of cell wall material and natural degradative processes such ripening. At present the only reported AFM studies of cellulase-cellulose interactions are low resolution imaging of the effect of adding either cellobiohydrolase I (CBH I), or CBH I inactivated at the catalytic site, to cotton fibres (Lee *et al* 1996). The surface morphology was reported to remain unchanged after addition of the inactivated enzyme whereas some disruption of the microfibrillar structure was seen as a result of adding the active enzyme. Another goal of this type of study would be to be able to image cell wall structure in intact, isolated plant cells under physiological conditions. If successful such work would allow investigation of structural changes in cell walls during biological processes such as growth or elongation, using plant cell cultures.

Starch

Starch is the major storage polysaccharide in plants. It is biosynthesised as large spheroidal semi-crystalline granules. Starch can be dissolved in solvents such as DMSO and fractionated into essentially two extreme types of polysaccharides: an essentially linear high molecular weight (about 10^6 D) polymer called amylose and very high molecular weight, highly branched polysaccharide called amylopectin. At present there are a few reported AFM studies on the starch polysaccharides but none of these yield new information on the structure of these molecules. There is also interest in imaging the intact starch granule. Starch granules range in size from μms to 100's μms depending on their botanical source. This raises problems in the use of the AFM. When such objects are deposited onto a flat surface they cause the effective surface roughness to become comparable to, or larger than the height of the probe (Fig. 4.27a). Currently AFMs are designed to image very flat surfaces at high resolution and the axial ratio of the normal pre-fabricated tips is quite small. One approach to imaging rough surfaces is to use higher aspect ratio probe tips. However, such tips are often brittle and can easily snap if they 'hit' objects on the surface during scanning.

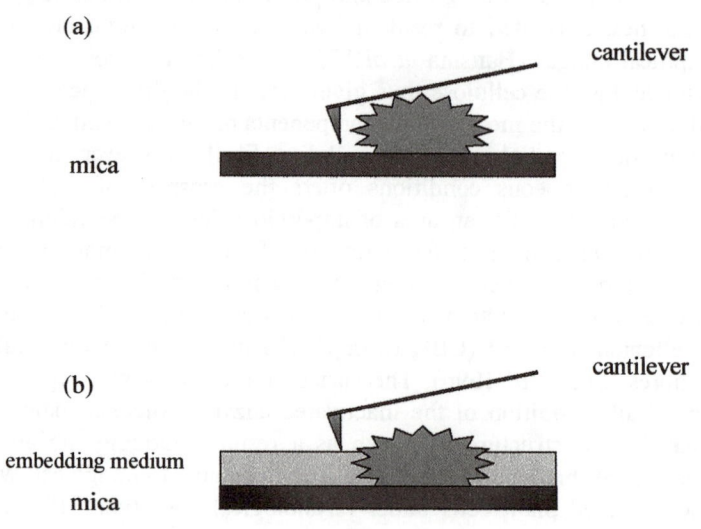

Figure 4.27. Schematic diagram illustrating the difficulties encountered in imaging large objects. (a) Large starch granules will contact the cantilever forcing the tip off the surface. (b) By embedding the starch granules the effective roughness of the surface is reduced and the exposed surface of the granule can be imaged.

The development of new 'flexible' probe tips, based on materials such as carbon nanotubes, may provide a solution to this problem. Even then the scan range available in the 'z' direction will be limited by the lateral scan size of the image, making high resolution imaging of rough surfaces very difficult. A practical solution to this problem is to embed the starch granules (Thomson *et al* 1994: Baldwin *et al* 1996; 1997) thus decreasing the effective roughness of the surface (Fig. 4.27b). This approach has been used in a real time investigation of the enzymatic degradation of wheat starch granules (Thomson *et al* 1994). The starch granules were dusted onto mica, metal coated and then further coated with carbon. After soaking in water it was possible to strip of the mica with cellotape exposing the surfaces of the granules. The preparation could then be imaged in an aqueous environment. Examination of the surface of the granule revealed various surface features ranging in size between 50- 450 nm. With granules with exposed surface pores, or those whose surfaces were cracked, possibly as a result of milling, it was possible to follow, in real time, breakdown of the starch granules due to the action of the enzyme α-amylase. The types of pits formed are consistent with the those previously observed by electron microscopy (Frannon *et al* 1992). Differences in surface morphology have been reported between wheat and potato starch granules and surface protrusions have been attributed to the ends of ordered clusters of amylopectin sidechains (Baldwin *et al* 1997).

4.4.5. Proteoglycans

Proteoglycans are not polysaccharides but show similarities to polysaccharides in their fibrous structures. This important class of biological material has been little studied by AFM. There are images of the cell aggregation factors (MAF) of the marine sponge *Microciona prolifera*, a proteoglycan system responsible for inter cell adhesion (Fritz *et al* 1997). The images reveal a complex star-like branched structure. The proteoglycan has been labelled with anti-MAF antibodies which have been enhanced with gold labelled secondary antibodies. Other AFM studies involving proteoglycans include investigation of their distribution on the surface of collagen fibres (Raspanti *et al* 1997) and studies of their binding strength through measurement of force-distance curves (Dammer *et al* 1995). These important materials deserve further study.

4.5. Proteins

There are several reasons for imaging proteins at surfaces. Firstly there is interest in the intrinsic structure of the protein. Here the surface is considered to act as an inert carrier. Information is required on the overall size, shape and subunit structure of the protein. At a larger scale it is possible to follow molecular interactions involved in processes such as bioassembly or protein gelation. A second area of interest is protein surface interactions. These are of interest in a wide variety of applications from surface biocompatibility, biofouling, the cleaning of surfaces, through to development of biosensors and immunoassays. Here the interest is in how the protein binds to the surface and how such binding affects its biological functionality.

4.5.1. Globular proteins

There is a large literature on the study of simple globular proteins by AFM. A wide variety of methods have been used to immobilise proteins onto a wide variety of surfaces. However, most proteins adsorb fairly easily to surfaces such as mica and can be imaged under alcohols or aqueous conditions using dc or Tapping methods. Care needs to be taken in eliminating adhesive forces and in minimising damage and displacement of the molecules. It is also very important to prevent formation of multilayers on the substrate. The upper layers are easily displaced and adsorbed by the probe resulting in poor imaging conditions. There is a tendency to increase the concentration of the sample in order to improve images if few proteins are seen. However, if multilayers are present, and if this is causing difficulties with retaining the sample on the substrate then, paradoxically, the solution may be to use lower rather than higher protein concentrations in order to obtain better images. In general the images of proteins show globular structures with dimensions which, when corrected for probe broadening, are consistent with either individual proteins or protein aggregates. A correlation has been shown between the volume of native or denatured proteins as measured by AFM and molecular weight (Schneider *et al* 1998). Here probe broadening effects are compensated for by measuring diameters at half-height and calculating volume by treating the molecules as segments of spheres. It is not clear why the same calibration should apply for native and denatured proteins. However, the approach does allow discrimination between individual molecules and molecular aggregates.

Deliberate induced aggregation of proteins can be followed by AFM. A number of plant and dairy proteins can be induced to aggregate or gel. Such processes can be followed by isolating multimers or gel precursors. Figure 4.28 shows a series of images following the association of the 7S soya protein β conglycinin. The sample has been heat treated at 100 °C for different lengths of time resulting in the formation of linear aggregates. The individual proteins are disc-like. Measurements of the height and length of the aggregates suggest they develop by the stacking of these discs into cylinders. By this sort of study of the behaviour of purified components it becomes possible to understand the functional behaviour of more complex protein isolates. A more dramatic example of protein aggregation is the classic and pioneering study of thrombin induced fibrin polymerisation (Drake *et al* 1989). This was one of the first demonstrations of the use of AFM to monitor the dynamics of a biologically important process and provided new information on the mode of assembly in this complex system.

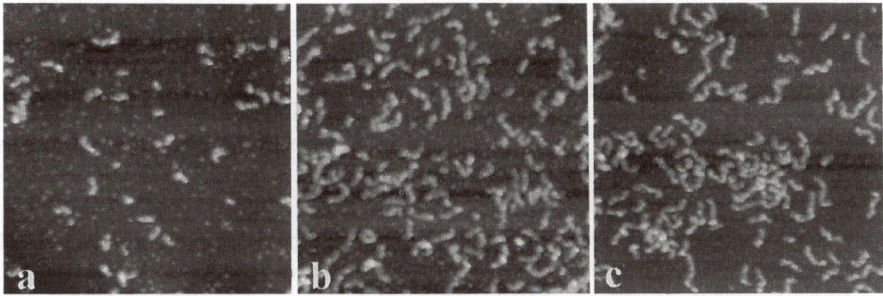

Figure 4. 28. Sequence of AFM images following the aggregation of 7S soya protein. Scan size 1.2 x 1.2 μm. The protein solution was heated to 100°C and the aggregation process was followed with time. The solution was diluted, deposited onto mica, air dried and then imaged under butanol. (a) After heating for 2 minutes. Both individual proteins and linear aggregates can be seen. (b) After heating for 5 minutes and (c) after heating for 60 minutes. These proteins associate to form linear soluble aggregates.

The AFM can be used in studies of proteins at surfaces or interfaces where the interest is in understanding how the proteins adsorb and also what effect the adsorption has on the subsequent functionality of the proteins. Typical applications of such studies can involve development of biosensors, biofouling and cleaning of surfaces or construction of biocompatible surfaces. AFM studies can be used to complement investigations by other methods, and to provide information on levels of protein adsorption, orientation of the proteins on the surface, aggregation and the formation of monolayers or multilayers. Information on protein aggregation and coverage does not require high resolution and can be inferred from measurements of surface roughness. Examples of studies on protein adsorption are investigations of ferritin adsorption (Caruso *et al* 1997; Davies *et al* 1994) and the binding of serum albumins to mica (Quist *et al* 1995; Mori and Imae, 1997) or more complex surfaces (Nakata *et al* 1996; Kowalczyk *et al* 1996). The nature and type of binding of enzymes to surfaces is of importance in the development and use of certain types of biosensors. In this case it should be possible to use AFM to study both the adsorption of the protein and the activity of the resulting surface. For example, AFM has been used to study the binding of glucose oxidase to gold surfaces (Quinto *et al* 1998) as a function of preparative conditions. It may also be possible to investigate the functional behaviour of such surfaces: conducting tips and the measurement of electron tunnelling could be used to study electron transfer processes, and the recently developed protein tracking methods, whereby enzymatic activity can be monitored by fluctuations in the height of enzymes (Radmacher *et al* 1994; Thomson *et al* 1996b), could be employed to map enzymatic activity on the surface in the presence of substrate.

This type of study is similar to investigations of the efficiency of bound antibodies in immunoassays (section 4.5.2).

The adsorption and interaction of proteins at air-water and oil-water interfaces is of importance in understanding the stability of foams and emulsions. In this case AFM provide the only direct method for visualising the structures formed. Examples of such studies are given later in section 5.7.

An important goal in the use AFM to study globular proteins would be to provide detailed information on protein shape and internal structure. Clearly AFM is never going to rival the atomic resolution obtainable by X-ray diffraction or modern nmr methods. Electron diffraction patterns contain information on the atomic structure of proteins. In very high resolution electron microscopy it is possible to reconstruct images showing the secondary structure and its connectiveness within the protein. Such studies are few and, at present, require data on 2D crystals. However, there is the real prospect in the future of generating such images on individual proteins. The information required to construct such high resolution images is not present in the AFM data. However, the aim of the use of AFM would be to improve on the resolution obtained through the use of conventional electron microscopy, or to achieve comparable resolution but under natural imaging conditions. To obtain high resolution images the proteins need to be immobilised onto a flat substrate. Air drying proteins onto substrates such as mica and imaging in air is difficult for several reasons. The deposition process itself may partially denature the protein, motion within the protein structure will tend to blur the image and, unless the imaging force is carefully controlled, then the probe will distort the protein during scanning. For passively adsorbed proteins it may be possible to obtain information on overall shape and size and the images can often be interpreted in terms of known models of the protein structure. An example of this type of study is the Tapping mode images of ribosomes deposited onto mica (Wu *et al* 1997). For studies on isolated proteins there is the added complication of probe broadening effects which may further complicate interpretation of the images. Probe broadening effects can be reduced by organising the deposited proteins into ordered arrays. This also appears to reduce distortion or displacement of the proteins.

If the proteins bind strongly to mica, and resist disruption by the probe, then quite high resolution images can be obtained on individual proteins. Perhaps the best example of such studies is work on pertussis toxin (Yang *et al* 1994b). The intact pertussis toxin and the B-oligomers were deposited onto mica and imaged in the dc contact mode under water without passing through a drying stage. Even in the raw data it was possible to resolve 5 (2 large and 3 small) subunits in the B-oligomer and, by noting that the central pore of the B-oligomer

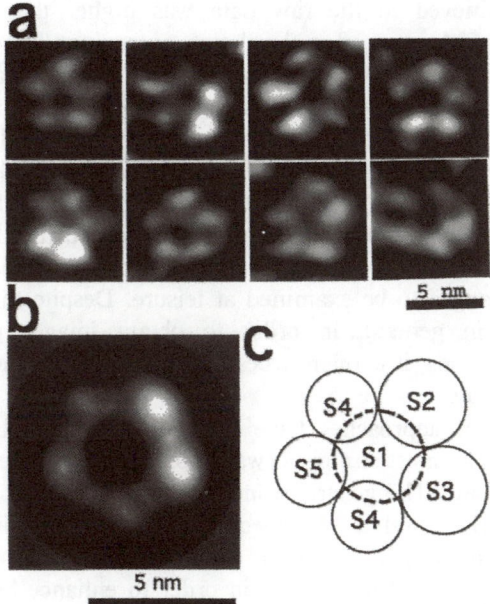

Figure 4.29. AFM images of pertussis toxin. (a) Eight representative images of the pertussis toxin B-oligomer. Two of the subunits appear brighter indicating that they are larger in size. (b) An averaged image formed from 300 individual images after alignment. (c) The proposed subunit structure of the molecule. Because it is not possible to distinguish between the subunits S2 and S3 it is not clear to which side the A-subunit is attached. Data reproduced from Yang *et al* 1994b with permission.

is absent in the intact toxin, to fix the central location of the A-oligomer. The raw images were enhanced by correlation averaging and the result of such treatment is illustrated in Fig. 4.29. In these studies it was claimed that features as small as 0.5 nm were resolvable in the images. The pentameric structure observed by AFM differs from the heptameric structure deduced from X-ray diffraction studies (Stein *et al* 1994). At present it is not clear why there is this discrepancy. However, this illustrates an important role which AFM can play in examining whether the protein structures found from X-ray data are appropriate for these molecules in solution. At present the only other method which could provide such information is nmr. It was also possible to use the AFM to investigate the stability of the toxin under different pH and temperature conditions (Yang *et al* 1994b).

Another example of high resolution images of physically adsorbed proteins on mica is the study of *E. coli* chaperonin proteins (Mou *et al* 1996a,b). High resolution AFM images of both GroEL and GroES proteins were obtained in solution and the images were improved by chemical fixation using glutaraldehyde.

The resolution achieved in the raw data was higher than that obtained by negatively stained EM, even after correlation averaging. The AFM was used to 'dissect' GroEL particles revealing information on the internal structure of this complex protein. In the case of these chaperonin proteins the structures obtained by AFM broadly agreed with those deduced from EM and X-ray diffraction studies, although combined EM and AFM data have been used to generate a new improved model for GroES. Chemical cross linking is a useful aid in imaging proteins. Not only does it improve the stability of the structure for imaging but, in addition, it may be used to trap a range of transient stages, allowing conformational changes to be examined at leisure. Despite the excellent studies described above, in general, in order to obtain images under aqueous or physiological conditions, it is often necessary to immobilise the proteins onto the substrate in some way.

A variety of approaches have been used to immobilise proteins for imaging by AFM. The most straightforward approach is to chemically attach the protein to the substrate. The general principles were discussed at the start of this chapter (section 4.1.6) and a few specific examples are given below. Many proteins contain thiols which can bind to gold coated mica. Proteins such as BSA and gelatin have been thiol-derivatised in order to enhance binding to gold for AFM studies (Nakata *et al* 1996). Photocrosslinkers have been employed to attach HPI layers to glass substrates for AFM studies (Karrasch *et al* 1994). More specific tagging has also been employed. Bacteriorhodopsin molecules have been genetically modified, replacing a serine residue with a cysteine residue, in order to allow covalent attachment to gold for imaging by AFM (Brizzolarra *et al* 1997). A nickel chelating dipeptide was attached to the carboxyl terminus of the heavy chains of an IgG antibody, which was then bound to nickel chloride treated mica (Ill *et al* 1993). Insertion of targeted binding sites allows the orientation of the bound molecule to be defined but the modifications are complex, may alter the natural structure or function of the molecules, and thus will probably only be used as a last resort. As mentioned earlier the formation of ordered arrays of proteins often stabilises the molecules. An interesting variation of this approach described by Shao and coworkers (Shao *et al* 1996) is to stabilise sparsely populated deposits of large molecules from lateral motion by packing the intervening space with smaller molecules. This approach has been found to be successful in imaging low density lipoprotein using the smaller cholera toxin B oligomer (Shao *et al* 1996). A particularly successful approach for imaging membrane proteins has been to reconstitute purified proteins into supported planar bilayers for study by AFM (Yang *et al* 1993a,b). Using laterally polymerised phosphatidylcholine bilayers it was possible to image membrane bound cholera toxin in low salt buffer to a resolution of 1-2 nm (Yang *et al* 1993a,b). For the B-oligomer (CTX-B) 5 subunits were resolved. In order to eliminate the possibility that the polymerisation may alter the function of the protein Mou and coworkers (Mou *et al* 1995b) have shown recently, through studies on cholera toxin, that AFM is fully capable of imaging membrane proteins on supported phospholipid bilayers of physiologically relevant species in solution. Surface features of the order of 1-2 nm can be resolved without the need for correlation averaging. In addition 2-D crystalline arrays can be grown directly on these model membranes and imaged by AFM (section 5.5). It is suggested that this methodology may be generally applicable for imaging membrane proteins, including integral membrane proteins, as long as supported bilayers can be made to incorporate the proteins (Mou *et al* 1995b). To this end these authors (Mou *et al* 1995b) suggest that direct fusion of protein-containing vesicles, and the spreading of these vesicles at air-water interfaces (Pearce *et al* 1992; Schindler, 1980; Schurholz and Schindler, 1991), may provide a route to achieving this goal.

It is possible to form 3D and 2D protein crystals and a number of membrane fragments are naturally occurring crystalline materials. The imaging of these types of material is described in chapter 6.

4.5.2. Antibodies

Antibodies are large, flexible multidomain proteins which have been well characterised by biophysical and biochemical methods. Because of their biological importance, and their use in immunolabelling and immunoassays, there is interest in characterising their structure and also their interactions with antigens and surfaces.

Most room temperature AFM studies have failed to match the resolution obtainable with electron microscopy (Parkhouse *et al* 1970). In general the 'molecules' appear to be globular, featureless and are variously ascribed to individual antibodies or aggregates (Lea *et al* 1992; Ill *et al* 1993, Yang *et al* 1994a; Fritz *et al* 1997; Harada *et al* 1997; Thimonier *et al* 1995). In the best images shapes can be identified which would be consistent with the expected structures (Fritz *et al* 1997; Harada *et al* 1997). The globular shapes are attributed to deformation of the molecules (Lea *et al* 1992) or molecular flexibility. Lowering the temperature should reduce molecular motion and higher quality images have been obtained by cryo-AFM (Han *et al* 1995; Zhang,Y *et al* 1996; Shao and Zhang, 1996). Y shaped IgG (human IgG1) molecules are clearly visible with structural heterogeneity consistent with a flexible hinge region (Fig. 4.30).

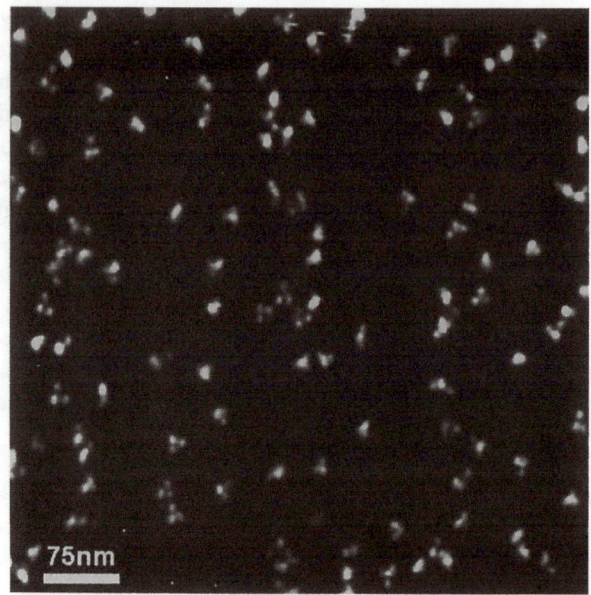

Figure 4.30. Cryo-AFM image of IgG (human IgG1), obtained at about 85°K. The characteristic Y shape of the antibody is clearly resolved. A range of conformations are visible in the images indicating the flexibility of the molecules. Data reproduced from Zhang, Y *et al* 1996 with permission of the authors and the Biophysical Society.

Images of monoclonal IgA (mouse) revealed monomers and various multimers instead of the expected dimers. Both J chains and Fab domains can be seen in many of the molecules. Studies on monoclonal IgM (mouse) antibodies by AFM have revealed (Han *et al* 1995; Zhang, Y *et al* 1996) a new conformation in addition to the accepted flat pentamer structure based on X-ray scattering data (Perkins *et al* 1991) and EM studies (Parkhouse *et al* 1970).

For immunoassays and biosensors there is considerable interest in the way in which antibodies are bound to surfaces, and how this affects antibody-antigen interactions. The adsorption of antibodies onto mica, modified mica, modified silicon oxide, gold and microtiter plates has been monitored by AFM (You and Lowe, 1996; Caruso *et al* 1996; Perrin *et al* 1997; Roberts *et al* 1995). After tip deconvolution it has been possible to discriminate between IgG and IgM antibodies on surfaces (Roberts *et al* 1995). Imaging can be used to assess the distribution of individual antibodies on the surface, assess problems due to antibody aggregation, or to study the stability of the interfacial structure. The nature of the binding mechanism, and the orientation of the antibody, will affect the efficiency with which the antibodies bind antigens, and this can be studied by observing the antigen binding directly. These types of experiments have been used to assess methods of preparing biosensors or immunoassays, and for assessing their sensitivity (Perrin *et al* 1997; Caruso *et al* 1996; You and Lowe, 1996; Davies *et al* 1994).

Antibodies provide specific labels for identification and mapping. Immunolabelling techniques are well established in both light and electron microscopy. It is possible to use antibodies to probe and map specific antigens on individual molecules (e.g. DNA section 4.2.4), macromolecular complexes such as chromosomes (section 4.2.5), cells (Putman *et al* 1993a) or tissue (Saoudi *et al* 1994). In the case of individual molecules, or simple molecular complexes, the antibody is large, and easily recognisable. For flat layered structures, such as bacterial S layers (Ohnesorge *et al* 1992), it may still be possible to recognise the antibody-antigen complex directly, although care must be taken to discriminate between specifically bound and passively adsorbed material. In more complex systems, particularly where the surface is rough, or where there is a need to ensure specificity, some additional form of labelling is necessary. Procedures can be adapted from established immunolabelling methods. Thus gold labelled antibodies can be prepared and used directly, or gold labelled secondary antibodies can be used to locate antibody-antigen complexes (Mulhern *et al* 1992; Putman *et al* 1993a; Saoudi *et al* 1994). Gold labelling is not as straightforward in AFM as it is for EM studies. The gold labels may be confused with similar sized surface protrusions (Putman *et al* 1993a) or even compressed into the surface by the probe, making them difficult to spot (Mulhern *et al* 1992). The labelling procedure can be improved by generating larger particulate deposits: examples used with AFM include silver enhancement (Neagu *et al* 1994; Putman *et al* 1993a), peroxidase labelled antibodies and their reaction with DAB (sections 4.2.5 and 4.3), or even fluorescently labelled complexes (McMaster *et al* 1996a,b) (section 4.2.5). There may, however, be more sensitive types of labelling which could be used with the AFM: magnetic labels or tips coated with antibodies may lead to enhanced interactions and improved contrast for the labels. Antigen coated tips have been used to study antibody-antigen interactions for antibodies used in immunoassay systems (Allen, S *et al* 1997). One area which does seem to require further research is the development of well established negative controls for immunolabelling studies by AFM.

4.5.3. Fibrous proteins

The structure and organisation of fibrous proteins is a good topic for investigation by AFM: here AFM offers high resolution and the prospects of studying bioassembly under natural conditions. A number of such studies have been reported in the literature and some examples of this type of work are given below:

Muscle proteins

The proteins myosin and titin are important structural components of muscle. Studies on myosin have shown that the glycerol-mica method, widely used for preparing proteins for electron microscopy, can be used for AFM studies (Hallett *et al* 1995; 1996). The protein solution in 50% glycerol is sandwiched between mica sheets, the sheets pulled apart and then dried under vacuum to remove glycerol and water. The molecules can be imaged under propanol or water/propanol mixtures by dc contact or Tapping. The resolution is similar to that obtained by electron microscopy. For example, it is possible to observe the periodicity of the coiled-coiled α-helical myosin tail. Too high an imaging force was shown to damage or displace the molecules but, on one occasion, the probe appears to have separated the helical strands within the tail (Hallett *et al* 1995). Under propanol the heads appear to aggregate but, in water/propanol mixtures, they are separated and the images obtained are in agreement with those seen by electron microscopy (Hallett *et al* 1996). Cryo-AFM images (Zhang, Y *et al* 1997) of smooth muscle myosin clearly resolved the motor domains of the head and the pitch of the α-helical coiled-coiled tail. In addition it has been possible to obtain new information on the effects of thiophosphorylation on the tail structure and the flexibility of the head-tail junction of myosin maintained in the physiologically relevant 6S conformation.

For titin, which would normally adopt a coiled configuration, the molecules in 50% glycerol were dropped onto mica, subjected to centrifugation to extend the molecules, and then vacuum dried before imaging under propanol by dc contact or Tapping. Elongated structures, similar to that seen by electron microscopy, with a globular head and extended tail were observed by AFM (Fig. 4.31). Use of Tapping mode in liquid reveals some substructure suggesting that the 'tadpole-like' structures may in fact be assemblies of individual titin molecules (Hallett *et al* 1996).

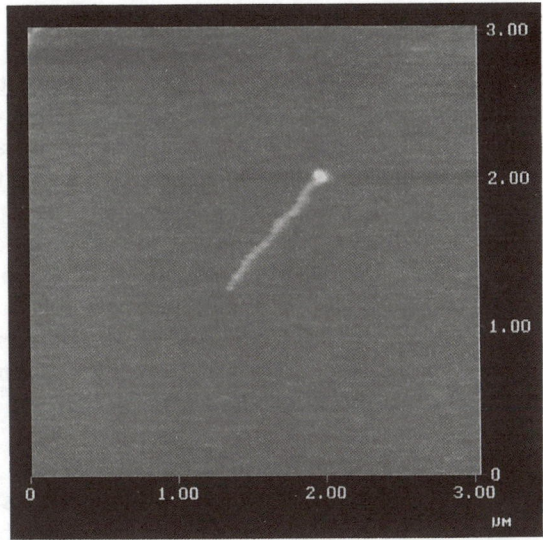

Figure 4.31. AFM image of titin. Data reproduced from Hallett *et al* 1996 with permission.

It is the elastic properties of the molecule which are important for its biological role in muscle. AFM studies permitted the mechanical properties of individual molecules to be examined. By adsorbing titin molecules onto gold, and then picking up a fraction of the molecule onto the tip, it has been possible to obtain force-extension curves in phosphate buffered saline on what are believed to be individual molecules. Studies on titin, and recombinant fragments, has allowed a determination of the forces required to unfold individual domains of the protein on extension, and also observation of protein refolding on relaxation (Rief *et al* 1997b). This type of study shows that the AFM is not just a microscope: it can also be used as a mini-laboratory for investigating mechanical properties at the molecular level.

Cytoskeleton proteins

Although there are reports of AFM studies on isolated actin filaments (Fritz *et al* 1995), spectrin molecules (Almqvist *et al* 1994; Zhang, P *et al* 1996), and microtubules (Fritz *et al* 1995; Vinckier *et al* 1995), the main area of interest with these materials is in the use of AFM to probe the cytoskeleton structures of cells, and to study dynamic changes of these structures in living cells (chapter 7).

Collagen

Collagen is the most abundant structural protein found in connective tissue. It exists in a variety of morphological forms, and is a good example of a complex self-assembling biological structure. Individual collagen molecules are stiff coils 280 nm in length and 1 nm diameter. Studies on collagen molecules by cryo-AFM (Shattuck *et al* 1994) and normal AFM studies of segment-long-spacing (SLS) crystals of collagen (Fujita *et al* 1997) are starting to reveal variations in the structure (diameter) along the molecules, which may be of importance in aspects of their assembly in calcified tissue. Collagen molecules assemble into a range of fibrous structures and networks. The most well studied assembly product is the native fibril, characterised by a periodic banding pattern with a repeat of ≈ 68 nm. AFM has been used to visualise the D banding (Fig. 4.32a) and the detailed surface structure of native (Raspanti *et al* 1996; Arogani *et al* 1995; Yamamoto *et al* 1997; Revenko *et al* 1994; Chernoff and Chernoff, 1992) and reconstituted collagen fibrils (Revenko *et al* 1994). The aggregation of monomers (Shattuck *et al* 1994), and the various stable transient states in the *in vitro* assembly into fibres have been quantified using AFM, in terms of the amounts and structures of the various intermediates present at different stages of fibrillogenesis (Gale *et al* 1995). Abnormal fibrillar structures such as fibrous long spacing collagen (FLS) are associated with various of types of pathogenic conditions such as Hodgkin's disease, athlerosclerotic plaques, myeloproliferative disorder and silicosis. It has been suggested that interactions with other molecules, such as glycoproteins, may influence formation of FLS fibres. *In vitro* AFM studies of FLS fibre assembly (Fig. 4.32b), in the presence of glycoprotein, suggests an unique assembly process, rather than formation of normal fibres and their conversion into FLS (Paige *et al* 1998). In bones and other calcified tissue the final structure is based on the interaction between the collagen and deposited apatite. AFM Tapping mode studies of *in vitro* and physiologically calcified tendon collagen have suggested that surface structure of the fibril induces nucleation of apatite crystals, and that their subsequent growth does not markedly alter the fibril structure (Bigi *et al* 1997). Tapping mode AFM on dried collagen fibrils from rat tail tendon have been used to visualise proteoglycan bound to the collagen surfaces (Raspanti *et al* 1997). The distribution of the proteoglycans was determined by comparing images obtained for the native structure, samples treated with chondroitinase, and samples incubated with Cupromeronic blue, a copper phthalocyanin specifically designed to stabilise the anionic glycosaminoglycan chains. Such studies contribute to an understanding of this complex and vital bioassembly process.

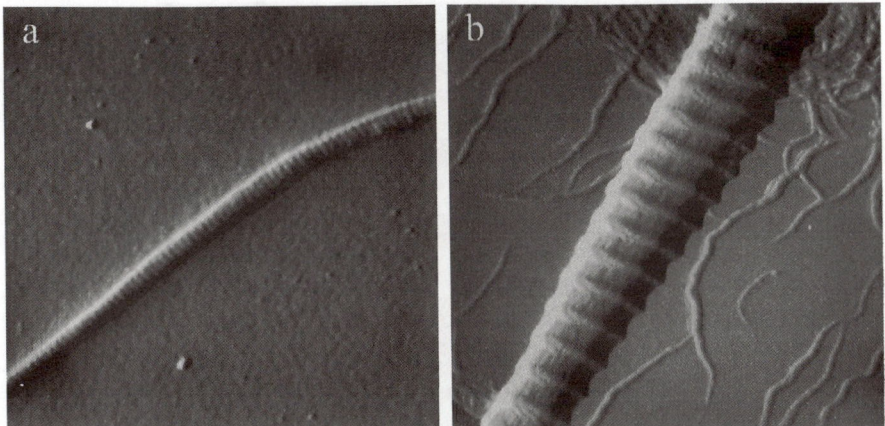

Figure 4.32. AFM images of (a) normal collagen fibril, scan size 3 x 3 μm showing the normal banding pattern, and (b) a fibrous long spacing collagen (FLS) fibril with a higher periodicity banding, scan size 3.5 x 3.5 μm. Data reproduced from Paige *et al* 1998 with permission of the authors and the Biophysical Society.

Gelatin is basically denatured collagen. It has a wide variety of industrial uses which involve its ability to form gels and films. These network structures are usually formed by cooling hot solutions. The gelation mechanism is believed to involve a coil-helix transition: formation of the triple helical structure is believed to lead to intermolecular aggregation and network formation. The initial gelation step is rapid, and is then followed by a slower stiffening of the network structure, usually attributed to a further level of aggregation, possibly involving reformation of a collagen fibre-like structure. AFM studies on gelatin films failed to reveal their molecular structure (Radmacher *et al* 1994; Haugstead *et al* 1993). This is generally thought to be because the films deform during scanning, blurring the image. By isolating gel precursors during the bulk gelation of gelatin it has been possible to use AFM to obtain clues about the gelation process (Mackie *et al* 1998). Fig. 4.33 shows a gelatin gel precursor which is believed to be the type of junction zone found in the gel. These appear to be aggregates of smaller fibres, believed to be reformed tropocollagen triple helical structures, and the aggregates do, on occasions, show periodicities reminiscent of those seen in collagen fibres. More information can be obtained on gelatin association and gelation by studying networks formed at air-water interfaces (Mackie *et al* 1998) and this is discussed in more detail in chapter 5.

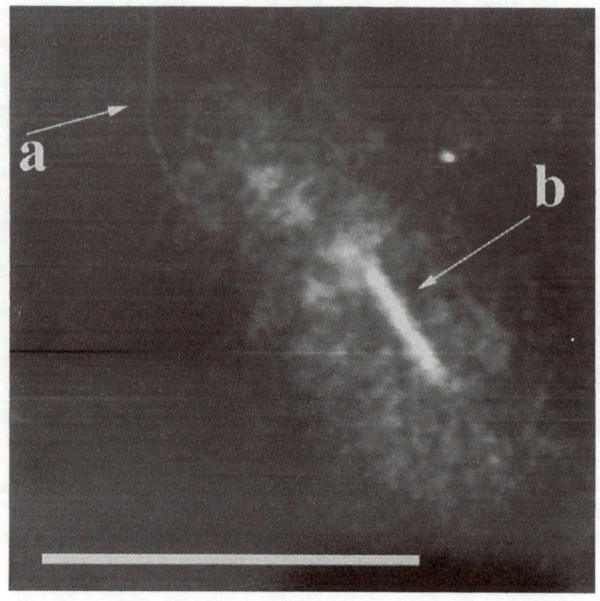

Figure 4.33. AFM image of a gelatin gel precursor. The junction zones labelled 'b' appear to be composed of filamentous structures labelled 'a', which are believed to be individual gelatin triple helices. Scale bar is 500 nm. Data is reproduced from Mackie *et al* 1998 with permission.

Neurofilaments

Neurofilaments are important cytoskeletal components. They are unusual in that the filament is coated with sidechains. When imaged by AFM (Brown and Hoh, 1997) under aqueous conditions the sidechains are in motion and cannot be imaged directly. However, they reveal themselves by creating a zone around the molecules from which 'debris' deposited from solution onto the mica is excluded (Fig. 4.34). Such an exclusion zone is not seen with filaments lacking sidechains. Measurement of force-distance curves at each image point (force mapping - section 7.1.2) has revealed a repulsive interaction due to these sidechains. This 'entropic' repulsion has been proposed as a new mechanism for determining inter-filament spacing in nerve axons. Clearly AFM can complement EM in the development of molecular models for the structure of nerves and the structural changes due to degenerative diseases.

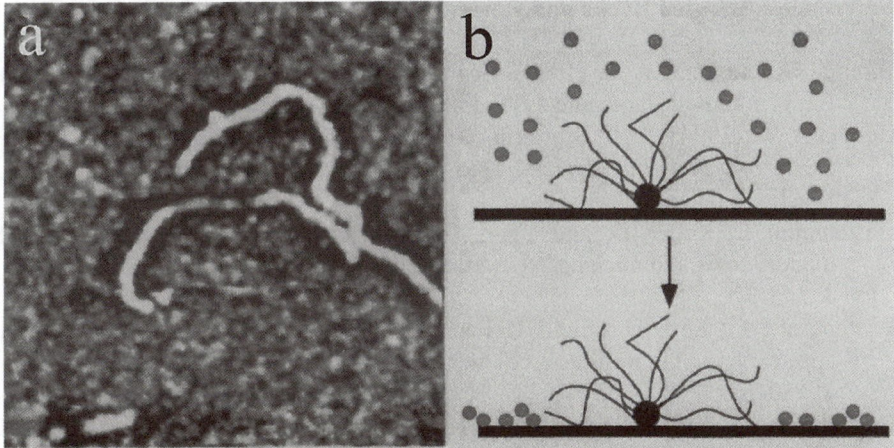

Figure 4.34. AFM Tapping mode image of (a) native neurofilaments adsorbed to mica and imaged in solution. The scan size is about 1.9 x 1.9 μm. The mobile sidearms create an exclusion region (black) around the molecule about 35-50 nm in width. (b) Schematic diagram illustrating how this exclusion zone is formed. Based on data reported by Brown and Hoh, 1997 with permission. Copyright (1997) American Chemical Society.

Amyloid fibres

The formation of fibrous particles is associated with certain diseases. For example, fibrous aggregates of the protein amyloid-β (Aβ) are major components of the neuritic plaques and vascular deposits that appear in Alzheimers disease. Neurotoxicity is related to aggregation of the Aβ protein but the interrelation between toxicity and structure is still unclear. AFM has been used to study self assembly and the surface structure of the resultant fibrils (Shivji *et al* 1996; Stine *et al* 1996). The high resolution attainable offers promise for understanding the structural origins of biological responses in this complex neurodegenerative disease.

References: Selected Books and Reviews

Binnig, G. and Rohrer, H. (1984). Scanning tunnelling microscopy. *Trends in Physics*, (eds. J. Janta and J. Pantollick) pp. 38-46.

Burchard, W. (1994). Light Scattering Techniques. In *Physical Techniques for the Study of Food Biopolymers*, (ed. S.B. Ross-Murphy), chap. 4. 151-213. Blackie, London.

Bustamante, C, Keller, D. and Yang, G. (1993). Scanning force microscopy of nucleic acids and nucleoprotein assemblies. *Current Opinion Structural Biol.* **3**, 363-372.

Bustamante, C, Eire, D.A. and Keller, D. (1994). Biochemical and structural applications of scanning force microscopy. *Current Opinion Structural Biol.* **4**, 750-760.

Bustamante, C. and Rivetti, C. (1996). Visualising protein-nucleic acid interactions on a large scale with the scanning force microscope. *Ann. Rev. Biophys. Biomol. Structure* **25**, 395-429.

Clemmer, C.R. and Beebe, T.P. (1992). A review of graphite and gold surface studies for use as substrates in biological scanning tunnelling microscopy studies. *Scanning Microscopy* **6**, 319-333.

Engel, A. (1991). Biological applications of scanning probe microscopes. *Ann. Rev. Biophys. Biophys. Chem.* **20**, 79-108.

Firtel, M. and Beveridge, T.J. (1995). Scanning probe microscopy in microbiology. *Micron.* **26**, 347-362.

Fritz, J, Anselmetti, D, Jarchow, J. and Fernandez-Busquets, X. (1997). Probing biomolecules with atomic force microscopy. *J. Structural Biol.* **119**, 165-171.

Fritzsche, W, Takac, L. and Henderson, E. (1997). Application of atomic force microscopy to visualisation of DNA, chromatin, and chromosomes. *Critical Rev. Eukaryotic Gene Expression* **7**, 231-240.

Hagerman, P.J. (1988). Flexibility of DNA. *Ann. Rev. Biophys. Chem.* **17**, 265-286.

Hansma, H.G. and Hansma, P.K. (1993). Potential applications of atomic force microscopy of DNA to the human genome project. *Proc. SPIE Int. Opt. Eng. (USA)* **1891**, 66-70.

Hansma, H.G. and Hoh, J.H. (1994). Biomolecular imaging with the atomic force microscope. *Ann. Rev. Biomed. Structure* **23**, 115-139.

Hansma, H.G, Laney, D.E, Bezanilla, M, Sinsheimer, R. L. and Hansma, P.K. (1995). Applications for atomic force microscopy of DNA. *Biophys. J.* **68**, 1672-1677.

Hansma, H.G. (1996). Atomic force microscopy of biomolecules. *IJ. Vac. & Technol. B* **14**, 1390-1394.

Hansma, H.G, Kim, K.J, Laney, D.E, Garcia, R.A, Argaman, M, Allen, M.J. and Parsons, S.M. (1997). Properties of biomolecules measured from atomic force microscopic images: a review. *J. Structural Biol.* **119**, 99-108.

Ikai, A. (1996). STM and AFM of bio/organic molecules and structures. *Surface Sci. Reports* **26**, 261-332.

Kasas, S, Thomson, N.H, Smith, B.L, Hansma, P.K, Miklossy, J. and Hansma, H.G. (1997). Biological applications of the AFM: from single molecules to organs. *Int. J. Imaging Syst. Technol.* **8**, 151-161.

Lal, R. and John, S.A. (1994). Biological applications of atomic force microscopy. *Amer. J. Physiol.* **266** (*Cell Physiol.* **35**), C1-C21.

Lyubchenko, Y.L, Jacobs, B.L, Lindsay, S.M. and Stasiak, A (1995). Atomic force microscopy of nucleoprotein complexes. Scanning *Microscopy* **9**, 705-727.

Morris, V.J. (1994). Biological applications of scanning probe microscopies. *Prog. Biophys. Mol. Biol*.**61**,131-185.

Shao, Z, Yang, J. and Somlyo, A.P. (1995). Biological atomic force microscopy: from microns to nanaometers and beyond. *Ann. Rev. Cell Dev. Biol.* **11**, 241-265.

Shao, Z and Zhang, Y. (1996). Biological cryo atomic force microscopy: a brief review. *Ultramicroscopy* **66**, 141-152.

Shao, Z, Mou, J, Czajkowski, D.M, Yang, J. and Yuan, J-Y. (1996). Biological atomic force microscopy: what is achieved and what is needed. *Adv. Phys.* 45, 1-86.

Yang, J. and Shao, Z. (1995). Recent Advances in Biological Atomic Force Microscopy. *Micron* **26**, 35-49.

Selected Research Papers

Allen, M.J, Dong, X.F, O'Neil, T.E, Yau, P, Kowalczykowski, S.C, Gatewood, J, Balhorn, R. and Bradbury, E.M. (1993a). Atomic force microscopic measurements of nucleosome cores assembled along defined DNA sequences. *Biochemistry* **32**, 8390-8396.

Allen, M.J, Lee, C, Lee, J.D, Pogany, G.C, Balooch, M, Siekhaus, W.J. and Balhorn, R. (1993b). Atomic force microscopy of mammalian sperm chromatin. *Chromosoma* **102**, 623-630.

Allen, S, Chen, X.Y, Davies, J. Davies, M.C, Dawkes, A.C, Edwards, J.C, Roberts, C.J, Sefton, J, Tendler, S.J.B. and Williams, P.M. (1997). Detection of antigen-antibody binding events with the atomic force microscope. *Biochemistry* **36**, 7457-7463.

Allison, D.P, Bottomley, L.A, Thundat, T, Brown, G.M, Woychik, R.P, Schrick, J.J, Jacobson, K.B. and Warmack, R.J. (1992a). Immobilization of DNA for scanning probe microscopy. *Proc. Natl. Acad. Sci.* **89**, 10129-10133.

Allison, D.P, Warmack, R.J, Bottomley, L.A, Thundat, T, Brown, G.M, Woychik, R.P, Schrick, J.J, Jacobson, K.B. and Ferrell, T.L. (1992b). Scanning tunnelling microscopy of DNA-a novel technique using radiolabelled DNA to evaluate chemically mediated attachment of DNA to surfaces. *Ultramicroscopy* **42**, 1088-1094.

Allison, D.P, Kerper, P.S, Doktycz, M.J, Thundat, T, Modrich, P, Larimer, F.W, Johnson, D.K, Hoyt, P.R, Mucenski, M.L. and Warmack, R.J. (1997). Mapping individual cosmid DNAs by direct AFM imaging. *Genomics* **41**, 379-384.

Almqvist, N, Backman, L. and Fredriksson, S. (1994). Imaging human erythrocyte spectrin with atomic force microscopy. *Micron* **25**, 227-232.

Aragano, I, Odetti, P, Altamura, F, Cavalleri, O. and Rolandi, R. (1995). Structure of rat tail collagen examined by atomic force microscopy. *Experientia* **51**, 1063-1067.

Argaman, M, Golan, R, Thomson, N.H. and Hansma, H.G. (1997). Phase imaging of moving DNA molecules and DNA molecules replicated in the atomic force microscope. *Nucleic Acids Res.* **25**, 4379-4384.

Baker, A.A, Helbert, W, Sugiyama, J. and Miles, M.J. (1997). High-resolution atomic force microscopy of native *Valonia* cellulose I microcrystals. *J. Structural. Biol.* **119**, 129-138.

Baker, A.A, Helbert, W, Sugiyama, J. and Miles, M.J. (1998). Surface structure of native cellulose microcrystals by AFM. *Appl. Phys. A* **66**, S559-S563.

Baldwin, P.M, Frazier, R.A, Alder, J, Glasbey, T.O, Keane, M.P, Roberts, C.J, Tendler, S.B.J, Davies, M.C. and Melia, C.D. (1996) Surface imaging of thermally sensitive particulate and fibrous materials with the atomic force microscope: a novel sample preparation method. *J. Microscopy* **184**, 75-80.

Baldwin, P.M, Davies, M.C. and Melia, C.D. (1997). Starch granule surface imaging using low-voltage scanning electron microscopy and atomic force microscopy. *Int. J. Biol. Macromolecules* **21**, 103-107.

Becker, J.C, Nikroo, A, Brabletz, T. and Resfeld, R.A. (1995). DNA loops induced by cooperative binding of transcriptional activator proteins and preinitiation complexes. *Proc. Natl. Acad. Sci. USA* **92**, 9727-9731.

Bezanilla, M, Drake, B, Nudler, E, Kashlev, M, Hansma, P.K. and Hansma, H.G. (1994). Motion and enzymatic degradation of DNA in the atomic force microscope. *Biophys. J.* **67**, 2454-2459.

Bigi, A, Gandolfi, M, Roveri, N. and Valdre, G. (1997). *In vitro* calcified tendon collagen: an atomic force and scanning electron microscopy investigation. *Biomaterials* **18**, 657-665.

Binnig, G. and Rohrer, H. (1984). Scanning tunnelling microscopy. *Trends in Physics*, (eds. J. Janta and J. Pantollick). pp. 38-46.

Bordas, J, Perez-Grau, L, Koch, M.H.J, Vega, M.C. and Nave, C. (1986). The superstructure of chromatin and its condensation mechanism. 1. Synchrotron radiation X-ray scattering results. *Eur. Biophys. J.* **13**, 157-173.

Bouchiat, C, Wang, M.D, Allemand, J-F, Strick, T, Block, S.M. and Croquette, V. (1999). Estimating the persistence length of a worm-like chain molecule from force-extension data. *Biophys. J.* **76**, 409-413.

Brant, D.A. and McIntire, T.M. (1996). Cyclic polysaccharides. In *Large Ring Molecules*, (ed. J.A. Semlyen), chapter 4, pp. 113-154. John Wiley & Sons, Oxford.

Brizzolara, R.A, Boyd, J.L. and Tate, A.E. (1997). Evidence for covalent attachment of purple membrane to a gold surface via genetic modification of bacteriorhodopsin. *IJ. Vac. & Technol. A* **15**, 773-778.

Brown, H.G. and Hoh, J.H. (1997). Entropic exclusion by neurofilament sidearms: a mechanism for maintaining interfilament spacing. *Biochemistry* **36**, 15035-15040.

Burchard, W. (1994). Light Scattering Techniques. In *Physical Techniques for the Study of Food Biopolymers*, (ed. S.B. Ross-Murphy), chap. 4. 151-213. Blackie, London.

Bustamante, C, Vesenka, J, Tang, C.L, Rees, W, Guthold, M. and Keller, R. (1992). Circular DNA molecules imaged in air by scanning force microscopy. *Biochemistry* **31**, 22-26.

Bustamante, C, Keller, D. and Yang, G. (1993). Scanning force microscopy of nucleic acids and nucleoprotein assemblies. *Current Opinion Structural Biol.* **3**, 363-372.

Bustamante, C, Eire, D.A. and Keller, D. (1994a). Biochemical and structural applications of scanning force microscopy. *Current Opinion Structural Biol.* **4**, 750-760.

Bustamante, C, Marko, J.F, Siggia, E.D. and Smith, S. (1994b). Entropic elasticity of λ-phage DNA. *Science* **265**, 1599-1600.

Bustamante, C. and Rivetti, C. (1996). Visualising protein-nucleic acid interactions on a large scale with the scanning force microscope. *Ann. Rev. Biophys. Biomol. Structure* **25**, 395-429.

Butt, H-J. (1992a). Electrostatic interaction in scanning probe microscopy when imaging in electrolyte solutions. *Nanotechnology* **3**, 60-68.

Butt, H-J. (1992b). Measuring local surface charge densities in electrolyte solutions with a scanning force microscope. *Biophys. J.* **63**, 578-582.

Capron, I, Alexander, S. and Muller, G. (1998). An atomic force microscopy study of the molecular organisation of xanthan. *Polymer* **39**, 5725-5730.

Caruso, F, Rodda, E. and Furlong, D.N. (1996). Orientational aspects of antibody immobilisation and immunological activity on quartz crystal microbalance electrodes. *J. Colloid & Interface Sci.* **178**, 104-115.

Caruso, F, Furlong, D.N. and Kingshott, P. (1997). Characterisation of ferritin adsorption onto gold. *J Colloid & Interface Sci.* **186**, 129-140.

Chernoff, E.A.G. and Chernoff, D.A. (1992). Atomic force microscope image of collagen fibers. *J Vac. Sci. & Technol. A* **10**, 596-599.

Chidsey, C.E.D, Loiacono, D.N, Sleator, T. and Nakahara, S. (1988). STM study of the surface morphology of gold on mica. *Surface Sci.* **200**, 45-66.

Clemmer, C.R. and Beebe, T.P. (1991). Graphite-a mimic for DNA and other biomolecules in scanning tunnelling microscopic studies. *Science* **251**, 640-642.

Clemmer, C.R. and Beebe, T.P. (1992). A review of graphite and gold surface studies for use as substrates in biological scanning tunnelling microscopy studies. *Scanning Microscopy* **6**, 319-333.

Cowman, M.K, Li, M. and Balazs, E.A. (1998a). Tapping mode atomic force microscopy of hyaluronan: extended and intramolecularly interacting chains. *Biophys. J.* **75**, 2030-2037.

Cowman, M.K, Lui, J, Li, M, Hittner, D.M. and Kim, J.S. (1998b). Hyaluronan interactions: self, water, ions. In *The chemistry, biology and medical applications of hyaluronan and its derivatives*, (ed.T. Laurent), pp.17-24. Portland Press, London.

Dammer,U, Popescu, O, Wagner, P, Anselmetti, D, Guntherodt, H-J. and Misevic, G.N. (1995). Binding strength between cell adhesion proteoglycans measured by atomic force microscopy. *Science* **267**, 1173-1175.

Davies, J, Roberts, C.J, Dawkes, A.C, Sefton, J, Edwards, J.C, Glasbey, T.O, Haymes, A.G, Davies, M.C, Jackson, D.E, Lomas, M, Shakessheff, K.M, Tendler, S.J.B, Wilkins, M.J. and Williams, P.M. (1994). Use of scanning probe microscopy and surface plasmon resonance as analytical tools in the study of antibody-coated microtiter wells. *Langmuir* **10**, 2654-2661.

DeGrooth, B.G. and Putman, C.A.J. (1992). High resolution imaging of chromosome-related structures by atomic force microscopy. *J Microscopy* **168**, 239-247.

Drake, B, Prater, C.B, Weisenhorn, A.L, Gould, S.A.C, Albrecht, T.R, Quate, C.M, Cannel, D.S, Hansma, H.G. and Hansma, P.K. (1989). Imaging crystals, polymers, and processes in water with the atomic force microscope. *Science* **243**, 1586-1589.

Engel, A. (1991). Biological applications of scanning probe microscopes. *Ann. Rev. Biophys. Biophys. Chem.* **20**, 79-108.

Erie, D.A, Yang, G, Schultz, H.C. and Bustamante, C. (1994). DNA bending by Cro protein in specific and non-specific complexes-implications for protein site recognition and specificity. *Science* **266**, 1562-1566.

Fang, Y. and Hoh, J.H. (1998). Surface-directed DNA condensation in the absence of soluble multivalent cations. *Nucleic Acids Res.* **26**, 588-593.

Fang, Y, Spisz, T.S, Wiltshire, T, D'Costa, N.P, Bankman, I.N, Reeves, R.H. and Hoh, J.H. (1998). Solid-state DNA sizing by atomic force microscopy. *Anal. Chem.* **70**, 2123-2129.

Fannon, J.E, Hauber, R.J. and BeMiller, J.N. (1992). Surface pores of starch granules. *Cereal Chem.* **69**, 284-288.

Finch, J.T. and Klug, A. (1976). Solenoidal model for superstructure in chromatin. *Proc. Natl. Acad. Sci. USA* **73**, 1897-1901.

Firtel, M. and Beveridge, T.J. (1995). Scanning probe microscopy in microbiology. *Micron.* **26**, 347-362.

Frank, B.P and Belfort, G. (1997). Intermolecular forces between extracellular polysaccharides measured using the atomic force microscope. *Langmuir* **13**, 6234-6240.

Frazier, R.C, Davies, M.C, Matthijs, G, Roberts, C.J, Schacht, E, Tendler, S.J.B. and Williams, P.M. (1997a). High-resolution atomic force microscopy of dextran monolayer hydration. *Langmuir* **13**, 4795-4798.

Frazier, R.A, Davies, M.C, Matthijs, G, Roberts, C.J, Schacht, E, Tendler, S.J.B. and Williams, P.M. (1997b). In situ surface plasmon resonance analysis of dextran monolayer degradation by dextranase. *Langmuir* **13**, 7115-7120.

Fritz, J, Anselmetti, D, Jarchow, J. and Fernandez-Busquets, X. (1997). Probing biomolecules with atomic force microscopy. *J. Structural Biol.* **119**, 165-171.

Fritz, M, Radmacher, M, Cleveland, J.P, Allersma, M.W, Stewart, R.J, Gieselmann, R, Janmey, P. Schmidt, C.F. and Hansma, P.K. (1995). Imaging globular and filamentous proteins in physiological buffer solutions with tapping mode atomic-force microscopy. *Langmuir* **11**, 3529-3535.

Fritzsche, W, Schaper, A. and Jovin, T.M. (1994). Probing chromatin with the scanning force microscope. *Chromosoma* **103**, 231-236.

Fritzsche, W. and Henderson, E. (1996a). Scanning force microscopy revealed ellipsoid shape of chicken erythrocyte nucleosomes. *Biophys. J.* **71**, 2222-2226.

Fritzsche, W. and Henderson, E. (1996b). Ultrastructural characterisation of chicken erythrocyte nucleosomes by scanning force microscopy. *Scanning* **18**, 138-139.

Fritzsche, W. and Henderson, E (1996c). Volume determination of human metaphase chromosomes scanning force microscopy. *Scanning Microscopy* **10**, 103-110.

Fritzsche, W, Martin, L, Dobbs, D, Jondle, D, Miller, R, Vesenka, J. and Henderson, E. (1996). Reconstruction of ribosomal subunits and rDNA chromatin by scanning force microscopy. *J. Vac. Sci. & Technol. B* **14**, 1405-1409.

Fritzsche, W, Takac, L. and Henderson, E. (1997). Application of atomic force microscopy to visualisation of DNA, chromatin, and chromosomes. *Critical Rev. Eukaryotic Gene Expression* **7**, 2311-240.

Fujita, Y, Kobayashi, K. and Hoshino, T. (1997). Atomic force microscopy of collagen molecules. Surface morphology of segment-long-spacing (SLS) crystallites of collagen. *J. Electron Microscopy* **46**, 321-326.

Gale, M, Pollansen, M.S, Markiewicz, P. and Goh, M.C. (1995). Sequential assembly of collagen revealed by atomic force microscopy. *Biophys. J.* **68**, 2124-2128.

Gunning, A.P, Kirby, A.R, Morris, V.J, Wells, B. and Brooker, B.E. (1995). Imaging bacterial polysaccharides by AFM. *Polymer Bull.* **34**, 615-619.

Gunning, A.P, Kirby, A.R. and Morris, V.J. (1996a) Imaging xanthan gum in air by ac "tapping" mode atomic force microscopy. *Ultramicroscopy* **63**, 1-3.

Gunning, A.P, Kirby, A.R, Ridout, M.J, Brownsey, G.J. and Morris, V.J. (1996b). Investigation of gellan networks and gels by atomic force microscopy. *Macromolecules* **29**, 6791-6796.

Gunning, A.P, Morris, V.J, Al-Assaf, S. and Phillips, G.O. (1996c). Atomic force microscopic studies of hylan and hyaluronan. *Carbohydr. Polym.* **30**, 1-8.

Gunning, A.P, Kirby, A.R, Ridout, M.J, Brownsey, G.J. and Morris, V.J. (1997). Investigation of gellan networks and gels by atomic force microscopy (erratum). *Macromolecules* **30**, 163-164.

Gunning, A.P, Cairns, P, Kirby, A.R, Bixler, H.J. and Morris, V.J. (1998). Characterising semi-refined iota-carrageenan networks by atomic force microscopy. *Carbohydr. Polym.*36, 67-72.

Guthold, M, Bezanilla, M, Erie, D.A, Jenkins, B, Hansma, H.G. and Bustamante, C. (1994). Following the assembly of RNA polymerase-DNA complexes in aqueous solution with the scanning force microscope. *Proc. Natl. Acad. Sci. USA* **91**, 12927-12931.

Hagerman, P.J. (1988). Flexibility of DNA. *Ann. Rev. Biophys. Chem.* **17**, 265-286.

Hallett, P, Offer, G. and Miles, M.J. (1995). Atomic force microscopy of the myosin molecule. *Biophys. J*. **68**, 1604-1606.

Hallett, P, Tskhovrebova, L, Trinick, J, Offer, G. and Miles, M.J. (1996). Improvements in atomic force microscopy protocols for imaging fibrous proteins. *J. Vac. Sci. & Technol. B* **14**, 1444-1448.

Han, W, Mou, J, Sheng, J, Yang, J. and Shao, Z. (1995). Cryo atomic force microscopy: a new approach for biological imaging at high resolution. *Biochemistry* **34**, 8215-8220.

Hanley, S.J, Giasson, J, Revol, J-F. and Gray, D (1992). Atomic force microscopy of cellulose microfibrils; comparison with transmission electron microscopy. *Polymer* **33**, 4639-4642.

Hanley, S.J, Revol, J-F, Godbout, L. and Gray, D.G. (1997). Atomic force microscopy and transmission electron microscopy of cellulose from *Micrasterias denticulata* ; evidence for a chiral helical microfibril twist. *Cellulose* **4**, 209-220.

Hansma, H.G, Sinsheimer, R.L, Li, M.Q. and Hansma, P.K. (1992a). Atomic force microscopy of single and double stranded DNA. *Nucleic Acids Res.* **20**, 3585-3590.

Hansma, H.G, Vesenka, J, Siegerist, C, Kelderman, G, Morrett, H, Sinsheimer, R.L, Elings, V, Bustamante, C. and Hansma, P.K. (1992b). Reproducible imaging and dissection of plasmid DNA under liquid with the atomic force microscope. *Science* **256**, 1180-1184.

Hansma, H.G. and Hansma, P.K. (1993) Potential applications of atomic force microscopy of DNA to the human genome project. *Proc. SPIE Int. Opt. Eng. (USA)* **1891**, 66-70.

Hansma, H.G, Bezanilli, M, Zenhausern, F, Adrian, M. and Sinsheimer, R.L. (1993a). Atomic force microscopy of DNA in aqueous solutions. *Nucleic Acids Res.* **21**, 505-512.

Hansma, H.G, Sinsheimer, R.L, Groppe, J, Bruice, T.C, Elings, V, Gurley, G, Bezanilla, M, Mastrangelo, I.A, Howe, P.V.C. and Hansma, P.K. (1993b). Recent advances in atomic force microscopy of DNA. *Scanning* **15**, 296-299.

Hansma, H.G. and Hoh, J.H. (1994). Biomolecular imaging with the atomic force microscope. *Ann. Rev. Biomed. Structure* **23**, 115-139.

Hansma, H.G, Brown, K.A, Bezanilla, M. and Bruice, T.C. (1994). Bending and straightening of DNA induced by the same ligand: characterisation with the atomic force microscope. *Biochemistry* **33**, 8436-8441.

Hansma, H.G, Laney, D.E, Bezanilla, M, Sinsheimer, R. L. and Hansma, P.K. (1995). Applications for atomic force microscopy of DNA. *Biophys. J.* **68**, 1672-1677.

Hansma, H.G. (1996). Atomic force microscopy of biomolecules. *J. Vac. Sci. & Technol. B* **14**, 1390-1394.

Hansma, H.G. and Laney, D.E. (1996). DNA binding to mica correlates with cationic radius: assay by atomic force microscopy. *Biophys. J.* **70**, 1933-1939.

Hansma, H.G, Revenko, I, Kim, K. and Laney, D.E. (1996). Atomic force microscopy of long short double-stranded, single stranded and triple stranded nucleic acids. *Nucleic Acids Res.* **24**, 713-720.

Hansma, H.G, Kim, K.J, Laney, D.E, Garcia, R.A, Argaman, M, Allen, M.J. and Parsons, S.M. (1997). Properties of biomolecules measured from atomic force microscopic images: a review. *J. Structural Biol.* **119**, 99-108.

Hansma, H.G, Golan, R, Hsieh, W, Lollo, C.P, Mullen-Ley, P. and Kwoh, D. (1998). DNA condensation for gene therapy as monitored by atomic force microscopy. *Nucleic Acids Res.* **26**, 2481-2487.

Harada, A, Yamaguchi, H. and Kamachi, M. (1997). Imaging antibody molecules at room temperature by contact mode atomic force microscope. *Chemistry Letts.* part 11, 1141-1142.

Haugstad, G, Gladfelter, W.L, Keyes, M.P. and Weberg, E.B. Atomic force microscopy of AgBr crystals and adsorbed gelatin films. (1993). *Langmuir* **9**, 1594-1600.

Heckl, W.M. (1992). Scanning tunneling microscopy and atomic force microscopy on organic and biomolecules. *Thin Solid Films* **210-211**, 640-647.

Henderson, E. (1992). Imaging and nanodissection of individual supercoiled plasmids by atomic force microscopy. *Nucleic Acids Res.* **20**, 445-447.

Heslop-Harrison, J.S, Leitch, A.R, Schwarzacher, T, Smith, J.B, Atkinson, M.D. and Bennett, M.D. (1989). The volume and morphology of human chromosomes in mitotic reconstructions. *Human Genet.* **84**, 27-34.

van Hulst, N.F, Garcia-Parajo, M.F, Moers, M.H.P, Veerman, J-A. and Ruiter, A.G.T. (1997). Near-field fluorescence imaging of genetic material: towards the molecular limit. *J. Structural Biol.* **119**, 222-231.

Ikai, A. (1996). STM and AFM of bio/organic molecules and structures. *Surface Sci. Reports* **26**, 261-332.

Ill, C.R, Kievens, V.M, Hale, J.E, Nakamura, K.K, Jue, R.A, Cheng, S, Melcher, E.D, Drake, B. and Smith, M.D. (1993). A COOH-terminal peptide confers regiospecific orientation and facilitates atomic force microscopy of an IgG1. *Biophys. J.* **64**, 919-924.

Jeffrey, A.M, Jing, T.W, DeRose, J.A, Vaught, A, Rekesh, D, Lu, F-X. and Lindsay, S.M. (1993). *Nucleic Acids Res.* **21**, 5896-5900.

Jondle, D.M, Ambrosio, L, Vesenka, J. and Henderson, E (1995). Imaging and manipulating chromosomes with the atomic force microscope. *Chromosome Research* **3**, 239-244.

Kalle, W.H.J, Macville, M.V.E, van der Corput, M.P.C, de Grooth, B.G, Tanke, H.J. and Raap, A.K. (1996). Imaging of RNA *in situ* hybridization by atomic force microscopy. *J. Microscopy* **182**, 192-199.

Karrasch, S, Hegerl, R, Hoh, J.H, Baumeister, W. and Engel, A. (1994). Atomic force microscopy produces faithful high-resolution images of protein surfaces in an aqueous environment. *Proc. Natl. Acad. Sci. USA* **91**, 836-838.

Kasas, S, Thomson, N.H, Smith, B.L, Hansma, P.K, Miklossy, J. and Hansma, H.G (1997a). Biological applications of the AFM: from single molecules to organs. *Int. J. Imaging Syst. Technol.* **8**, 151-161.

Kasas, S, Thomson, N.H, Smith, B.L, Hansma, H.G, Zhu, X, Guthold, M, Bustamante, C, Kool, E.T, Kashlev, M. and Hansma, P.K. (1997b). *Escherichia coli* RNA polymerase activity observed using atomic force microscopy. *Biochemistry* **36**, 461-468.

Kirby, A.R, Gunning, A.P. and Morris, V.J. (1995a). Imaging xanthan gum by atomic force microscopy. *Carbohydr. Res.* **267**, 161-166.

Kirby, A.R, Gunning, A.P, Morris, V.J. and Ridout, M. J. (1995b). Observation of the helical structure of the bacterial polysaccharide acetan by atomic force microscopy. *Biophys. J.* **68**, 360-363.

Kirby, A.R, Gunning, A.P. and Morris, V.J. (1996a). Imaging polysaccharides by atomic force microscopy. *Biopolymers* **38**, 355-366.

Kirby, A.R, Gunning, A.P, Waldron, K.W, Morris, V.J. and Ng, A. (1996b). Visualisation of plant cell walls by atomic force microscopy. *Biophys. J.* **70**, 1138-1143.

Kowalczyk, D, Marsault, J-P. and Slomkowski, S. (1996). Atomic force microscopy of human serum albumin (HSA) on poly(styrene/ acrolein) microspheres. *Colloid & Polymer Sci.* **274**, 513-519.

Kuutti, L, Peltonen, J, Pere, J. and Teleman, O. (1995). Identification and surface structure of crystalline cellulose studied by atomic force microscopy. *J. Microscopy* **178**, 1-6.

Lal, R. and John, S.A. (1994). Biological applications of atomic force microscopy. *Amer. J. Physiol.* **266** (*Cell Physiol.* **35**), C1-C21.

Lea, A.S, Pugnor, A, Hlady, V, Andrade, J.D, Herron, J.N. and Voss Jnr, E.W. (1992). Manipulation of proteins on mica by atomic force microscopy. *Langmuir* **8**, 68-73.

LeCam, E, Frechon, D, Barry, M, Fourcade, A. and Delain, E. (1994). Observation of binding and polymerization of Fur repressor onto operator-containing DNA with electron and atomic force microscopes. *Proc. Natl. Acad. Sci. USA* **91**,11816-11820.

Lee, I, Evans, B.R, Lane, L.M. and Woodward, J. (1996). Substrate-enzyme interactions in cellulase systems. *Bioresource Technol.* **58**, 163-169.

Leuba, S.H, Yang, G, Robert, C, Samori, B, van Holde, K, Zlatanova J. and Bustamante, C. (1994). Three-dimensional structure of extended chromatin fibers as revealed by tapping-mode scanning force microscopy. *Proc. Natl. Acad. Sci. USA* **91**, 11621-11625.

Li, H, Rief, M, Oesterhelt, F. and Gaub, H.E. (1998). Single-molecule force spectroscopy on xanthan by AFM. *Adv. Mater.* **3**, 316-319.

Li, J.W, Tian, F, Wang, C, Bai, C.L. and Cao, E.H. (1997). Possible multistranded DNA induced by acid denaturation-renaturation. *J. Vac. Sci. & Technol. B* **15**, 1637-1640.

Li, M-Q, Xu, L. and Ikai, A (1996). Atomic force microscope imaging of ribosome and chromosome. *J. Vac. Sci. & Technol. B* **14**, 1410-1412.

Luginbuehl, R. and Sigrist H. (1998). Light-dependent substrate functionalization and biomacromolecule immobilization. In *Procedures in Scanning Probe Microscopy*, (eds. R.J.Colton, A. Engel, J.E. Frommer, H.E. Gaub, A.A. Gewirth, R. Guckenberger, J. Rabe, W. Heckl and B Parkinson), module 7.12.2, pp 488-492. J. Wiley & Sons, New York.

Lyubchenko, Y.L, Gall, A.A, Shlyakhtenko, L.S, Harrington, R.E, Oden, P.I, Jacobs, B.L. and Lindsay, S.M. (1992a). Atomic force microscopy imaging of double stranded DNA and RNA. *J. Biomolecular Structural Dynamics* **9**, 589-606.

Lyubchenko, Y.L, Jacobs, B.L. and Lindsay, S.M. (1992b). Atomic force microscopy of reovirus dsRNA: a routine technique for length measurements. *Nucleic Acids Res.* **20**, 3983-3986.

Lyubchenko, Y.L, Oden, P.I, Lampner, D, Lindsay, S.M. and Dunker, K.A. (1993). Atomic force microscopy of DNA and bacteriophage in air, water and propanol: the role of adhesion forces. *Nucleic Acids Res.* **21**, 1117-1123.

Lyubchenko, Y.L, Jacobs, B.L, Lindsay, S.M. and Stasiak, A (1995). Atomic force microscopy of nucleoprotein complexes. *Scanning Microscopy* **9**, 705-727.

Lyubchenko, Y.L. and Shlyakhtenko, L.S. (1997). Visualization of supercoiled DNA with atomic force microscopy *in situ* . *Proc. Natl. Acad. Sci. USA* **94**, 496-501.

Maaloum, M, Pernodet, N. and Tinland, B. (1998). Agarose gel structure using atomic force microscopy: gel concentration and ionic strength effects. *Electrophoresis* **19**, 1606-1610.

Mackie, A.R, Gunning, A.P, Ridout, M.J. and Morris, V.J. (1998). Gelation of gelatin: observation in the bulk and at the air-water interface. *Biopolymers* **46**, 245-252.

Marszalek, P.E, Oberhauser, A.F, Pang, Y-P. and Fernandez, J.M. (1998). Polysaccharide elasticity governed by chair-boat transitions of the glucopyranose ring. *Nature* **396**, 661-664.

Marsh, T.C, Vesenka, J. and Henderson, E. (1995). A new DNA nanostructure, the G-wire, imaged by scanning probe microscopy. *Nucleic Acids Res.* **23**, 696-700.

Martin, L.D, Vesenka, J.P, Henderson, E. and Dobbs, D.L. (1995). Visualization of nucleosomal structure in native chromatin by atomic force microscopy. *Biochemistry* **34**, 4610-4616.

McIntire, T.M, Penner, R.M. and Brant, D.A. (1995). Observations of a circular, triple helical polysaccharide using noncontact atomic force microscopy. *Macromolecules* **28**, 6375-6377.

McIntire, T.M. and Brant, D.A. (1997a). Imaging carbohydrates polymers with noncontact mode atomic force microscopy. In *Techniques in Glycobiology*, (eds. R.R. Townsend and A.T. Hotchkiss, Jnr), chapter 12, 187-206. Marcel Dekker, New York.

McIntire, T.M. and Brant, D.A. (1997b). Imaging of individual biopolymers and supramolecular assemblies using noncontact atomic force microscopy. *Biopolymers* **42**, 133-146.

McIntire, T.M. and Brant, D.A. (1998). Observation of the $(1\rightarrow3)$-β-D glucan linear triple helix to macrocycle interconversion using noncontact atomic force microscopy. *J. Amer. Chem. Soc.* **120**, 6909-6919.

McKeown, T.A, Moss, S.T. and Jones, E. B. G. (1996). Atomic force and electron microscopy of sporangial wall microfibrils in *Linderina pennispora*. *Mycol. Res.* **100**, 821-826.

McMaster, T.J, Hickish, T, Min, T, Cunningham, D. and Miles, M.J. (1994). Application of scanning force microscope to chromosome analysis. *Cancer Genet. Cytogenet.* **76**, 93-95.

McMaster, T.J, Winfield, M.O, Baker, A.A, Karp, A. and Miles, M.J. (1996a). Chromosome classification by atomic force microscopy volume measurement. *J. Vac. Sci. & Technol. B* **14**, 1438-1443.

McMaster, T.J, Winfield, M.O, Karp, A. and Miles, M.J. (1996b). Analysis of cereal chromosomes by atomic force microscopy. *Genome* **39**, 439-444.

Meyer, A, Rouquet, G, Lecourtier, L. and Toulhoat, H. (1992). Characterisation by atomic force microscopy of xanthan in interaction with mica. In *Physical Chemistry of Colloids and Interfaces in Oil Production*, (eds. H. Toulhoat and J. Lecourtier), pp. 275-278. Editions Techniq., Paris..

Mori, O. and Imae, T. (1997). AFM investigation of the adsorption process of bovine serum albumin on mica. *Colloids & Surfaces B Biointerfaces* **9**, 31-36.

Morris, V.J. (1994). Biological applications of scanning probe microscopies. *Prog. Biophys. Mol. Biol*. **61**,131-185.

Morris, V.J, Gunning, A.P, Kirby, A.R, Round, A.N, Waldron, K. and Ng, A. (1997). Atomic force microscopy of plant cell walls, plant cell wall polysaccharides and gels. *Int. J. Biol. Macromolecules*. **21**, 61-66.

Morris, V.J. (1998a). Atomic Force Microscopy. *The European Food and Drink Review*. Spring, 17- 21.

Morris, V.J. (1998b). Applications of atomic force microscopy in food science. In *Gums and Stabilisers for the Food Industry 9*, (eds. P.A. Williams and G.O. Phillips), pp. 361-370. Royal Society Chemistry, Cambridge, Special Publication no. 218.

Morris, V.J. (1998c). Gelation of polysaccharides. *In Functional Properties of Food Macromolecules*, (eds. S.E. Hill, D.A. Ledward and J.R. Mitchell), pp. 143-226. Aspen Publishers, Gaithersburg, USA, chapter 4.

Morris, V.J, Mills, E.C.N, Mackie, A.R, Wilde, P, Kirby, A.R. and Gunning, A.P. (1999). Probing biopolymer functionality in foods with the atomic force microscope. *Food Industry J.* **2**, 11-35.

Mosher, C, Jondle, D, Ambrosio, L, Vesenka, J. and Henderson, E. (1995). Microdissection and measurements of polytene chromosomes using the atomic force microscope. *Scanning Microscopy* **8**, 491-497.

Mou, J, Czajkowsky, D.M, Zhang, Y. and Shao, Z. (1995a). High resolution atomic-force microscopy of DNA: the pitch of the double helix. *FEBS Letters* **371**, 279-282.

Mou, J, Yang, J. and Shao, Z. (1995b). Atomic force microscopy of cholera toxin B-oligomers bound to bilayers of biologically relevant lipids. *J. Mol. Biol.* **248**, 507-512.

Mou, J, Sheng, S.J, Ho, R. and Shao, Z. (1996a). Chaperonins GroEL and GroES: views from atomic force microscopy. *Biophys. J.* **71**, 2213-2221.

Mou, J, Czajkowsky, D.M, Sheng, S.J, Ho, R. and Shao, Z. (1996b). High resolution surface structure of *E. coli* GroES oligomer by atomic force microscopy. *FEBS Letters* **381**, 161-164.

Mulhern, P.J, Blackford, B.L, Jericho, M.H, Southam. G. and Beveridge, T.J. (1992). AFM and STM studies of the interaction of antibodies with S-layer sheath of the archaeobacterium *Methanospiririllum hungatei*. *Ultramicroscopy* **42**, 1214-1224.

Muller, D.J. and Engel, A. (1997). The height of biomolecules measured with the atomic force microscope depends on electrostatic interactions. *Biophys. J.* **73**, 1633-1644.

Muller, D.J, Fotiaddis, D, Scheuring, S, Muller, S.A. and Engel, A. (1999). Electrostatically balanced subnanometer imaging of biological specimens by atomic force microscope. *Biophys. J.* **76**, 1101-1111.

Murray, M.N, Hansma, H.G, Bezanilla, M, Sano, T, Ogletree, D.F, Kolbe, W, Smith, C.L, Cantor, C.R, Spengler, S, Hansma, P.K. and Salmeron, M. (1993). *Proc. Natl. Acad. Sci. USA* **90**, 1037-1038.

Musio, A, Mariani, T, Frediani, C, Sbrana, I. and Ascoli, C. (1994). Longitudinal patterns similar to G-banding in untreated human chromosomes: evidence from atomic force microscopy. *Chromosomes* **103**, 225-229.

Nakata, S, Kido, N, Hayashi, M, Hara, M, Sasabe, H, Sugawara, T .and Matsuda, T. (1996). Chemisorption of proteins and their thiol derivatives onto gold surfaces: characterisation based on electrochemical non-linearity. *Biophys. Chem.* **62**, 63-72.

Neagu, C, van der Werf, K.O, Putman, C.A.J, Kraan, Y.M, de Grooth, B.G, van de Hulst, N.F. and Greve, J. (1994). Analysis of immunolabelled cells by atomic force microscopy, optical microscopy and flow cytometry. *J. Structural Biol.* **112**, 32-40.

Nettikadan, S, Tokumasu, F. and Takeyasu, K. (1996). Quantitative analysis of the transcription factor Ap2 binding to DNA by atomic force microscopy. *Biochem. Biophys. Res. Commun.* **226**, 645-649.

Ng, J.D, Kuznetsov, Y.G, Malkin, A.J, Keith, G, Giege, R. and McPherson, A. (1997). Visualiazation of RNA crystal growth by atomic force microscopy. *Nucleic Acids Res.* **25**, 2582-2588.

Oberleithner, H, Schneider, S. and Henderson, R.H. (1996). Viewing the renal epithelium with the atomic force microscope. *Kidney & Blood Pressure Res.* **19**, 142-147.

Ohnesorge, F, Heckl, W.M, Häberle, W, Pum, D, Sara, M, Schindler, H, Schilcher, K, Kiener, A, Smith, D.P.E, Sleytr, U.B. and Binnig, G. (1992). Scanning force microscopy studies of the S layers from *Bacillus coagulans* E38-66, *Bacillus sphaericus* CCM2177 and of an antibody binding process. *Ultramicroscopy* **42**, 1236-1242.

Ohta, T, Nettikadan, S, Tokumasu, F, Ideno, H, Abe, Y, Kuroda, M, Hayashi, H and Takeyasu, K. (1996). Atomic force microscopy proposes a new model for stem-loop structure that binds a heat shock protein in the *Staphylococcus aureus* HSP70 operon. *Biochem. Biophys. Res. Commun.* **226**, 730-734.

Paige, M.F, Rainey, J.K. and Goh, M.C. (1998). Fibrous long spacing collagen ultrastructure elucidated by atomic force microscopy. *Biophys. J.* **74**, 3211-3216.

Parkhouse, R.M, Askonas, B.A. and Dourmashkin, R. R. (1970). Electron microscopic studies of mouse immunoglobulin M structure and reconstruction following reduction. *Immunology* **18**, 575-584.

Pearce, K.H, Hiskey, R.G. and Thompson, N.H. (1992). Surface binding kinetics of prothrombin fragment 1 on planar membranes measured by total internal reflection fluorescence microscopy. *Biochemistry* **31**, 5983-5995.

Perkins, S.J, Nealis, A.S, Sutton, B.J. and Feinstein, A. (1991). Solution structure of human and mouse immunoglobulin M by synchrotron X-ray scattering and molecular graphics modeling. *J. Mol. Biol.* **221**, 1345-1366.

Pernodet, N, Maaloum, M. and Tinland, B. (1997). Pore size of agarose gels by atomic force microscopy. *Electrophoresis* **18**, 55-58.

Perrin, A, Lanet, V. and Theretz, A. (1997). Quantification of specific immunological reactions by atomic force microscopy. *Langmuir* **13**, 2557-2563.

Pietrasanta, L.I, Schaper, A. and Jovin, T.M. (1994). Probing specific molecular conformations with the scanning force microscope. Complexes of plasmid DNA and anti-Z-DNA antibodies. *Nucleic Acids Res.* **22**, 3288-3292.

Power, D, Larsen, I, Hartley, P, Dunstan, D. and Boger, D.V. (1998a). Molecular images of gels formed under shear using atomic force microscopy. In *Gums and Stabilisers for the Food Industry 9*, (eds. P.A. Williams and G.O. Phillips), pp. 388-394. Royal Society Chemistry, Cambridge, Special Publication no. 218.

Power, D, Larsen, I, Hartley, P, Dunstan, D. and Boger, D.V. (1998b). Atomic force microscopy studies on hydroxypropylguar gels formed under shear. *Macromolecules* **31**, 8744-8748.

Puppels, G.J, Putman, C.A.J, De Grooth, B.G. and Greve, J. (1992). Raman microspectroscopy and atomic force microscopy of chromosomal banding patterns. *SPIE Proc.* **1922**, 145-155.

Putman, C.A.J, de Grooth, B.G, Hansma, P.K, van Hulst, N.F. and Greve, J. (1993a). Immunogold labels: cell-surface markers in atomic force microscopy. *Ultramicroscopy* **42**, 177-182.

Putman, C.A.J, de Grooth, B.G, Wiegrant, J, Raap, A.K, van der Werf, K.O, van Hulst, N.F. and Greve, J. (1993b). Detection of in situ hybridisation to human chromosomes with the atomic force microscope. *Cytometry* **14**, 356-361.

Quinto, M, Ciancio, A .and Zambonin, P.G. (1998). A molecular resolution AFM study of gold-adsorbed glucose oxidase as influenced by enzyme concentration. *J. Electroanalytical Chem.* **448**, 51-59.

Quist, A.P, Bjorck, L.P, Reimann, C.T, Oscarsson, S.O. and Sundqvist, B.U.R. (1995). A scanning force microscopy study of human serum albumin and porcine pancreas trypsin adsorption on mica surfaces. *Surface Sci.* **325**, L406-L412.

Radmacher, M, Fritz, M, Cleveland, J.P, Walters, D.R. and Hansma, P.K. (1994). Imaging adhesion forces and elasticity of lysozyme adsorbed on mica with the atomic force microscope. *Langmuir* **10**, 3809-3814.

Radmacher, M, Fritz, M. and Hansma, P.K. (1996). Imaging soft samples with the atomic force microscope-gelatin in water and propanol. *Biophys. J.* **69**, 264-270.

Rampino, N.J. (1992.) Cisplatin induced alterations in oriented fibers of DNA studies by atomic force microscopy. *Biochem. Biophys. Res. Commun.* **182**, 201-207.

Rasch, P, Wiedemann, U, Wienberg, J. and Heckl, W.M. (1993). Analysis of banded human chromosomes and in situ hybridisation patterns by scanning force microscopy. *Proc. Natl. Acad. Sci. USA* **90**, 2509-2511.

Raspanti, M, Alessandrini, A, Gobbi, P. and Ruggeri, A. (1996). Collagen fibril surface: TMAFM, FEG-SEM and freeze-etching observations. *Microscopy Res. & Technique* **35**, 87-93.

Raspanti, M, Alessandrini, A, Ottani, V. and Ruggeri, A. (1997). Direct visualisation of collagen-bound proteoglycans by Tapping-mode atomic force microscopy. *J. Structural Biol.* **119**, 118-122.

Rees, W.A, Keller, R.W, Vesenka, J.P, Yang, G. and Bustamante, C. (1993). Scanning force microscopy imaging of transcription complexes: evidence for DNA bending in open promoter and elongation complexes. *Science* **260**, 1646-1649.

Revenko, I, Sommer, F, Minh, D.T, Garrone, R. and Franc, J-M. (1994). Atomic force microscopy of the collagen fibril. *Biol. Cell* **80**, 67-69.

Ridout, M.J, Brownsey, G.J, Gunning, A.P. and Morris, V.J. (1998). Characterisation of the polysaccharide produced by *Acetobacter xylinum* strain CR1/4 by light scattering and atomic force microscopy. *Int. J. Biol. Macromolecules* **23**, 287-293.

Rief, M, Oesterhelt, F, Heymann, B. and Gaub, H.E. (1997a). Single molecule force spectroscopy on polysaccharides by atomic force microscopy. *Science* **275**, 1295-1297.

Rief, M, Gautel, M, Oesterhelt, F, Fernandez, J.M. and Gaub, H.E. (1997b). Reversible unfolding of individual titin immunoglobulin domains by AFM. *Science* **276**, 1109-1112.

Rivetti, C, Guthold, M. and Bustamante, C. (1996). Scanning force microscopy of DNA deposited onto mica: equilibrium versus kinetic trapping studied by statistical polymer chain analysis. *J Mol. Biol.* **264**, 919-932.

Roberts, C.J, Williams, P.M, Davies, J, Dawkes, A.C, Sefton, J, Edwards, J.C, Haymes, A.G, Bestwick, C, Davies, M.C. and Tendler, S.J.B. (1995). Real space differentiation of IgG and IgM antibodies deposited on microtiter wells by scanning force microscopy. *Langmuir* **11**, 1822-1826.

Round, A. N, Kirby, A.R. and Morris, V.J. (1996) Collection and processing of AFM images of plant cell walls. *Microscopy & Analysis* **55**, 33-35.

Round, A.N, MacDougall, A.J, Ring, S.G. and Morris, V.J. (1997). Unexpected branching in pectins observed by Atomic Force Microscopy. *Carbohydr. Res.* **303**, 251-253.

Samori, B, Nigro, C, Armentano, V, Cimieri, S, Zuccheri, G. and Quagliariello, C. (1993). DNA supercoiling imaged in 3 dimensions by scanning force microscopy. *Angew. Chem. Int. Ed. Engl.* **32**, 1461-1463.

Saoudi, B, Lacapere, J-J, Chatenay, D, Pepin, R, Derpirre, C. and Sartre, A. (1994). Imaging surface of gold-immunolabelled thin sections by atomic force microscopy. *Biol. Cell* **80**,63-66.

Schindler, H. (1980). Formation of planar bilayers from artificial or native membrane vesicles. *FEBS Letters* **122**, 77-79.

Schurholz, Th. and Schindler, H. (1991). Lipid-protein surface films generated from membrane vesicles: self-assembly, composition, and film structure. *Eur. Biophys. J.* **20**, 71-78.

Schleif, R. and Hirsh, J. (1980). Electron microscopy of proteins bound to DNA. *Method Enzymol.* **65**, 885-896.

Schneider, S.W, Larmer, J, Henderson, R.M. and Oberleithner, H. (1998). Molecular weights of individual proteins correlate with molecular volumes measured by atomic force microscopy. *Pflugers Arch-Eur. J. Physiol.* **435**, 362-367.

Shaiu, W.L, Larson, D.D, Vesenka, J. and Henderson, E. (1993a). Atomic force microscopy of oriented linear DNA molecules labelled with 5 nm gold spheres. *Nucl. Acids Res.* **21**, 99-103.

Shaiu, W.L, Vesenka, J, Jondle, D, Henderson, E. and Larson, D.D. (1993b). Visualization of circular DNA molecules labelled with colloidal gold spheres using atomic force microscopy. *J. Vac. Sci. & Technol. A* **11**, 820-823.

Shao, Z, Yang, J. and Somlyo, A.P. (1995). Biological atomic force microscopy: from microns to nanometers and beyond. *Ann. Rev. Cell Dev. Biol.* **11**, 241-265.

Shao, Z. and Zhang, Y. (1996) Biological cryo atomic force microscopy: a brief review. *Ultramicroscopy* **66**, 141-152.

Shao, Z, Mou, J, Czajkowski, D.M, Yang, J. and Yuan, J-Y. (1996). Biological atomic force microscopy: what is achieved and what is needed. *Adv. Phys.* **45**, 1-86.

Shattuck, M.B, Gustafsson, M.G.L, Fisher, K.A, Yanagimoto, K.C, Veis, A, Bhatnagar, R.S. and Clarke, J. (1994). Monomeric collagen imaged by cryogenic force microscopy. *J. Microscopy* **174**, RP1-RP2.

Shivji, A.P, Davies, M.C, Roberts, C.J, Tendler, S.J.B. and Wilkinson, M.J. (1996). Molecular surface morphology studies of beta-amyloid self-assembly: Effect of pH on fibril formation. *Protein & Peptide Letts.* **3**, 407-414.

Smith, B.L, Gallie, D.R, Le, H. and Hansma, P.K. (1997). Visualization of poly(A)-binding protein complex formation with poly(A) RNA using atomic force microscopy. *J. Structural Biol.* **119**, 109-117.

Soon, L.L.L, Bottema, C. and Breed W.G. (1997). Atomic force microscopy and cytochemistry of chromatin from marsupial spermatozoa with special reference to *Sminthopsis crassicaudata* . *Mol. Reproduction & Development* **48**, 367-374.

Stein, P.E, Boodhoo, A, Armstrong, G.D, Cockle, S.A, Klein, M.H. and Reid, R.J. (1994). The crystal structure of pertussis toxin. *Structure.* **2**, 45-57.

Stine Jnr, W.B, Snyder, S.W, Ladror, U.S, Wade, W.S, Miller, M.F, Perun, T.J, Holzman, T.F. and Krafft, G.A. (1996). The nanometer-scale structure of amyloid-β visualised by atomic force microscopy. *J. Protein Chem.* **15**, 193-203.

Suzuki, A, Yamazaki, M. and Kobiki, Y. (1996). Direct observation of polymer gel surfaces by atomic force microscopy. *J. Chem. Phys.* **104**, 1751-1757.

Suzuki, A, Yamazaki, M, Kobiki, Y. and Suzuki, H. (1997). Surface domains and roughness of polymer gels observed by atomic force microscopy. *Macromolecules* **30**, 2350-2354.

Suzuki, A, Yamazaki, M, Kobiki, Y. and Suzuki, H. (1998). Surface roughness of polymer gels. In *The Wiley Polymer Networks Group Review Series, Volume 1*, (eds. T. Nijenhuis and W.J. Mijs), pp 489-503.

Tanigawa, M. and Okada,T. (1998). Atomic force microscopy of supercoiled DNA structure on mica. *Anal. Chem.* **365**, 19-25.

Tasker, S, Matthijs, G, Davies, M.C, Roberts, C.J, Schacht, E.H. and Tendler, S.J.B. (1996). Molecular resolution imaging of dextran monolayers immobilized on silica by atomic force microscopy. *Langmuir* **12**, 6436-6446.

Thalhammer, S, Stark, R.W, Muller, S, Wienberg, J. and Heckl W.M. (1997). The atomic force microscope as a new microdissecting tool for the generation of genetic probes. *J. Structural Biol.* **119**, 232-237.

Thimonier, J, Chauvin, J.P, Barber, J. and Rocca-Serra, J. (1995). Preliminary studies of an immunoglobulin M by near field microscopies. *J. Trace & Microprobe Techn.* **13**, 353-359.

Thomson, N.H, Miles, M.J, Ring, S.G, Shewry, P.R. and Tatham, A.S. (1994). Real-time imaging of enzymatic degradation of starch granules by atomic force microscopy. *J. Vac. Sci. & Technol. B* **12**, 1565-1568.

Thomson, N.H, Kasas, S, Smith, B, Hansma, H.G. and Hansma, P.K. (1996a). Reversible binding of DNA to mica for AFM imaging. *Langmuir* **12**, 5905-5908.

Thomson, N.H, Fritz, M, Radmacher, M, Cleveland, J.P, Schmidt, C.F. and Hansma, P.K. (1996b). Protein tracking and detection of protein motion using atomic force microscopy. *Biophys. J.* **70**, 2421-2431.

Thundat, T, Allison, D.P, Warmack, R.J, Doktycz, M.J, Jacobson, K.B. and Brown, G.M. (1993). Atomic force microscopy of single and double stranded deoxyribonucleic acid. *J. Vac. Sci. & Technol. A* **11**, 824-828.

Thundat, T, Allison, D.P. and Warmack, R.J. (1994). Stretched DNA observed with atomic force microscopy. *Nucleic Acids Res.* **22**, 4224-4228.

Valle, M, Valpuesta, J.M, Carrascosa, J.L, Tamayo, J. and Garcia, R. (1996). The interaction of DNA with bacteriophage ϕ29 connector: a study by AFM and TEM. *J. Structural Biol.* **116**, 390-398.

Vesenka, J, Guthold, M, Tang, C.L, Keller, D, Delanie, E. and Bustamante, C. (1992a). A substrate preparation for reliable imaging of DNA molecules with the scanning force microscope. *Ultramicroscopy* **42**, 1243-1249.

Vesenka, J, Hansma, H.G, Siegerist, C, Siligardi, G, Schabtach, E. and Bustamante, C. (1992b). Scanning force microscopy of circular DNA and chromatin in air and propanol. *SPIE Proc.* **1639**, 127-137.

Vesenka, J, Mosher, C, Schaus, S, Ambrosio, L. and Henderson, E. (1995). Combining optical and atomic force microscopy for life sciences research. *Biotechnique* **19**, 240-253.

Vinckier, A, Heyvaert, I, D'Hoore, A, McKittrick, T, Van Haesendonck, C, Engelborghs, Y. and Hellemans, L. (1995). Immobilizing and imaging microtubules by atomic force microscopy. *Ultramicroscopy* **57**, 337-343.

Wagner, P, Kernen, P, Hegner, M, Ungewickell, E. and Semenza, G. (1994). Covalent anchoring of proteins onto gold-directed NHS-terminated self-assembled monolayers in aqueous buffers-SFM images clathrin cages and triskelia. *FEBS Letts.* **356**, 267-271.

Wagner, P, Hegner, M, Kernen, P, Zaugg, F. and Semenza, G. (1996). Covalent immobilization of biomolecules onto Au (111) via N-hydroxysuccinimide ester functionalized self-assembled monolayers for scanning probe microscopy. *Biophys. J.* **70**, 2052-2066.

Van der Wel, N.H, Putman, C.A.J, Van Noort, S.J.T, de Grooth, B.G. and Emons, A,M.C. (1996). Atomic force microscopy of pollen grains, cellulose microfibrils, and protoplasts. *Protoplasma* **194**, 29-39.

Wiegrabe, W, Monajembashi, S, Dittmar, H, Greulich, K-O, Hafner, S, Hildebrandt, M, Kittler, M, Lochner, B. and Unger, E. (1997). Scanning near-field optical microscope: a method for investigating chromosomes. *Surface & Interface Analysis* **25**, 510-513.

Wilkins, M.J, Davies, M.C, Jackson, D.E, Mitchell, J.R, Roberts, C.J, Stokke, B.T. and Tendler, S.J.B. (1993). Comparison of scanning tunnelling microscopy and transmission electron microscopy image data of a microbial polysaccharide. *Ultramicroscopy* **48**, 197-201.

Winfield, M, McMaster, T.J, Karp, A. and Miles, M.J. (1996). Atomic force microscopy of plant chromosmes. *Chromosome Res.* **3**, 128-131.

Woodcock, C.L.F, Frado, L.L.Y. and Rattner, J.B. (1984). The higher-order structure of chromatin-evidence for a helical ribbon arrangement. *J. Cell Biol.* **99**, 42-52.

Woodcock, C.L, Grigoryev, S.A, Horowitz, R.A. and Whitaker, N. (1993). Chromatin folding model that incorporates linker variability generates fibers resembling the native structures. *Proc. Natl. Acad. Sci. USA* **90**, 9021-9025.

Wu, M. and Davidson, N. (1975). Use of gene 32 protein staining of single-strand polynucleotides for gene mapping by electron microscopy: application to the ϕ80d$_3$*ilvsu*+7 system. *Proc. Natl. Acad. Sci. USA* **72**, 4506-4510.

Wu, X. and Liu, W. (1997). Secondary structure of rat ribosomal RNAs studied by atomic force microscope. *Prog. Biochem. Biophys. (China)* **24**, 430-435.

Wu, X, Liu, W, Xu, L. and Li, M. (1997). Topography of ribosomes and initiation complexes from rat liver as revealed by atomic force microscopy. *Biol. Chem.* **378**, 363-372.

Wyman, C, Grotkoop, E, Bustamante, C. and Nelson, H.C,M. (1995). Determination of heat-shock transcription factor 2 stochiometry at looped DNA complexes using scanning force microscopy. *EMBO J.* **14**, 117-123.

Yamamoto, S, Hitomi, J, Shigeno, M, Sawaguchi, S, Abe, H. and Ushiki, T. (1997). Atomic force microscopic studies of isolated collagen fibrils of the bovine cornea and sclera. *Arch. Histol. Cytol.* **60**, 371-378.

Yang, J, Takeyasu, K. and Shao, Z. (1992). Atomic force microscopy of DNA molecules. *FEBS Letters* **301**, 173-176.

Yang, J, Tamm, L.K, Somlyo, A.P. and Shao, Z. (1993a). Promises and problems of biological atomic force microscopy. *J. Microscopy* **171**, 183-198.

Yang, J, Tamm, L.K, Tillack, T.W. and Shao, Z. (1993b). New approach for atomic force microscopy of membrane proteins: the imaging of cholera toxin. *J. Mol. Biol.* **229**, 286-290.

Yang, J, Mou, J. and Shao, Z. (1994a). Molecular resolution atomic force microscopy of soluble proteins in solution. *Biochim. Biophys. Acta* **1199**, 105-114.

Yang, J, Mou, J. and Shao, Z. (1994b). Structure and stability of pertussis toxin studied by in situ atomic force microscopy. *FEBS Letters* **338**, 89-92.

You, H.X. and Lowe, C. (1996). AFM studies of protein adsorption 2. Characterisation of immunoglobulin G adsorption by detergent washing. *J. Colloid & Interface Sci.* **182**, 586-601.

Zenhausern, F, Adrian, M, Heggeler-Bodrier, B, Emch, R, Jobin, M, Taborelli, M. and Descouts, P. (1992a). Imaging of DNA by scanning force microscopy. *J. Structural Biol.* **108**, 69-73.

Zenhausern, F, Adrian, M, Heggeler-Bordier, B, Eng, L.M. and Descouts, P. (1992b). DNA and RNA polymerase/DNA complexes imaged by scanning force microscopy: influence of molecular-scale friction. *Scanning* **14**, 212-217.

Zhang, P, Bai, C, Cheng, Y, Fang, Y, Feng, L. and Pan, H. (1996). Direct observation of uncoated spectrin with atomic force microscope. *Science in China (Series B)* **39**,378-385.

Zhang, Y, Sheng, S.J. and Shao, Z. (1996). Imaging biological structures with the cryo atomic force microscope. *Biophys. J.* **71**, 2168-2176.

Zhang, Y, Shao, Z, Somlyo, A.P. and Somlyo, A.V. (19970. Cryo-atomic force microscopy of smooth muscle myosin. *Biophys. J.* **72**, 1308-1318.

Zlatanova, J, Leuba, S.H, Yang, G, Bustamante, C. and van Holde, K. (1994). Linker DNA accessibility in chromatin fibers of different conformation-a reevaluation. *Proc. Natl. Acad. Sci. USA* **91**, 5277-5280.

CHAPTER 5

INTERFACIAL SYSTEMS

5.1. Introduction to interfaces

The term interfacial system covers a vast array of biological samples ranging from the phospholipid bilayer membranes which encase animal cells, self-assembly of proteins into 2D crystalline arrays on bacterial surfaces, to the interfacial layers in colloidal dispersions such as emulsions or foams in foods. The major use of AFM in this area is the investigation of the structure of such interfaces. However, before discussing what has been done with the AFM it is useful to introduce certain quantities used to define and characterise surfaces and interfaces.

5.1.1. Surface activity

A fundamental property of any interface between two phases is the existence of a definite quantity of free energy associated with every unit of interfacial area. Interfacial energy is related to interfacial or surface tension and sometimes these terms are used interchangeably. Interfacial tension in the simplest case of liquids, can be understood by considering intermolecular interactions. Within the bulk phase van der Waals forces which hold the molecules in a liquid form, act uniformly on each molecule in every direction, hence the net force on each molecule is on average zero. However, this does not apply to the molecules located at the interface, as they only interact with molecules underneath them in the bulk of the material. Fig. 5.1 illustrates this point schematically. These unbalanced forces at the surface can be equated to a free energy per unit area of surface. The surface free energy results in a contracting force, due to the net inward attraction from the molecules below the surface. Surface tension (γ) is defined as the amount of work required to increase the surface area by a unit amount (Fig. 5.2), therefore the surface tension (γ) and surface free energy are related. The interfacial or surface tension (γ) is defined as the force (F) per unit length required to expand the surface isothermally (Fig. 5.2). This means that the surface tension can also be described in terms of the work required to expand the surface area (A) by a unit amount.

Since reversible work and free energy (G) are equivalent then:

$$\gamma = \left(\frac{\partial G}{\partial A} \right)_{V_1 T_1}$$ (5.1)

where V is the volume and T the temperature.

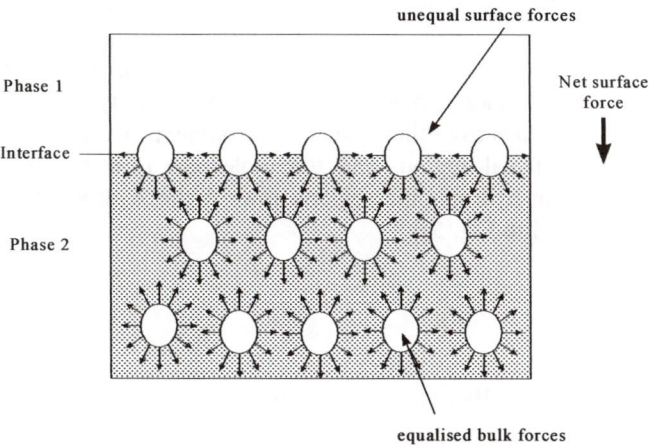

Figure 5.1. Surface tension: molecules in the bulk of phase 2 have equal van der Waals forces acting on all the molecules. Molecules at the surface only interact with neighbours at the surface or in the bulk, resulting in a net inward force. This has the effect of making a liquid behave as if its' surface is enclosed in an elastic skin.

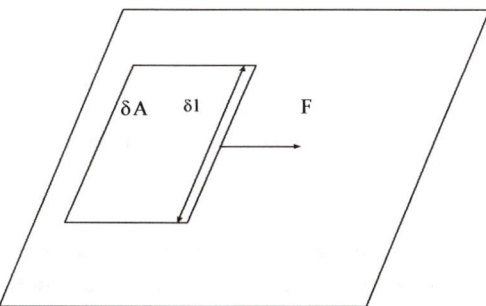

Figure 5.2. Surface tension γ is the force F per unit length δl required to expand the surface isothermally. Increasing the surface area by an amount δA leads to an increase in free energy δG. γ is the work done per unit area leading to equation 5.1.

Certain solutes, even when present in only very low concentrations, alter the surface energy of their solvents to an extreme degree. Generally the effect is to lower the surface tension, and such substances are known as *surface active* agents or *surfactants*. Although the term surfactant can be used to describe any surface active molecule it is generally used to describe only relatively small surface active molecules such as lipids and not large ones such as proteins. In the following pages the term surfactant will be applied according to this convention. The most familiar surfactant is soap. The principle of action of a surface active species is self-assembly at the interfacial region to form a 'film'. Most surface active molecules contain hydrophilic and hydrophobic regions, i.e. they are amphiphilic. In general, the assembly at the interface is driven by expulsion of the hydrophobic regions of the amphiphilic molecules from the aqueous environment. A concentration gradient is setup at the interfacial region, and the molecules near the interface orient themselves with their hydrophobic regions toward the hydrophobic side of the interface, as illustrated in Fig. 5.3.

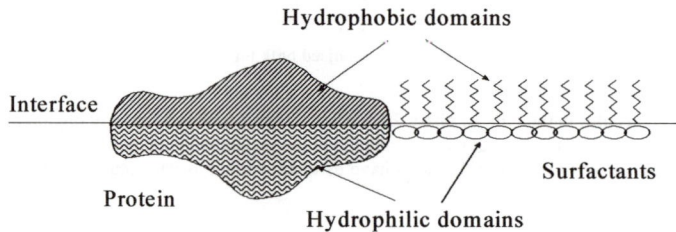

Figure 5.3. Assembly of surface active molecules at an interface by exclusion of hydrophobic regions of the molecules from the hydrophilic (aqueous) phase.

The amount of surface active molecules present at the interface can be quantified by measuring the change in surface tension. Interfacial films are not purely of intellectual curiosity, they do actually confer useful physical qualities to the interface, the most important of them are rheological characteristics. To illustrate this it is instructive to consider how foams are formed and stabilised by surfactants.

Obviously the formation of a foam increases the interfacial area greatly and, from a thermodynamic point of view, this can be explained simply by the fact that surfactants lower the surface free energy, meaning that the increase in surface area is energetically possible. However, lowering of interfacial energy is not the whole story since if all that was required for a stable foam was low interfacial energy, then pure liquids with low surface tensions such as alcohols would be able to form stable foams, which in reality they cannot. In fact it is the nature of the surfactant molecules themselves which enable the foam to be stabilised. They do this by the so-called Gibbs-Marangoni mechanism. The surfactants are highly mobile in the interfacial region; that is they can diffuse laterally with relative ease (Marangoni mechanism) and, if soluble, there is also a reasonably high rate of exchange of molecules between the bulk liquid and the surface (Gibbs mechanism). Thinning of the interfacial layer between gas bubbles in a foam, if not checked would quickly lead to its collapse. In a surfactant stabilised foam such thinning causes localised depletion of the concentration of the surfactant molecules. Because of their lateral mobility within the interface, and their ability to adsorb from the bulk liquid, other surfactant molecules quickly move into the depleted region to restore equilibrium (Fig. 5.4).

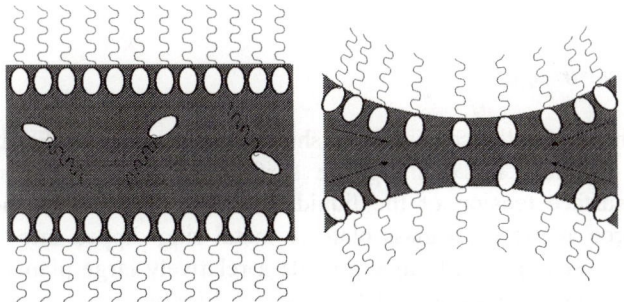

Figure 5.4. When a film is stretched, restoration of interfacial film thickness occurs by the migration of surfactant molecules with adherent water into the depleted (thin) region of the film.

As the surfactant molecules move into the depleted region they bring an attached water layer with them so that the film thickness is restored, the thin spot is 'healed' and the bubble stabilised. It is easy to see how this effect confers a real physical attribute to the interfacial film. In fact it is defined as the interfacial elasticity and the elastic modulus (E) is defined by the rate of change of surface tension with the rate of change of area, thus:

$$E = \frac{d\,\gamma}{d \ln A} \tag{5.2}$$

Interfacial rheology plays an important role in the biological functions of the cell membrane, for instance membrane fluidity affects cell fusion.

5.1.2. AFM of interfacial systems

Interfacial regions are often inaccessible making direct imaging with the AFM problematic. For example, in an oil in water emulsion oil droplets are soft, spherical and vary in size over a range of several microns. Similarly, the phospholipid bilayer which surrounds animal cells are not only curved but also are very soft meaning that molecular resolution AFM images of cell membranes are extremely difficult to obtain *in-situ*. As a final example, the interfacial films in a protein foam are difficult to image by AFM because the bubbles themselves are not rigid enough to scan directly. All of these examples present problems which need to be overcome in order to allow high resolution examination of interfaces by AFM. The most convenient way to study interfacial systems is to recreate them in the controlled environment of a Langmuir trough.

5.1.3. The Langmuir trough

Langmuir troughs generally consist of a shallow rectangular PTFE bath with moveable PTFE barriers for defining the surface area, and some means of monitoring the surface tension of the liquid, usually a delicate balance which quantifies the force acting on a glass plate or metal ring dipped into the liquid surface. Perhaps most importantly they provide a relatively large planar interface which has distinct advantages for AFM imaging. Firstly, the interface can be imaged directly in the 'submarine' AFM (section 2.9.2) or the interfacial film can be transferred onto a suitable substrate without distortion of its shape, and then imaged by AFM. The interfacial film can be formed in two ways, and these are partially dependant on the solubility of the material of interest. Firstly, by carefully adding the sample solution dropwise to the surface of the liquid in the trough an interfacial film can be formed; this technique is known as spreading. Alternatively for some soluble surface active molecules such as proteins, the interfacial film can be formed at the interface by simply filling the trough with a solution of the sample, and allowing the molecules to adsorb to the interface by self-assembly; a technique known as adsorption.

So far, the basic principles of interfacial tension and elasticity have been discussed. Another parameter, directly linked to the surface or interfacial tension, and one which is particularly relevant to the Langmuir trough, is the *surface pressure* (Π). The surface pressure defines the tendency of a surface to spread out in a way analogous to the molecules in a gas, but is confined to a two dimensional area rather than a three dimensional volume. The surface pressure of a given interface is defined as the difference between the surface tension of the bare interface (γ_0) and the surface tension of the interfacial film (γ_1) and is given by

$$\Pi = \gamma_0 - \gamma_1 \qquad (5.3)$$

Since there is this direct relationship between surface tension and surface pressure it can be seen that the more molecules which are present in an interfacial film the higher will be the surface pressure. This reinforces the gas analogy mentioned earlier, meaning that surface pressure is a useful concept because it can describe the state of packing of the molecules in an interfacial film. For this reason it is often quoted instead of surface tension. Fig. 5.5 illustrates how measurement of surface pressure versus interfacial area (in a graph known as a Π-A isotherm) can define the various states of an interfacial film.

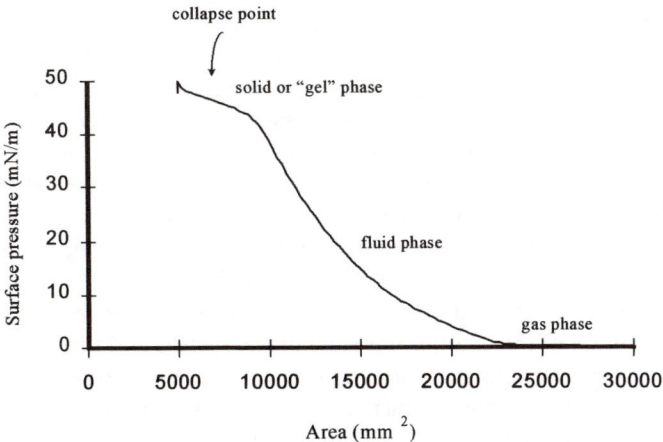

Figure 5.5. Typical Π-A isotherm of a phospholipid film showing gas, liquid, solid phase and collapse point.

At very low surface pressure (high surface tension, Fig 5.5) the interfacial film is composed of a very low density of amphiphilic molecules, and in for example a lipid interfacial layer, they would be described as being in the expanded 'gas phase', and as in a gas the molecules can diffuse freely, but within the surface. The packing density can be increased by adding more lipid to the surface or, more commonly, by reducing the interfacial area by moving the barriers inwards on the Langmuir trough. In either case the molecules are forced to pack more closely and the film reaches what is termed the 'fluid phase'; the molecules can still diffuse at this stage but are more confined (Fig. 5.5). If the area is reduced still further the molecules are close packed into what is known as the 'solid' or 'gel' phase. Now there is very little room for molecular diffusion (Fig. 5.5). Finally if the area of the interface is reduced beyond this point then the film will quickly be forced to collapse with the lipid molecules leaving the surface and entering the bulk or 'sub-phase' (Fig. 5.5).

Now that the interfacial film has been defined, how is it transferred to a substrate for AFM imaging?

5.1.4. Langmuir-Blodgett film transfer

Generally speaking the interfacial film is very soft and so must be transferred from the Langmuir trough onto a substrate in order to support it mechanically so that AFM imaging may be carried out. The principle is to pull the desired substrate through the interfacial film in a controlled manner. For a planar interface this is quite straightforward and can be done using Langmuir-Blodgett (LB) techniques. An excellent and detailed review of LB film structures can be found elsewhere (Shwartz, 1997). Measurement of surface pressure (or surface tension) can be used to monitor transfer of films to the substrate (Fig. 5.6). Because the surface tension of a given system is inversely related to the amount of the adsorbate at the interface a convenient measure of film transfer is surface pressure. The principle is that when some of the adsorbate is transferred onto a substrate during a dipping cycle, the surface pressure will also change. The bigger the substrate to trough ratio the larger the observed change in surface pressure. This change in surface pressure implies a disruption of the equilibrium state of the system and therefore a feedback loop is often employed to maintain constant surface pressure as the transfer progresses, by moving the trough barriers inwards reducing the surface area of the trough, and hence maintaining the packing of the interfacial layer. In this case a plot of the feedback correction signal during the dip will confirm film transfer. However, since the substrate area required for an AFM sample is usually small compared to the trough area such feedback control is not always necessary.

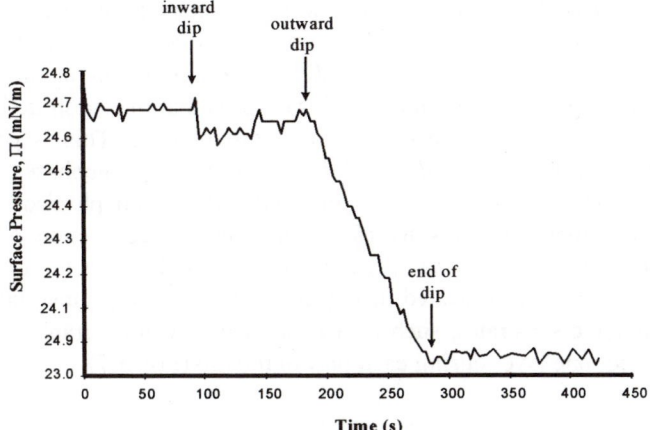

Figure 5.6. Film transfer during dipping is verified by a change in surface pressure. Data shown here is for a protein film transferring onto mica. Note that in this case that film transfer, as indicated by a change in surface pressure, occurs only on the outward portion of the dip. See Fig. 5.7. for further details.

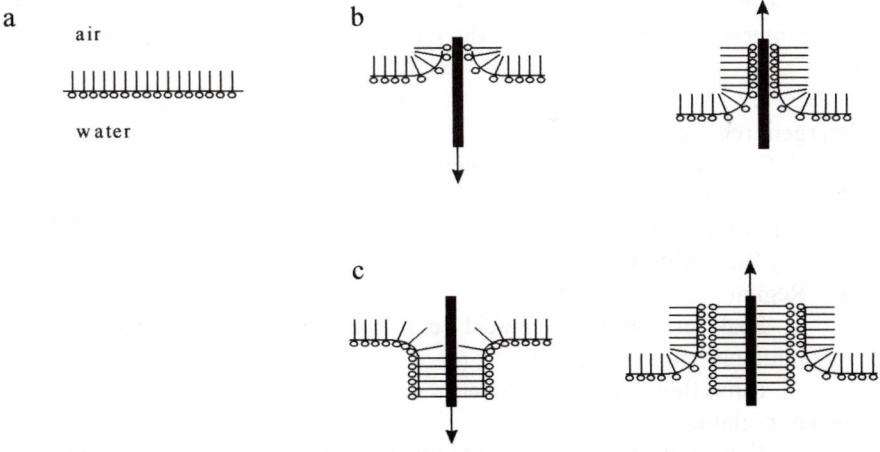

Figure 5.7. Langmuir-Blodgett film transfer from an air-water interface. (a) Molecules oriented with their polar head groups downward mean that upon dipping a hydrophilic substrate (b) transfer occurs only on the upward stroke producing monolayer coverage. With an hydrophobic substrate (c) transfer occurs during both strokes producing bilayer coverage.

The nature of the transferred film depends to some extent on the substrate that is dipped through the interface. It is easier for a number of reasons to start the dip with the substrate above the liquid and dip in and back out again. Principle amongst the advantages of doing things in this order is that the substrate is exposed to the bulk solution for only the minimum length of time. This is particularly important when dealing with *adsorbed* films, where the subphase may have significant concentrations of the sample molecule which can physisorb onto the substrate causing misleading results (for more details see section 5.7.1). For hydrophilic substrates such as mica only a monolayer will be transferred during the upward part of the dip, as illustrated in Fig. 5.7 and shown experimentally in Fig. 5.6. For hydrophobic substrates, such as acid treated silicon or graphite, a bilayer will be transferred, one layer during each part of the dip (Fig. 5.7).

5.2. Sample preparation

5.2.1. Cleaning protocols: glassware and trough

The following protocols can be used to clean glassware and the trough prior to forming interfacial films. One of the three cleaning regimes described below can be used.

1. Detergent regime:
   i. Clean thoroughly with free rinsing detergent and hot water.
   ii. Rinse with hot water.
   iii. Rinse with ethanol.
   iv. Rinse with water.
2. Acid Regime:
   i. Soak in concentrated acid (e.g. chromic acid, hydrofluoric acid or ammonium persulphate).
   ii. Rinse thoroughly with water.
3. Solvent regime:
   i. Rinse stepwise in solvents of increasing, then decreasing hydrophobicity. e.g. water-acetone-chloroform/methanol (2:1)-acetone water.

Note: Hydrofluoric acid (HF) is an extremely hazardous reagent and its use should not be undertaken lightly. This advice is particularly relevant when it is used to wash apparatus since relatively large volumes of HF are required, making it especially dangerous. Also it should be noted that contrary to expectation it actually becomes *more* dangerous with dilution since dissociation liberates more F⁻ ions.

To verify that the Langmuir trough is clean and that the water being used is 'surface pure' the surface tension should be measured before spreading the sample at the air-water interface. At 20°C the surface tension value for water is 72.6 mN m^{-1}. Surface active species act to reduce the surface tension and thus, in practice, a value of surface tension lower than 72.5 mN m^{-1} (at 20°C) indicates contamination. Remember though that surface tension varies with temperature and ionic strength (corrected values can be found in most data books). Finally, a good way of ensuring that there is no surface contamination, even at very low levels, is to reduce the trough area to its smallest extent by moving the barriers together whilst measuring the surface tension. This will close-pack any surface contaminants making them easier to detect by a drop in the surface tension. If the surface tension remains constant at 72.5-72.6 mN m^{-1} then the surface is clean.

5.2.2. Substrates

Interfacial films are of course highly sensitive to any surface active species which may come into contact with them so, in addition to cleaning the Langmuir trough itself and all glassware used during sample preparation, the whole of the substrate used for film transfer must be clean, not simply the face to be imaged.

Mica

In the case of mica this means cleaving *both* sides of the sheet being used and also trimming edges which may have been handled prior to dipping it through the interface. A rectangular piece of mica is cut from a larger sheet using clean scissors (rinsed with acetone and allowed to dry) which is then held at the top and bottom edges only whilst both sides are cleaved using two pairs of fine pointed tweezers. This rather acrobatic task is really not that difficult with a little practice. Alternatively, the mica can be cleaved on both sides by sandwiching it between two lengths of adhesive tape and then carefully peeling them apart. This second method, although easier, does make it more difficult to verify complete cleavage of the mica and good quality adhesive tape, i.e. tape with an even coating of adhesive, is a must. Once cleaved the mica should only be handled with clean tweezers.

Polished silicon wafers

The other useful AFM substrate for LB film transfer, and incidentally one which is more commonly used for LB film formation, is polished silicon. As is the case for mica rectangular sections of substrate make the dipping process easier. These can be broken out of a wafer by scribing the back of the silicon using a diamond glass cutter, lining up the scratch with the edge of a glass slide, and then pressing gently downward to snap off the chord of silicon (Fig. 5.8). Alternatively the scribed silicon can be broken by placing it face up on a pad of compliant material (such as a Gel-Pak™, Vichem Corporation, USA) and then broken by carefully placing a straight metal edge such as the end of a steel ruler opposite the scribed marks and tapping sharply on the other end. The trick when breaking the silicon is that thinner wafers (thickness $\approx$ 0.4 mm) are much easier to break than thicker ones, which can be anything up to 1 mm in thickness, and often break into about twenty tiny shards! The free piece can now be carefully scribed near the end to be dipped, so that a suitable sized piece which fits into the liquid cell of the AFM can be broken off after dipping.

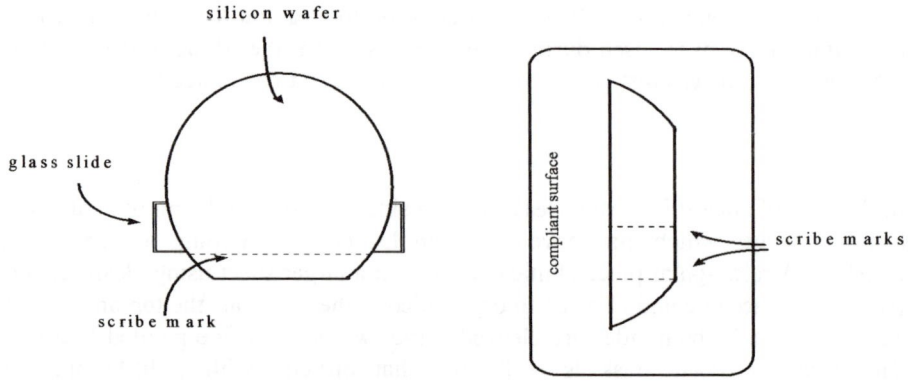

Figure 5.8. Sectioning a silicon wafer.

The silicon piece should be cleaned prior to dipping in a hot mixture of hydrogen peroxide and sulphuric acid (3:7), followed by a water rinse. This will remove any organic contamination leaving the native (hydrophilic) oxide surface. A further 'processing' step in 10% hydrofluoric acid will remove the oxide layer from the silicon leaving a hydrogen passivated hydrophobic silicon surface.

5.2.3. Performing the dip

After cleaning, a simple and convenient way to hold the substrate during dipping is to use a glass microscope slide fitted with a paper clip (Fig. 5.9). The substrate can simply be slid under the paper-clip. If using mica, the bottom edge should be trimmed off with the scissors since this may be contaminated by finger grease. Now one should have a perfectly clean substrate ready to dip.

Figure 5.9. Mica substrate being dipped through an air-water interface.

The advantage of using the glass slide now becomes clear: it can easily be attached to the dipping mechanism associated with the LB trough using a crocodile clip, thus eliminating the fiddly problems encountered if one tries attach the clean substrate directly to the dipper. The substrate can then be dipped through the interface of the liquid in the trough and out again, taking care not to let the bottom of the glass slide contact the liquid surface. The rate of the dip should be slow enough to ensure that minimal distortion of the interfacial film occurs during transfer. In practice we have found a rate of 8 mm per minute to be fine for protein films but for lipids an slower rate of about 2 mm per minute should be used. After this the substrate can be slid out from under the paper-clip and the dipped section cut to a suitable size to fit into the liquid cell of the AFM. For polished silicon the handling procedure is more difficult and the cutting must be done as described above (section 5.2.2).

5.3. Phospholipids

Phospholipids are a class of amphiphilic molecules of enormous importance in biology since phospholipid bilayers form the basic structure of the outer envelope, or plasma membrane, of all animal cells. Phospholipids are complex lipids consisting of polar (hydrophilic) head-groups containing glycerol with a phosphate group and two non-polar (hydrophobic) hydrocarbon tails. The tails are generally made up of fatty acids of varying length with one tail having one or more unsaturated *cis* double bonds which introduce kinks in the tail. Variation in the length and saturation of the tails affects the packing ability of the phospholipids, and thus confers rheological properties, such as fluidity to the membranes. Fluidity of the cell membrane is biologically important; for example certain transport processes and enzyme activities are inhibited when the bilayer viscosity increases beyond a threshold level. Due to their cylindrical shape phospholipid molecules spontaneously form bilayers in aqueous solution with the non-polar tails in the middle of the 'sandwich'. The bilayers eliminate free edges where the hydrophobic tails would be in water by forming compartments, and this tendency also means that they can re-seal if torn. The phospholipid bilayer acts as a two dimensional fluid in which membrane proteins are 'dissolved' and the individual lipid molecules are free to diffuse laterally and exchange places with their neighbours. Motion of phospholipid molecules between the monolayers on either side of the bilayer, known as 'flip-flop' motion, is less common, and this confinement creates a problem for the synthesis of the bilayers, since phospholipid molecules are synthesised in only one of the monolayers of a membrane, normally the cytosolic monolayer of the endoplasmic reticulum. If transfer of the molecules to the other side of the membrane could not occur no bilayer could be formed. Transfer is in fact mediated by a special class of membrane proteins called the phospholipid translocators, which catalyse the 'flip-flop' of the phospholipid molecules (Alberts *et al* 1994).

There are a large number of different phospholipids which are defined according to the substitution on the phosphorous atom. For example two of the most abundant in the plasma membranes of mammalian cells, phosphatidylcholine, also known as lecithin, and phosphatidylethanolamine have a choline group and an ethanolamine group attached via the phosphate respectively. The nature of the headgroup confers various properties to the molecules, such as charge and solubility, which have important bearing on the reactions of proteins and enzymes which reside within the membrane, or which interact with it. This specificity of the headgroup means that very often the lipid composition of the two halves of the bilayers (also known as leaflets) are different, for reasons of functionality, and this variability is controlled by the phospholipid translocators during synthesis in the endoplasmic reticulum. For example, the enzyme protein kinase C, which responds to extracellular signals, binds to the cytoplasmic face of the plasma membrane which is rich in phosphatidylserine, whose negative charge is required for enzymic catalysis. As mentioned above, as well as the chemical makeup of the plasma membrane, its physical characteristics also play an important role. The bilayer can change from being in a liquid like state to a rigid crystalline or gel-like 'frozen' state as the temperature is lowered. The temperature at which this happens is determined by the length and saturation of the hydrocarbon chains in the phospholipid molecules. Longer hydrocarbon chains allow neighbouring molecules to interact more strongly and so raise the 'freezing' temperature. *Cis*-double bonds (which represent an unsaturated structure) cause kinks in the chains which makes packing of molecules more difficult, and so lowers the 'freezing' temperature. This fact is used by the cells of primitive organisms such as yeasts, whose temperature is determined by their environment, and which increase the production of unsaturated phospholipids in their plasma membranes to preserve membrane fluidity when drops in temperature occur (Alberts *et al* 1994).

It can be seen that the detailed nature of the composition of the plasma membrane of cells, both in terms of its physical and chemical characteristics, determines functionality and so this is an area ideal for study by atomic force microscopy, offering the ability to examine both attributes at a highly localised level. As mentioned in the earlier section on sample preparation, the best way to image these materials is to prepare synthetic bilayers which can then be transferred onto suitable substrates for the AFM, since imaging of cell membranes *in situ* produces no resolvable detail and, even under optimised imaging conditions, damages the membrane structure (Schaus and Henderson, 1997). There are two ways of forming synthetic phospholipid bilayers experimentally. The first method is to create them in bulk aqueous solution in the form of spherical vesicles known as liposomes, the size of which can vary from 25 nm to 1 µm in diameter. The liposomes can then be adsorbed onto mica for AFM imaging (Shibata-Seki *et al* 1996) and, if high resolution is required, they can then be collapsed down to make them flat (Singh and Keller, 1991). The second method is to create a planar phospholipid bilayer at an air water interface on a Langmuir trough and then transfer this to the substrate by Langmuir-Blodgett dipping (Zasadzinski *et al* 1991). Artificially prepared model cell membranes which have been adsorbed to a planar solid substrate in this manner are known as supported bilayers.

5.3.1. AFM studies

In an early AFM study, membrane bilayers of dipalmitoyl phosphatidylcholine (DPPC) and dipalmitoyl phosphatidylethanolamine (DPPE) were adsorbed to a mica substrate and imaged after air-drying (Singh and Keller, 1991). Two methods were used for deposition of DPPC bilayers. In the first method the mica was incubated in a mixed solution of detergent and the lipid whilst the detergent was slowly removed by dialysis, resulting in the nucleation of small lipid islands which grew radially outwards to produce circular bilayer patches. The second method of deposition involved incubating mica in a suspension of lipid vesicles for 15-30 minutes, then blotting the excess liquid away with filter paper producing a nearly uniform coverage of lipid bilayer with a few cracks and salt crystals. It was noted that bilayer formation was sometimes enhanced by rolling the drop of lipid suspension around over the mica surface. Despite the use of relatively high forces of around 15 nN the bilayers were reproducibly imaged, although the researchers concluded that this force was near the limit above which damage to the bilayers occurred, as shown by areas which were deliberately damaged by scanning with high forces. Thickness measurements of the DPPC bilayers at 6.3 nm were in reasonable agreement with the expected values of 4-5.5 nm (Sackmann, 1983). When lipid vesicles formed from DPPE were deposited using the second method large circular discs up to 2 μm in diameter and 100-200 nm thick resulted, suggesting that, in this case, the vesicles remained intact and did not collapse. However, the samples had been stored at 4°C for several months prior to adsorption to the mica favouring the formation of very large vesicles in the suspension. The final point arising from the study is that rinsing of the adsorbed vesicles with water removed the vesicles, leaving only 'footprints' of inconsistent height where they had been adsorbed to the mica (Singh and Keller, 1991).

5.3.2. Modification of phospholipid bilayers with the AFM

A later study utilised the ability of the AFM to modify phospholipid bilayers composed of 1,2-bis (10,12-tricosadinoyl)-*sn*-glycerol-3-phosphocholine (DC$_{8,9}$PC) (Brandow *et al* 1993). In this case lipid tubules formed by cooling the lipid-solvent dispersion slowly through its transition temperature (Yager and Schoen, 1984) were deposited onto freshly cleaved highly oriented pyrolytic graphite (HOPG) and imaged in air after drying, whereupon they collapsed to form flat bilayers. Cutting of the lipid tubules was achieved by repeatedly scanning (some 2,500 times) over the same scan line until all of the material had been removed. A force of 12.9 ±1.0 nN was found to be the threshold needed to cut the tubules, irrespective of the angle of cut with respect to the tube axis. Above this force value the cutting process was quicker. When the cuts made to the tubules were narrow (< 80 nm) a process of self-annealing was observed over a period of 24 hours, indicating that the lipid molecules in the bilayer were mobile even in air, and that residual lipids were probably left on the graphite surface after cutting. However, for wider cuts this process could be halted. The AFM tip could, however, be used to heal such wider cuts, by shovelling lipid molecules into the cut, by scanning perpendicular to it with forces approximately 14% smaller than the threshold cutting force. The modification of the lipid tubules is shown in Fig. 5.10.

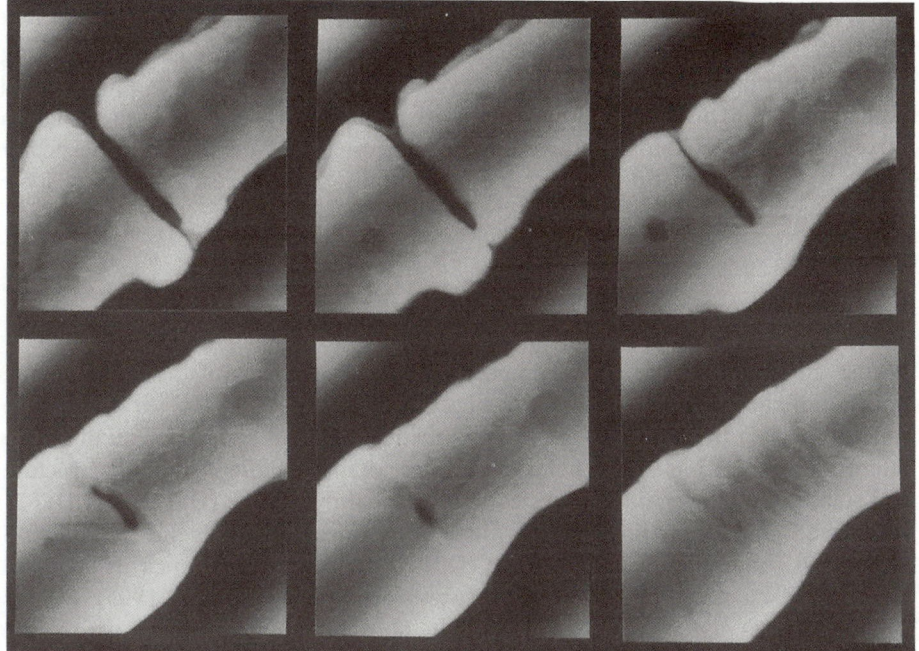

Figure 5.10. Time series of the annealing process of an 80 nm cut through a collapsed lipid tubule which occurs due to repeated scanning of the AFM tip perpendicular to the cut. Reproduced with permission from Brandow *et al* 1993 and the Biophysical Society.

Following polymerisation of the acyl chains of the $DC_{8,9}PC$ lipid tubules with UV irradiation, cutting of the tubules with the AFM tip proved to be impossible. The study was probably the first to demonstrate the ability of the AFM to image, manipulate and therefore potentially to measure and alter some fundamental properties of a lipid bilayer system, such as the intermolecular interactions which determine the viscosity in the bilayer (Brandow *et al* 1993).

Following this work several studies have examined the effect of the scanning tip of the AFM on phospholipid bilayers (Hui *et al* 1995; Knapp *et al* 1995), and intact liposomes (Shibati-Seki *et al* 1996). Knapp and co-workers treated AFM tips by glow discharge, in hexafluoropropene (HFP) to make them hydrophobic, and in air to make them hydrophilic. The interaction of the treated tips with the sample was then compared by scanning phospholipid bilayers composed of a bottom layer of DPPE and a top layer of DPPC supported on mica substrates (Knapp *et al* 1995). In addition to topography, friction and elasticity of the layers was measured proving that with modified tips and combined imaging modes the nature of the bilayer surface, whether hydrophobic or hydrophilic, could be determined. This raised the possibility that the AFM could be used to map a mixed lipid bilayer. Finally, the study demonstrated that dissection of particular layers with the AFM tip was possible by carefully selecting an appropriate force window in order to expose the middle of the bilayer 'sandwich'.

5.3.3. Studying intrinsic bilayer properties by AFM

Whilst the above studies concentrated mainly on the effects of the AFM on the imaging of phospholipid bilayers, there have been several other studies which have concentrated on the nature of the bilayers themselves, and so are potentially more interesting from a biologists' point of view. In a fairly detailed study the structure and stability of bilayer films of distearoyl phosphatidylcholine (DSPC) and DPPE, transferred as Langmuir-Blodgett films onto mica in the solid phase, and dilinoleoyl phosphatidylethanolamine (DLPE) transferred onto mica in the fluid phase, were examined by AFM under aqueous conditions (Hui *et al* 1995). Bilayer stability is an issue due the fact that phospholipid molecules which are asymmetrical tend to form curved monolayers, as illustrated schematically in Fig. 5.11. When these asymmetric lipids are incorporated into a bilayer each monolayer is constrained to a common curvature, giving rise to a bending or frustration energy which reduces the stability of the bilayer. Indeed, supplied with sufficient external energy, this stored energy can lead to the destruction of the bilayer (Seddon, 1990). The bending energy in phospholipid bilayers is believed to play an active role in the function of many membrane proteins (Hui and Sen, 1989).

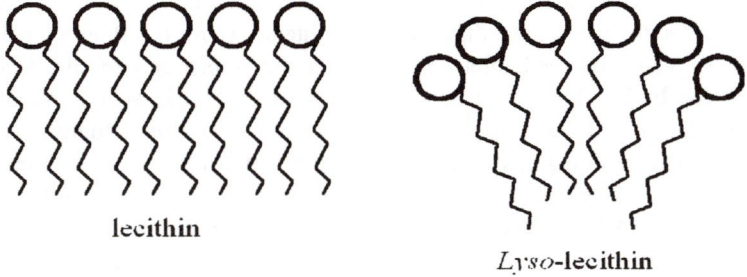

lecithin

Lyso-lecithin

Figure 5.11. Whereas symmetrical lipids such as lecithin form planar monolayers, asymmetric lipids such as *lyso*-lecithin pack in a wedge manner and tend to form curved monolayers.

The effect of headgroup on the structure of the bilayer was determined by comparing high magnification AFM images of bilayers composed of DPPE/DPPE and DPPE/DSPC. The bilayer composed entirely of DPPE revealed a periodic structure of ridges 0.49 nm apart, which is close to the hexagonal pattern seen for mica of spacing 0.52 nm, but this possibility was discounted since it agreed well with the crystallographic structure of a similar phospholipid, dilauroyl phosphatidylethanolamine. The periodicity seen on the DPPE/DPPE bilayer was therefore attributed to rows of aligned headgroups (Hui *et al* 1995). Furthermore, high magnification AFM images of the bilayer composed of DPPE/DSPC failed to reveal any lattice details, probably confirming that the periodicity seen for the pure DPPE bilayer was not simply due to an inadvertent imaging of the underlying mica lattice. If the results are taken at face value then they represent a demonstration that AFM imaging is capable of determining the effect of the phospholipid headgroup on the packing of molecules within a bilayer, since pure DPPE bilayers would be expected to pack into an ordered crystalline arrangement when in the 'solid' phase, whereas phosphatidylcholine exists in a less ordered gel-like solid phase.

Factors such as packing density, ordering of molecules and the number of layers when increased are all known to enhance the stability of supported phospholipid layers, but the role of bending energy upon stability had not been studied before. In the second part of the work Hui and coworkers studied this effect directly by creating bilayers with a natural tendency to form highly curved surfaces under experimentally controllable conditions (Hui *et al* 1995). By varying the pH of the solution in which the samples were imaged the charge on the headgroups of the phospholipid molecules was manipulated to produce high or low bending energy systems. For example, a monolayer of the unsaturated phospholipid DLPE has a high curvature below neutral pH so a bilayer composed of DPPE/DLPE was created in order to investigate this unstable system. At high values of pH (>11) some defects were observed after the water in the liquid cell of the AFM was exchanged for buffer, but no gross changes were observed. When the imaging buffer was exchanged for one of a low pH (pH 5.0) this situation changed dramatically. The initial scan revealed many more defects than were present at pH 11, and these defects grew in the period of scanning, eventually resulting in the removal of almost all of the top layer of DLPE, correlating with the increase in bending energy which occurs at low pH. These results are reproduced in Fig. 5.12.

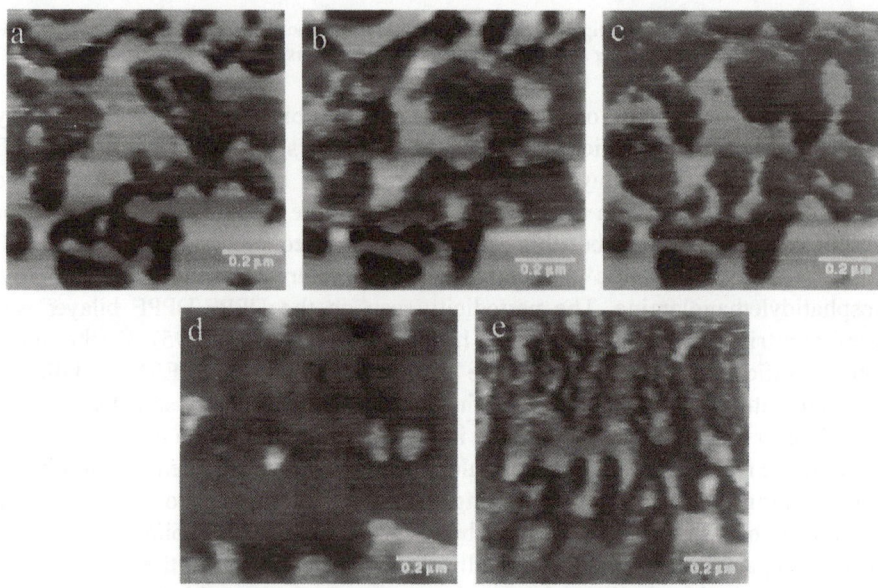

Figure 5.12 Sequence of AFM images of a DPPE/DLPE bilayer at pH5. (a-d) The top leaflet of unsaturated, fluid DLPE is systematically removed, indicating low cohesion between the leaflets at this value of pH. (e) After repeated scanning a regular pattern, perpendicular to the scanning direction appears. This is an artefactual phenomenon occurring when a soft material is rubbed by a hard material. The features are known as Shallamach waves. Reproduced with permission from Hui *et al* 1995 and the Biophysical Society.

Another interesting observation, and one which should serve as a cautionary reminder of the fact that AFM can affect soft samples, was that after removal of most of the top layer of DLPE, the residual material formed a pattern of roughly parallel lines in response to repeated scanning of the tip (Hui *et al* 1995). These lines are not real structural features, but rather an artefact known as Schallamach waves, which arise when hard objects slide over soft surfaces (Schallamach, 1971).

The effect of surface pressure on the molecular structure of monolayers of DPPC transferred onto quartz substrates has been examined by AFM (Zhai and Kleijn, 1997). At high and intermediate surface pressures, after the onset of a condensed liquid state, Langmuir-Blodgett films of the DPPC on quartz gave AFM images revealing molecular order, but films transferred from monolayers in the less packed liquid expanded state, as expected, revealed no order in the AFM images (Zhai and Kleijn, 1997).

5.3.4. Ripple phases in phospholipid bilayers

An interesting phenomenon which can occur in phospholipid bilayers is the so-called ripple phase which happens below the first order phase (main) transition temperature (Tardieu *et al* 1973). In stacks of bilayers it has been shown that in this ripple phase the surface of the membrane becomes wrinkled in a periodic manner (Rand *et al* 1975). Until recently the ripple phase had only been attributed to the temperature and hydration of the stacked bilayers. In an AFM study by Mou and co-workers a new ripple phase induced by a commonly used buffer compound, Tris (hydroxymethyl) aminomethane ($C_4H_{11}NO_3$), has been observed in single phospholipid bilayers of phosphatidylcholine (Mou *et al* 1994). Supported phospholipid bilayers were prepared by deposition of a sonicated suspension of vesicles in 20 mM NaCl onto freshly cleaved mica, followed by a short heating step above the main transition temperature (33 °C). The mica was then rinsed with water prior to imaging under appropriate buffer solution, with the bilayers never being exposed to air. Initially the bilayers were imaged in pure NaCl solution and shown to be flat and featureless, even after temperature cycling above the main transition temperature. By scraping a patch of the phospholipid layer off the mica with the AFM tip the depth of the layer ($\sim$ 6 nm) was measured in order to confirm that only a single bilayer was present. Upon the addition of 20 mM Tris in 50 mM NaCl the surface of the bilayer began to wrinkle and, after 16 hours, a pronounced ripple structure was observed with a periodicity of 18$\pm$ 2 nm, and amplitude of 0.3 nm. The sample was then heated above its transition temperature (42-50 °C) for half an hour and a second, wider, ripple phase formed with a periodicity of 32 $\pm$ 2 nm and an amplitude of 1.2 nm, coexisting with the narrower original phase. These results are reproduced in Fig. 5.13. Interestingly the thickness of the bilayers changed after ripple formation; for the narrow ripple phase the bilayer thickness went up to about 7 nm and, for regions of the wide ripple phase, the thickness went up to some 9-10 nm.

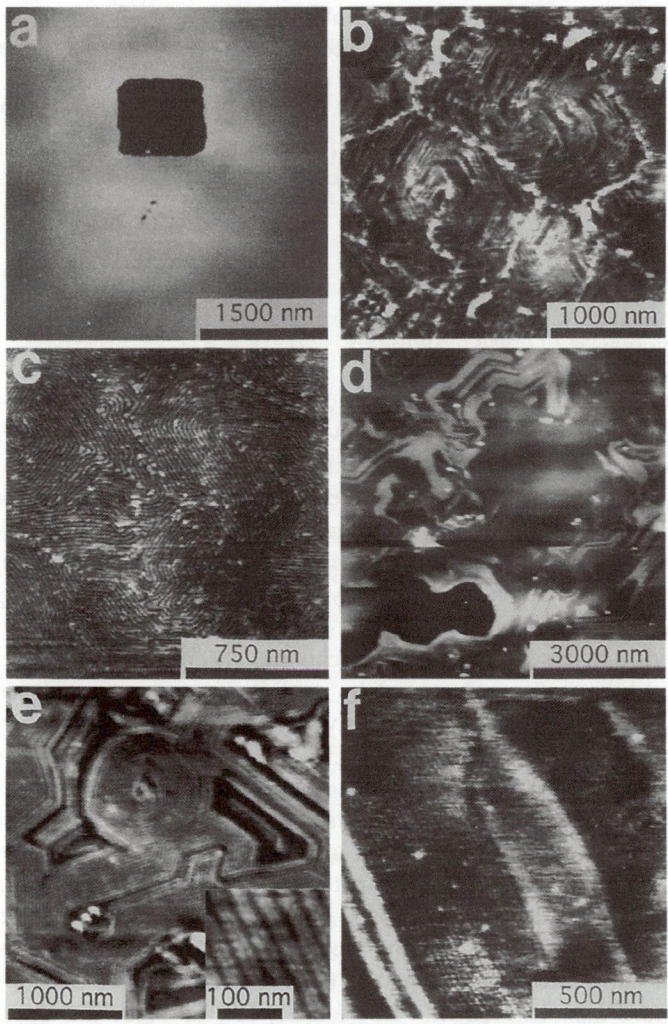

Fig 5.13. Ripple phases in bilayers of diC15-PC imaged by AFM. (a) diC15-PC bilayer imaged in pure NaCl solution (20 mM) displays no ripples. A square hole has been made in the bilayer by scanning at high speed and force so that the thickness of the bilayer could be measured (6 nm). (b) After 2 h incubation at 20 °C in 20 mM Tris and 50 mM NaCl some ripple-like features have started to form and the thickness of the bilayer increased (7 nm). (c) By 16 h incubation in the mixed buffer, the ripple structures are more pronounced, but the bilayer thickness was unchanged (7 nm). (d) Following heating of this sample above the main transition temperature (42-50 °C) two distinct ripple phases were seen, thick domains and thin domains. The thicker domains were 2-3 nm taller than the thin domains. (e) Higher magnification image of the thick domain showing very clearly the ripple structure. At even higher magnification (inset) each ripple can be seen to consist of two ridges. (f) High magnification image of the thinner domain. Reproduced with permission from Mou *et al* 1994, copyright (1994) American Chemical Society.

At high magnification the thicker ripple phase was seen to be made up of two ridges with a separation of around 11 nm. The different phases appeared to be relatively stable; at room temperature no conversion between these two phases was observed. When the Tris buffer was replaced by pure NaCl solution however, the ripple phases began to disappear, with the wider phase initially converting to the narrow one, and then the narrow phase disappearing, a process which took about 6 hours to complete at room temperature. The whole process was fully reversible and the ripple phases could be induced again by the addition of Tris buffer to the same sample and, furthermore, this process did not appear to damage the bilayers. The AFM with its inherent high resolution was able to reveal localised subtleties in the structure of the ripples, such as domain boundaries and edges where there were defects in the bilayer. Some examples of this are shown in Fig. 5.14.

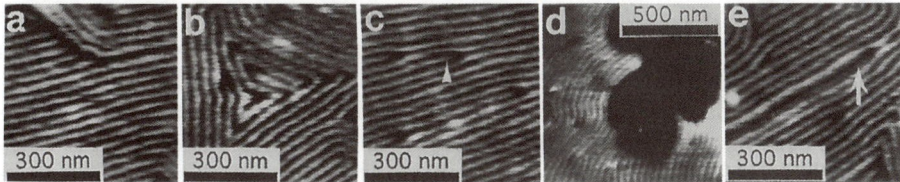

Figure 5.14. High magnification images of boundary regions in the ripple domains. (a) At a junction one domain is stopped and neither connects with nor crosses the other. (b) The triangular void in the centre of a region where three linear domains cross has been filled with a spiral ripple. (c) New ripples were sometimes observed to form in perfectly ordered domains without affecting the structure. (d) At a natural defect in the bilayer the ripple structure remained nearly flawless, both perpendicular and parallel to the edge: the ripple structure appears very stable and insensitive to defects. (e) Where two parallel domains have grown to meet, a thin gap was often seen which was too small to incorporate a complete ripple. However, in such gaps a weak line was occasionally observed (arrowed). Reproduced with permission from Mou *et al* 1994, copyright (1994) American Chemical Society.

The mechanism by which the Tris molecules cause ripples in the bilayers is not understood. Nevertheless the results demonstrated that the ripple phase can occur in single bilayers, and hence inter-bilayer interaction is not required. In a later study this work was extended to examine ripple phases in *asymmetric* phospholipid bilayers composed of either DPPC or DSPC in one leaflet and 1-palmitoyl-2-oleoylphosphatidylglycerol (POPG), 1-stearoyl-2-oleoylphosphatidylglycerol (SOPG), or 1- palmitoyl-2-oleoylphosphatidylethanolamine (POPE) in the other leaflet (Czajkowski *et al* 1995). For these systems the ripple phase was induced by imaging in phosphate buffered saline (PBS), the sodium and phosphate ions being the necessary components. The asymmetric phospholipid bilayers were formed on mica by sequential Langmuir-Blodgett film transfer of a monolayer of each phospholipid. The supported bilayer samples were then rinsed with the imaging solutions before being imaged with the AFM. Through experiments with differing concentrations of the component ions in PBS (K^+, Na^+, Cl^-, PO_4^{2-}) it was established that there was a specific requirement for sodium and phosphate if the ripple phase was to occur. It was noted that the presence of 0.5 mM potassium lowered the threshold level of sodium required for the onset of ripple phase by an order of magnitude. As before more than one periodicity was observed in the ripple phase. Most of the studies were made on bilayers of DPPC/POPG but the composition of each leaflet was also studied for different systems in order to determine whether there was any specificity of the lipid for the formation of ripple phase. Ripple phases were readily formed when the bottom layer was made up of saturated phosphatidylcholine and the top layer made up of a mixture of saturated and unsaturated lipids with small negatively charged headgroups, such as phosphatidylglycerol at pH 7.5, and phosphatidylethanolamine at pH 11.0. For bilayers with DSPC in the bottom layer the ripple phase formation was much slower, and no ripple phase was observed for symmetric bilayers of DPPC. So with PBS as the initiator only the asymmetric bilayers displayed ripple phases, indicating that the nature of the inter-leaflet coupling has some bearing on the ion-induced occurrence of the ripple phase. This result is an example of an AFM study which has genuinely revealed new information on a biological system and these observations cannot at present be accounted for by theories of ripple formation in phospholipid bilayers.

5.3.5. Mixed phospholipid films

In real cell membranes the phospholipid bilayers are not only asymmetric but generally composed of many different phospholipid molecules in each layer or leaflet. Therefore it would be nice if the AFM could map this distribution within a single layer. To date there are few papers on heterogeneous phospholipid films, but this is an area which is rife for study. In an interesting paper Langmuir-Blodgett films of the non-miscible phospholipids DPPE and 1,2 Di[(*cis*)-9-octadecanoyl]-sn-glycero-3-phosphoethanolamine) (DOPE) were formed on mica and HOPG and examined by AFM in air (Soletti *et al* 1996). The different substrates were used in order to examine the effect of the nature of the substrate on the transferred films and, in addition, the effect of surface pressure on film structure. Phase separation into distinct domains was clearly observed for the non-miscible phospholipids, since there was a difference in height of about 0.5 nm between the two species. The shape of the domains was dependent upon the molar ratio of the molecules. As the molar ratio was changed the shape of each domain varied from multi-lobed, to twin lobed, and then to discrete circular domains of one within the other. The relative areas of each domain corresponded well with the expected molecular area and molar ratio allowing identification of the phases in the AFM images. Another property measured was the coefficient of friction over the mixed films utilising the by friction force imaging mode of the AFM. The friction maps produced a clear difference in contrast, with the DOPE domains having a higher coefficient of friction than the DPPE domains, and appearing liquid-like. This is a significant observation, indicating that even if there were no size differences enabling topographic distinction of the lipid phases, one may be able to differentiate between different phospholipids by their frictional characteristics. The value of the surface pressure during film transfer of mixtures with a molar ratio of 0.75/0.25 of DPPE and DOPE effected the nature of the DPPE domains. As the collapse pressure for DPPE (45 mN m^{-1}) was approached cracks in the DPPE domains appeared accompanied by the some bilayer formation. As the surface pressure was increased to even higher values, 55 mN m^{-1}, bilayers were very clearly observed. All of the above measurements were made on films transferred onto mica. The effect of substrate on the structure seen in the transferred films was demonstrated with images of DPPE/DOPE mixtures on HOPG. The domains were no longer circular but irregularly shaped (Soletti *et al* 1996).

In a similar study mixed LB films, of distearoylphosphatidylethanolamine (DSPE) and DOPE formed on mica, were studied as monolayers in air and as bilayers in water (Dufrene *et al* 1997). The films were formed by LB dipping, the second leaflet of the bilayers being created on the downstroke, and the sample then kept under water by means of submerged beakers in the Langmuir trough. This allowed the LB films to be transferred to the liquid cell of the microscope without being exposed to the air, in which the bilayer was unstable. Once again phase separation was seen but, in this case, as well as in phospholipid monolayers (Fig 5.15), phase separation was observed in phospholipid bilayers (Fig. 5.16). Also observed was good contrast in friction and adhesion maps of the lipid domains, despite the fact that both have identical headgroups. Adhesion maps were obtained by performing force-distance curves at every eighth point during imaging (to produce an array of 64 x 64 pixels) and displaying the force at which the tip detaches from the sample surface at each of these points. The differences in topographic contrast seen in all three data sets was attributed to three factors; (I) the length of the phospholipid molecules, (II) the tilt assumed by the molecules in each phase and most importantly (III) the mechanical response of the different layers. This last factor is responsible for the contrast seen in the friction and adhesion images. The DOPE phases being fluid-like were inelastically deformed by the AFM tip. Furthermore, in the bilayers imaged under water, an additional contrast factor was observed; namely a short range repulsive force which was seen over the DSPE domains and which was speculated to arise from hydration or steric effects since the headgroup is identical in both phospholipids. This force manifested as a large height difference between the domains in the topography images (4.8± 0.7 nm!), which could not be accounted for by molecular length, tilt, or deformation. The origins of this large difference was discovered by close examination of the tip-sample force interactions that were taking place in the wet system. A large repulsive component was observed at a tip-sample separation of 3 nm in force versus distance curves recorded over the DSPE domains, suggesting that the tip hovered above the surface of these regions whilst imaging. By contrast, the force versus distance curves obtained over the DOPE regions suggested that the tip actually penetrated to a depth of some 1 nm whilst imaging.

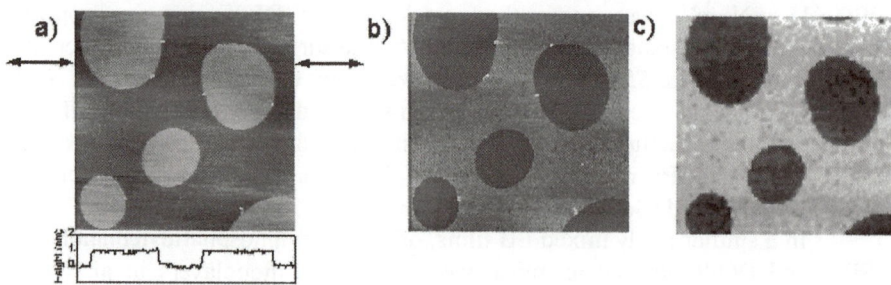

Figure 5.15. Phase separation of mixed monolayers of DSPE/DOPE on mica, imaged in air. (a) Topography image with line profile beneath, (b) friction image, (c) adhesion image (re-sampled to 512 x 512 pixels). In the images lighter areas correspond to larger height, friction and adhesion respectively. Scan size: 15 x 15 μm. Reproduced with permission from Dufrene *et al* 1997 and the Biophysical Society.

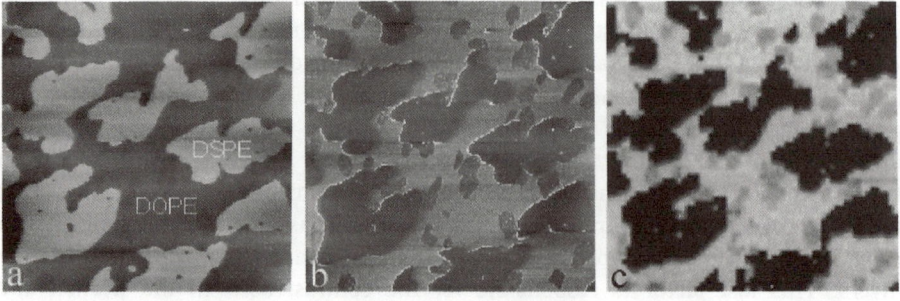

Figure 5.16. Phase separation observed in mixed phospholipid bilayers, imaged in water. The bottom leaflet of the bilayer (which is attached to the mica substrate) is composed of pure DSPE and the upper leaflet is composed of DSPE/DOPE. (a) Topography, (b) friction, (c) adhesion images (re-sampled to 512 x 512 pixels). As in Fig. 5.15 lighter areas correspond to larger height, friction or adhesion. Scan sizes: 15 x 15μm. Reproduced with permission from Dufrene *et al* 1997 and the Biophysical Society.

In a refinement of these studies on mixed films a chemically modified AFM tip was used to differentiate between phospholipids and glycolipids in a mixed bilayer, although at the time of writing these results were only reported in the form of an abstract (Dufrene *et al* 1998). However, such subtle chemical sensitivity is a highly promising step for the application of AFM to the study of real biological plasma membranes.

5.3.6. *Effect of supporting layers*

If more realistic plasma membrane models are to be studied, such as those incorporating proteins and enzymes, then simple LB film transfer directly onto a substrate is often not straightforward, due to factors such as the insolubility of membrane proteins and instability of the bilayers themselves. A more useful method in this case is the monolayer fusion technique (Kalb *et al* 1992). The first step in this technique involves LB transfer of a lipid monolayer onto a hydrophilic substrate. After this step vesicles containing reconstituted membrane proteins (Racker, 1985) are deposited onto the coated substrate, whereupon they form a supported lipid bilayer with incorporated proteins. This has the advantages that only small amounts of the protein are required, the native membrane environment of the proteins is preserved, and their orientation within the membrane will also be preserved. In addition to their application as model membranes, such lipid-protein vesicles are promising candidates for use in drug delivery systems, where they may provide a highly specific means of targeting particular cell or tissue types.

Supported phospholipid monolayers and bilayers, generated using the monolayer fusion technique, have been examined by AFM (Vikholm *et al* 1995) and a combination of AFM with the related SPM technique of scanning near field optical microscopy (SNOM - see section 8.3.) (Tamm *et al* 1996). In this study a phospholipid (either DPPC or DPPE) was transferred by LB methods onto a glass substrate to form the first supporting layer (Tamm *et al* 1996). Fluorescent lipid analogue probe molecules were mixed with these monolayers so that the structure of the layer could be characterised by SNOM, as an alternative to simple AFM examination of the surfaces. The SNOM images revealed that, under the transfer conditions used (30 mN m^{-1}), DPPC formed monolayers with numerous sub-micron sized crystallites which could easily be visualised because the fluorescent probe molecules were forced into the grain boundaries. The crystallisation of the layer would probably not have been obvious by AFM examination alone. By contrast, the monolayers produced after transfer of DPPE in the solid condensed phase had much larger domains, the excluded fluorescent probe molecule patches were much more diffuse and no regularity was seen. AFM imaging under water of subsequently fused bilayers from vesicles of POPC, POPE, cholesterol and the disialo-ganglioside G_{D1a} revealed that the structure of the supporting lipid monolayer had a profound effect upon the resultant 'biomembrane' bilayer, the results of which are shown in Fig. 5.17. The bilayer fused to the DPPC coated substrate was highly irregular and highly corrugated (Fig. 5.17a,c), whereas the bilayer fused onto the DPPE coated substrate (Fig. 5.17b,d) was relatively uniform.

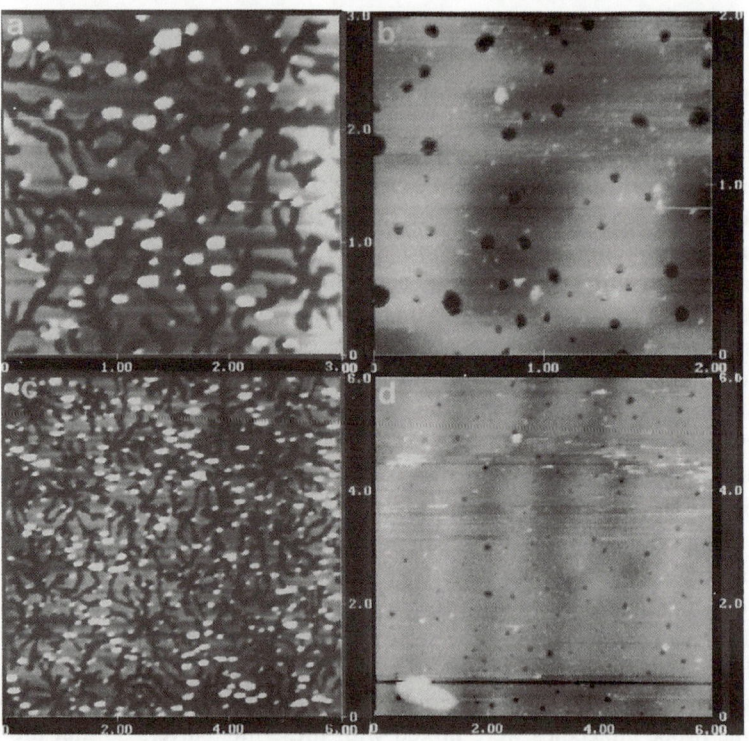

Figure 5.17. AFM images, obtained under water, showing the dramatic effect that a supporting phospholipid monolayer has on the resultant structure of an adsorbed bilayer (POPC:POPE:cholesterol:G_{D1a}). (a,c) DPPC and (d,e) DPPE. The scan sizes are indicated on the bottom of each image (units are in μm). Reproduced with permission from Tamm *et al* 1996.

It appeared that the supporting lipid layer determined the structure of the adsorbed bilayer by some form of epitaxy. Thus it was demonstrated that the combined use of scanning probe microscopes could characterise these complex systems in biologically relevant environments.

5.3.7. Dynamic processes of phospholipid layers

One of the most exciting possibilities of the use of AFM imaging is the ability to follow dynamic processes *in-situ,* and there are a few examples of real time imaging of processes on phospholipid bilayers.

In what was one of the first example of real-time imaging in a phospholipid bilayer system, the binding of streptavidin to bilayers, composed of DPPE in the first layer and a second layer of mixed dimyristoyl phosphatidylethanolamine (DMPE) and biotinylated DPPE, was followed by AFM (Weisenhorn *et al* 1992). The study found that streptavidin bound almost exclusively to biotin which was in fluid-phase lipid domains, but individual proteins could not be resolved due to the lipid fluidity. The structure of the few regions were streptavidin did bind to biotin in crystalline-phase lipid domains was found to be sensitive to the imaging force but, in the optimum range (< 200 pN), individual proteins were resolved. The conclusion was that at excessive force the AFM tip simply 'ploughs' through the lipid layer rendering visualisation of bound protein molecules impossible (Weisenhorn *et al* 1992).

Two studies have used AFM to examine the enzymatic modification of phospholipid layers (Turner *et al* 1996; Grandbois *et al* 1998). In the first instance AFM was combined with several other techniques (X-ray photoelectron spectroscopy, secondary-ion mass spectrometry, X-ray reflectivity and ellipsometry) to characterise the effect of free phospholipase C on a film of dimyristoyl phosphatidylcholine (DMPC). The DMPC layer was immobilised by chemical attachment to a silicon wafer, and the film examined before and after enzyme treatment. The study wasn't therefore a true dynamic study, but is worthy of mention as one of only two examples of AFM investigations of enzymatic breakdown of phospholipid films. The results showed that the enzyme was active even against an immobilised lipid layer, removing all of the phosphate from the lipid (Turner *et al* 1996).

The second study followed the process of enzymatic degradation of a phospholipid bilayer of DPPC by phospholipase A_2 as it actually took place in the liquid cell of the AFM (Grandbois *et al* 1998). The bilayer was formed on mica by double LB dipping, with the first film allowed to dry in air for 15 minutes prior to the second dip. The surface pressure was maintained at 35 mN m^{-1} for both dips, so that the DPPC layer was in the highly packed gel-phase, and the transfer speed was 3 mm min^{-1}. Sharpened tips can be grown by the deposition of carbon using the electron beam in an EM focused onto the apex of a standard silicon nitride tip, and these are often referred to as 'e-beam tips' or 'supertips'. Such a tip was used to minimise the effects of adhesion and tip contamination by the products of hydrolysis and, in this way, the imaging force was kept below 0.5 nN. The AFM imaging was carried out under buffer solution and then bee venom phospholipase A_2 was added to a concentration of 0.26 µM. The subsequent sequence of AFM images obtained are reproduced in Fig. 5.18 and illustrate that the degradation of the DPPC bilayers by the enzyme is nucleated by defects. This is because the rate of degradation of DPPC in the gel-phase is very slow, due to the close-packing of the

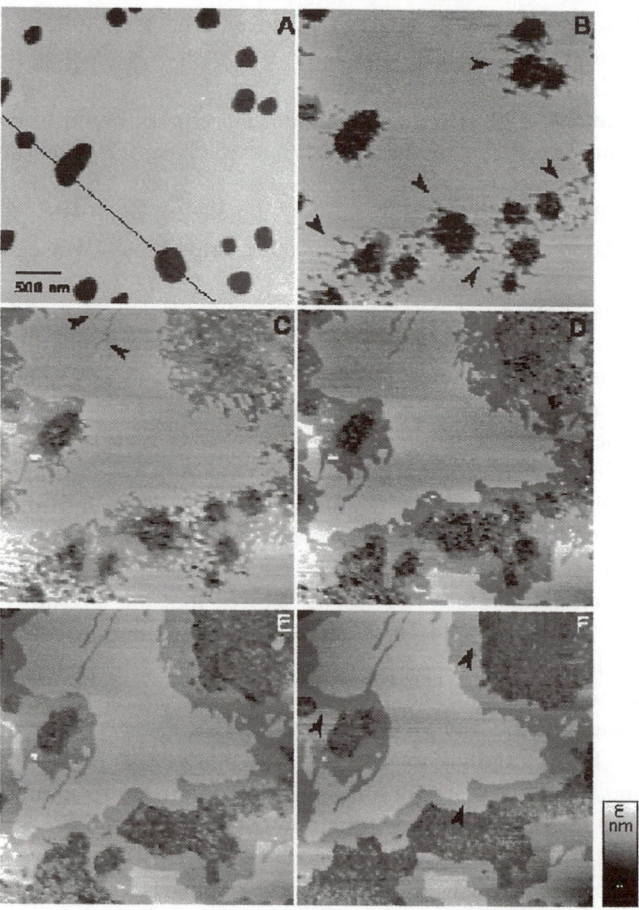

Figure 5.18. part 1. AFM time sequence of the hydrolysis of a DPPC bilayer by phospholipase A_2. (A) Before addition of enzyme, and (B) 2 min. (C) 4min. (D) 6min. (E) 9 min. (F) 12 min. after addition of the enzyme. Scale bar 500 nm. Sequence continued overleaf. Reproduced with permission from Grandbois *et al* 1998 and the Biophysical Society.

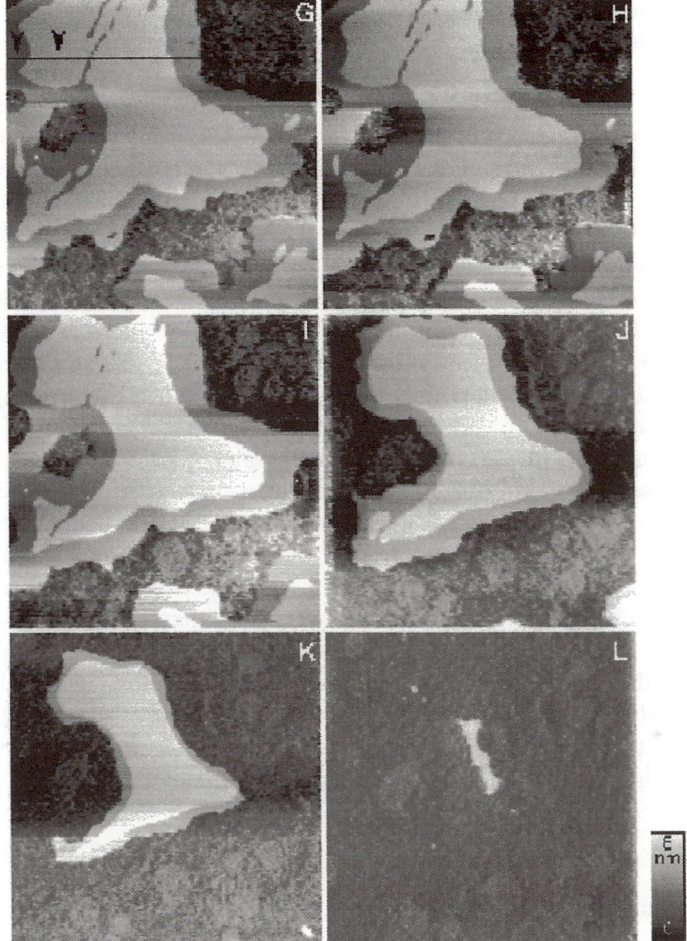

Figure 5.18. part 2. AFM time sequence continued of the hydrolysis of a DPPC bilayer by phospholipase A_2. (G) 14 min. (H) 17 min. (I) 30min. (J) 50min. (K) 80 min. and, (L) 140 min. after addition of the enzyme. Reproduced with permission from Grandbois *et al* 1998 and the Biophysical Society.

lipids, whereas at defects the lipids are not well packed and, indeed, no degradation was observed in defect free regions of the bilayers. This fact was proven by deliberately disturbing a defect free region of the bilayer by scanning at slightly higher force and scan rate (5 nN and 20 Hz) in the presence of enzyme and noting that a grove resulted. When this procedure was repeated in the absence of enzyme no damage occurred. The gross pattern of degradation is strikingly clear, with the enzyme eroding the bilayer around the holes and 'eating' its way into the defect free areas of the film (Fig. 5.18D-J) until virtually none of the film remains (Fig. 5.18L). Upon closer inspection of the image in Fig. 5.18B many small channels can be seen leading away from the holes, indicating that the erosion of the bilayer by the enzyme did not occur uniformly in every direction. At higher magnification these channels were seen to be confined, in the main, to discrete angles centred around 120°, indicating that the enzyme may be sensitive to the crystal orientation of the DPPC molecules since, in the gel phase, they are hexagonally packed. The AFM images of the degradation patterns were carefully analysed in terms of channel area, to estimate numbers of lipid molecules hydrolysed, and with time in order to estimate reaction rates, thus determining what level of hydrolysis was imaged in the sequence. For the narrow channels, which are only some 15 nm in width and are arrowed in Fig. 5.18B, the remarkable conclusion was that these channels represented hydrolysis by single enzyme molecules, since the figures derived from image analysis agreed well with calorimetric data obtained by others (Lichtenberg *et al* 1986). Thus the activity of single enzyme molecules may be followed by AFM and, if activity is really sensitive to the phospholipid organisation in bilayers, then it may be possible to examine the effect of factors believed to modulate bilayer organisation by AFM (Grandbois *et al* 1998).

Other dynamic studies on phospholipid bilayers of DPPC include lipid loss from mica surfaces *ex-situ* at elevated temperatures (Fang and Yang, 1997), domain motion under conditions of high relative humidity *in-situ* (Shiku and Dunn, 1998) and fillipin induced lesions in mixed DPPE/cholesterol bilayers *ex-situ* (Santos *et al* 1998).

5.4 Liposomes and intact vesicles

All of the AFM studies described in the previous sections have been made on flat bilayers. There are a few examples of AFM studies carried out on intact liposomes and vesicles. Shibata-Seki and coworkers imaged intact liposomes of DPPC/cholesterol by AFM under water after binding them to an antigen covered surface of gold on mica (Shibata-Seki *et al* 1996). The AFM images of the liposomes produced spherical features, with measured diameters which were in reasonable agreement with sizes obtained from light scattering measurements. Measurements of the heights of the liposomes indicated that the AFM tip compressed these soft structures by about 40% and, therefore, the image quality was found to depend quite strongly on the force set-point used (acceptable image quality was only obtained with forces <1 nN), and, of course, loading pressure which is determined by tip sharpness. In general the blunter standard silicon nitride tips produced better images. From rudimentary calculations it was determined that the critical loading force (for a standard tip) and pressure which the liposomes could withstand was 4.5 nN and 0.9×10^6 N m^{-2} respectively (Shibata-Seki *et al* 1996).

In a very recent and fascinating study the process of liposome collapse during adsorption to mica has been imaged by AFM (Egawa and Furusawa, 1999). Initially AFM scans were made of a freshly cleaved mica surface under pure buffer (10 mM MgCl$_2$) in the liquid cell of the microscope. Then the buffer was replaced with a suspension of liposomes at various salt concentrations in order to assess the effect of electrolyte concentration on the interaction between the mica and liposomes. The electrostatic interactions between the liposomes and mica were quantified by measurement of the zeta potential of the systems. Phosphatidylcholine (PC) and phosphatidylethanolamine (PE) were studied and the mechanism of bilayer formation on the substrate compared. The results obtained for PC are shown as a time sequence in Fig. 5.19.

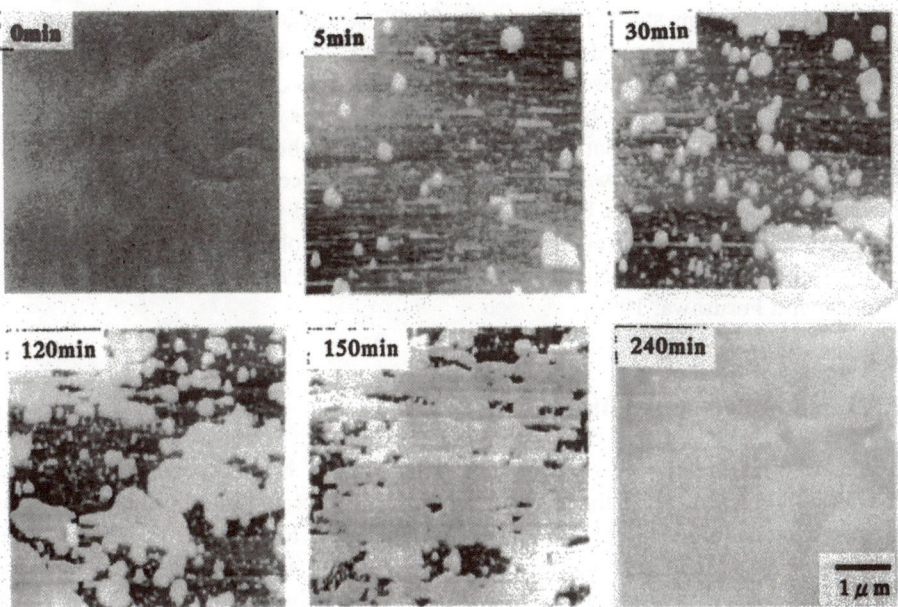

Fig 5.19. Time sequence of AFM images captured during the adsorption of a PC liposome onto mica in 10 mM MgCl$_2$ solution. Reproduced with permission from Egawa and Furusawa, 1999, copyright (1999) American Chemical Society.

Five minutes after injection of the PC suspension spherical features appear on the mica surface (Fig 5.19) which grow in number (Fig. 5.19, 30 min) until they overlap and begin to flatten (Fig 5.19, 120-150 min), and then finally form a continuous bilayer (Fig. 5.19, 240 min). In addition to the visual evidence for liposome collapse provided by the AFM images, namely spherical features flattening out, the collapse upon adsorption to mica was also monitored using a fluorimeter to measure the release of a fluorescent probe molecule encapsulated in the liposomes. The rate of PC bilayer coverage of the substrate was quantified by area measurement in the AFM images under three different ionic strengths (10^{-4} M, 10^{-3} M, 10^{-2} M $MgCl_2$) and compared to fluorescence measurements of PC liposome/mica mixtures at the same ionic strengths. The results from this experiment were in very close agreement and, when displayed graphically, the form of the curves were almost identical. The rate of adsorption and deformation of PC vesicles on mica was shown to increase with increasing electrolyte concentration due to the decreasing zeta potential of the vesicle/mica system, meaning that less electrostatic repulsion was present between them. Finally with PC only one bilayer was ever formed on the mica, no additional bilayers were seen under any of the electrolyte conditions studied. The results for PE were rather different in the respect that a second bilayer formed but the initial bilayer formation was similar to that of PC. The second bilayer however, was relatively easy to displace with the AFM tip, indicating that the interaction between bilayers (that is between the ethanolamine headgroups) was weaker than the interaction between the bilayer and mica surface. These differences, which were observed in the absence of electrostatic repulsion between lipid and mica in both cases, were attributed to the different degrees of hydration in the headgroups of the two phospholipids, demonstrating the importance of headgroup chemistry in determining the nature of bilayer formation (Egawa and Furusawa, 1999).

5.5. Lipid-protein mixed films

In the plasma membrane of cells the phospholipid bilayer contains numerous membrane proteins and so the study of proteins incorporated into lipid bilayers is of great relevance to biological AFM. This area overlaps to a certain degree with the section on 2D protein crystals described (section 6.2) since most such systems are membrane proteins which naturally occur in phospholipid bilayers. There are however, a few examples of studies where proteins have been deliberately incorporated into phospholipid bilayers created on Langmuir troughs or into lipid vesicles for reasons of improved protein stability, and therefore improved resolution by AFM imaging (Yang *et al* 1993; Mou *et al* 1995; Czajkowski *et al* 1998), or to examine intrinsic interactions between the components (von Nahmen *et al* 1997) or finally, as a model system for generating material contrast in friction AFM images (Sommer *et al* 1997).

In the first example a bilayer of 1,2-dipentacosa-10,12-diynoyl-phosphatidylcholine (DAPC) containing 20 mol% ganglioside GM1 was transferred from an air-water interface onto mica using LB dipping. Both cholera toxin and the its B-subunit were bound to the ganglioside in the bilayer by incubation of solutions of the toxin with the mica supported bilayer. This enabled high resolution (better than 2 nm) AFM images to be obtained under pure water and buffer as long as the imaging force was kept around 0.3 - 0.5 nN. The intact cholera toxin molecules did not pack as closely as the B-subunits, with the result that the resolution achieved for the former was lower, providing direct evidence of the importance of molecular packing for AFM imaging at high resolution. The whole cholera toxin molecules were imaged only as spherical globules but, by contrast, the B-subunits' pentameric structure was clearly resolved (Yang *et al* 1993). In an effort to increase the stability of the bilayer and protein contained within it, cross-linking of the leaflets was carried out by exposing the sample to ultraviolet light. In a later study by the same group cholera toxin B-subunits were re-examined, this time bound to bilayers of DPPE / PC, but without any cross-linking steps (Mou *et al* 1995). Bilayers were produced by both LB-dipping and the vesicle fusion technique. Slightly clearer AFM images were obtained, although the resolution was comparable to the first study (1-2 nm), and interestingly no dependence of image quality on lipid state was found; the AFM images of the oligomers were just as good whether they were bound in a gel-phase (Fig. 5.20b) or in a fluid-phase (Fig. 5.20c) lipid layer. As before the proteins were bound to ganglioside incorporated into the lipid layer and, if this step was omitted, none were bound (Fig. 5.20d). Since proteins will bind randomly and non-specifically to bare mica, this specific binding into a phospholipid bilayer provides a useful alternative method for protein deposition to a substrate which enables a far more ordered protein layer to be obtained. A further advantage of specific binding of proteins would appear to be that less tip contamination is encountered, probably because the molecules are far harder to displace. In this regard it was noted that preparation of these protein-lipid 'membranes' using the small scale lipid-coated droplet technique (Kornberg and Darst, 1991- see section 6.2.3.) was unsuitable producing very small 'membrane' fragments which quickly contaminated the AFM tip. Finally, in this study similar results were obtained when the protein was bound to bilayers composed of several other phospholipids with differing headgroups and acyl chains (Mou *et al* 1995). Staphylococcal α-hemolysin (α-HL), a small (33 kD) water soluble protein, converts into a pore forming oligomer upon binding to the plasma membrane. For some years there had been a degree of uncertainty over the structure adopted by the α-HL in membranes. Electron microscopy studies had indicated that it formed a hexamer (Gouaux *et al* 1994) but later X-ray crystallography studies demonstrated a heptameric shape (Song *et al* 1996). The

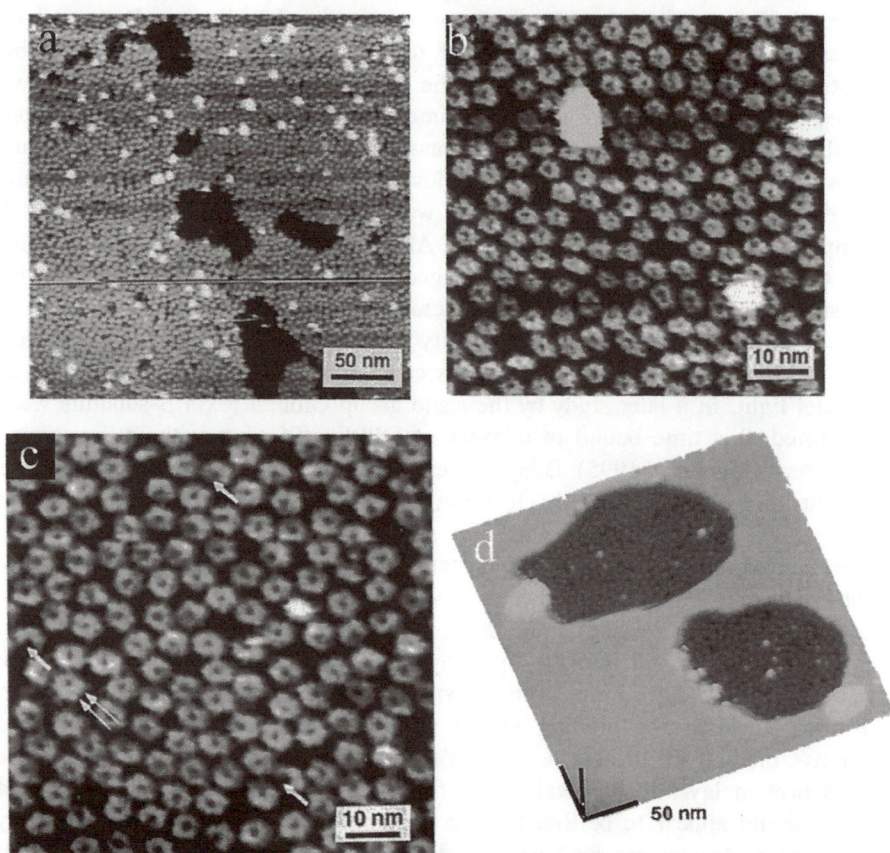

Figure 5.20. (a) Cholera toxin bound into a gel-phase phospholipid (DPPC) bilayer which incorporates a ganglioside (GM1) to bind the protein molecules. (b) At higher resolution individual toxin molecules are clearly seen, and in many the central cavity and pentameric structure is visible. (c) High resolution image of cholera toxin molecules bound to a fluid-phase phospholipid (DPPC/POPG) bilayer containing GM1. The image quality is just as good as in the gel-phase layer, indicating that membrane fluidity does not degrade AFM image resolution. The clarity of this image allows variation in the structure of individual cholera toxin molecules to be seen. In some cases (single arrows) subunits were missing, while others displayed six-fold symmetry (double arrows) instead of the expected five-fold symmetry. (d) When no ganglioside is incorporated into the bilayers cholera toxin molecules do not bind. However, many cholera toxin molecules are visible in the defect regions, where the bare substrate (mica) is exposed. Reproduced with permission from Mou *et al* 1995.

crystallographic study argued that the image processing required in the EM study may have led to an artefactual result (Song *et al* 1996). AFM was applied in order to try to resolve this problem (Czajkowski *et al* 1998) since it has already proved capable of resolving directly the shape of several membrane proteins (chapter 6) and soluble proteins incorporated into phospholipid bilayers, without the need for image processing and in realistic environments. A lipid bilayer was formed on mica by LB dipping but on this occasion no ganglioside was required because oligomerisation of the α-HL occurs spontaneously upon contact with suitable amphipathic substrates. The AFM images obtained, in contact mode with oxide sharpened silicon nitride tips under a selection of aqueous buffers, demonstrated that the α-HL could indeed form a hexamer, confirming the earlier EM study. This AFM study provided proof that α-HL can exist in two different oligomeric forms, the reasons for which will require further investigation. Thus the study provided a further and encouraging example of AFM being used tackling a real biological problem (Czajkowski *et al* 1998).

The interplay between a recombinant pulmonary surfactant associated protein (SP-C) and a phospholipid bilayer composed of DPPC and dipalmitoyl phosphatidylglycerol (DPPG) was studied by AFM (von Nahmen *et al* 1997). This system was used as a model for the surface active combination of phospholipids and proteins present in lungs, which act to keep the surface tension of the fluid layer on the aveoli at low levels. Low surface tension of this fluid layer is an essential feature of the 'mechanics' of breathing, enabling the lungs to re-expand after exhalation. To model the effects found in the lungs, mixed lipid-protein films drawn at various surface pressures to simulate inhalation (high surface pressure due to reduced interfacial area) and exhalation (low surface pressure due to increased interfacial area) were studied. LB films were formed on mica on the upstroke of the dip, which was carried out a rate of 2 mm min^{-1} . The transferred films were then examined in air by AFM. At low surface pressures (30 mN m^{-1}), AFM images of the mixed film showed that both protein and lipid were present in the interfacial film; some regions were entirely composed of lipid, some had a mixture of lipid and protein, and others were composed almost entirely of protein. AFM images of films transferred at higher surface pressures (50 mN m^{-1}), showed by contrast raised lamellar islands of protein-fluidic lipid complex sitting on top of a continuous phospholipid bilayer. As the compression of the film was increased, by reducing the area on the Langmuir trough, the phase separation between the lipid and protein-lipid complex became more distinct, in that the lamellae of the complex reduced in area but increased in height. The effect was reproducible and fully reversible, so that once the surface pressure was lowered again, the film structure seen by AFM reverted to a flat, partially mixed film. The conclusion was that the AFM images revealed that a protein-lipid complex moves in and out of the phospholipid bilayer depending upon the surface pressure, but never detaches from it completely, thus it acts as a sort of reservoir of surface active material which can quickly re-spread onto the interfacial layer upon reduction of surface pressure (expansion of the lungs), ensuring no rupture of this lubricating layer occurs (von Nahmen *et al* 1997). This is an important property because the main phospholipid constituent in the lungs is DPPC which enters the gel-phase readily and will not re-spread after it has exceeded its collapse pressure. A final and important check carried out in the study was a validation of the LB film transfer procedure. By incorporating a fluorescent dye it was possible to obtain fluorescence micro-graphs of the interfacial films at the air-water interface and after transfer onto mica. These proved that no gross changes to film structure were induced by the transfer procedure (von Nahmen *et al* 1997).

Finally, in a very recent paper Mueller and co-worker performed force-distance measurements on phospholipid bilayers, bare mica, and on bilayers in the presence of the nerve sheath protein myelin (Mueller *et al* 1999). Force versus distance curves recorded during stretching of the myelin away from the mica, or lipid coated mica surface indicated that, when in contact with the lipid layer, the protein adopts a different conformation to that observed when it is simply adsorbed to mica, a result which may not be surprising in view of the normal interfacial behaviour of proteins. In these force spectroscopy studies, the researchers reported that imaging was not possible and so the system could not be fully characterised.

5.6. Miscellaneous lipid films

As well as AFM studies on phospholipids there have been numerous investigations on other lipid films. These studies could justify a book in their own right but, as they are less biologically significant, they have not been included in this text. However, there are a couple of studies which are worth mentioning briefly here since their findings may impact upon the future study of phospholipids, particularly on more complex (mixed) or dynamic systems. Chemical identification of mixed amphiphile LB films has been achieved by measurement of surface potential during scanning using a hybrid mode AFM with conducting tips (Inoue and Yokoyama, 1994; Fujihara and Kawate, 1994; Chi *et al* 1996). Secondly *in-situ* growth of an adsorbed film of a cationic surfactant, cetyltrimethylammonium bromide (CTAB), on mica has been imaged by AFM (Li *et al* 1998).

5.7. Interfacial protein films

For proteins the process of assembly at an interface involves not just a simple reorientation of the molecules but also a degree of unfolding to expose hydrophobic regions. This means that unlike simple amphiphilic molecules, such as phospholipids, the assembly (or more strictly the *adsorption*) of proteins at an interface is a relatively slow process, which is influenced by many factors. As such interfacial protein films are an interesting candidate for study by AFM, although surprisingly little work has been carried out to date. In principle it should be possible to investigate many fundamental problems which have so far eluded explanation, such as how unfolding of protein molecules at an interface effects their functional behaviour, and the characteristics of the networks they can form in interfacial regions. For example, why do some proteins form better, more stable foams or emulsions than others, and what effect do other amphiphilic molecules, which compete for space at the interface, have on protein networks in the mixed systems one might expect to find in real environments?

5.7.1. Specific precautions

There are certain experimental problems which are specific to the study of interfacial protein films and which need to be considered before imaging by AFM. Obviously the objective of such studies is to examine the nature of the interfacial protein film but, because the interface is not imaged directly, it is necessary to ensure that when the film is transferred onto a substrate it is *only* the interfacial film that is deposited. For protein films this is not as straightforward as it might seem. The problem is twofold, firstly amphiphilic proteins have a high affinity for many substrates under most conditions. A second and more specific problem is that in realistic systems the process of interfacial protein film formation is by *adsorption* of the proteins from bulk solution to the interfacial region: think of the foam produced by beating egg whites, or the foam formed as the head on a glass of beer. This means that the usual technique for creating interfacial films on a Langmuir trough involving spreading the amphiphiles at the air-water interface, is not the most realistic way of creating interfacial protein films. Studies of the surface rheology of protein films have demonstrated that *spread* films possess different characteristics to *adsorbed* films (Krägel *et al* 1999; Mackie *et al* 1999). If the interfacial film is created by adsorption of the protein molecules from the bulk, then as well as an interfacial layer of proteins there will also be free protein in the bulk liquid or *subphase*. During dipping the substrate is exposed to the subphase and the possibility exists of non-specific protein adsorption in addition to transfer of the interfacial film.

During a study (Mackie *et al*) on the displacement of protein films from the air-water interface by surfactant, which is described in detail in the following section, it was possible to demonstrate that such 'parasitic' adsorption really does take place. Displacement of the interfacial protein films with surfactant has allowed production of an interfacial protein film with definite structural features which allow it to be distinguished from simple passive adsorption of protein onto the mica. Passive adsorption of the protein from the subphase simply produces fairly uniform monolayer coverage of the substrate. The passive adsorption can be eliminated by the following procedure. The interfacial protein film was created by filling a Langmuir trough with a 0.5 μM solution of the milk protein β-lactoglobulin and allowing it to adsorb to the air-water interface until a pseudo-equilibrium surface pressure was obtained (10 mN m^{-1} after 30 minutes). By partially displacing the interfacial β-lactoglobulin film using a surfactant (Tween-20) an interfacial protein network was produced which contains holes. Not all of the protein from the bulk solution has adsorbed to the interface, meaning that if LB dipping is done on the system as it stands parasitic adsorption of proteins from the subphase will occur as well as interfacial film transfer. This is illustrated by the AFM images in Fig. 5.21 which were obtained after LB film transfer onto mica and air-drying.

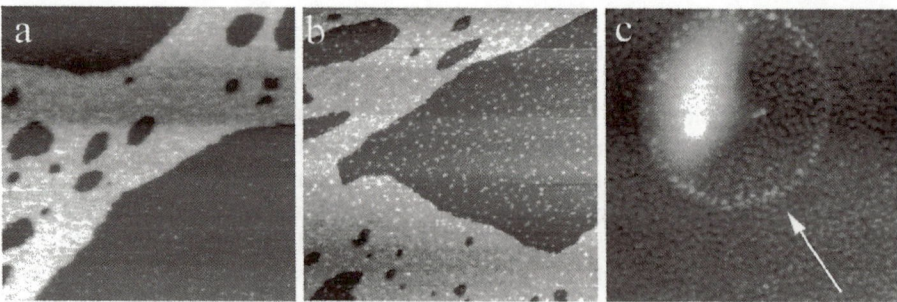

Figure 5.21. AFM images showing the effect of perfusion of the subphase prior to LB transfer, of an *adsorbed* mixed interfacial film of protein and surfactant. (a) After perfusion of the subphase of the Langmuir trough with 2L of surfactant solution most of the non-interfacial protein has been removed. This can be seen because the surfactant rich regions from the interface have little adsorbed protein in the AFM image. (b) If perfusion of the subphase is carried out less rigorously (in this image with only 1L of surfactant solution) then more 'parasitic' adsorption of proteins to the substrate occurs-the 'dark' regions are peppered with proteins (c) If no perfusion of the subphase is carried out, the substrate is completely covered by 'parasitically' adsorbed protein from the subphase and the interfacial film structure (arrowed) is almost completely masked. Scan sizes: (a) 3 x 3 μm, (b) 2 x 2 μm, (c) 1.6 x 1.6 μm. Imaged in dc mode under butanol.

The samples were imaged in contact mode and under butanol which dissolves the surfactant, leaving only the protein network attached to the mica. Therefore the bright regions are protein and the dark regions (should be) bare mica. The two AFM images shown in Fig. 5.21 are from LB dips after different levels of perfusion of the subphase with surfactant solution to remove free protein. Fortunately interfacial protein films are immobile because they form an elastic network, and this means that once the film has formed there is little exchange of the proteins at the interface with proteins in the bulk, as happens with films of smaller amphiphiles such as lipids and surfactants. This allows perfusion of the bulk solution or subphase with pure water or pure buffer solution without affecting the interfacial protein film, as long as the flow rate is kept low enough to prevent disruption of the surface (1-2 mL min^{-1} for a trough of 450 mL volume). For the situation shown in Fig. 5.20 the subphase must be perfused with the surfactant solution in order to maintain the equilibrium concentration of surfactant molecules at the interface. By transferring the interfacial films after different levels of perfusion it was possible to quantify the parasitic adsorption of protein molecules to the substrate, seen as bright specs in dark regions. Even a quick inspection of the AFM images confirms that parasitic adsorption of proteins onto the mica from the subphase is reduced as the perfusion volume increases (compare Fig. 5.20a,b which had 2L and 4L of perfusion respectively). The images prove that the protein molecules attached to the mica in the dark regions have not come from the interfacial film but from the subphase, so it is easy to see that if no perfusion is carried out when adsorbed protein films are being transferred by LB dipping, then the mica surface could potentially be completely saturated with passively adsorbed protein before any transfer takes place, leading to highly misleading AFM images.

5.7.2. AFM studies of interfacial protein films

Early AFM studies of interfacial protein films tended to simply demonstrate the capability of AFM to image such systems (Birdi *et al* 1994; Gunning *et al* 1996). More recently however, there are a couple of examples of studies which have focused on the behaviour of interfacial protein films, in particular how they react to the presence of competing surface active species such as surfactants or lipids (Alexandre *et al* 1994, 1997; Sommer *et al* 1997; Mackie *et al* 1999). Different LB film transfer mechanisms were studied for a system of mixed fatty acid (behenic acid)/protein (glucose oxidase) which is of interest in the production of biosensors (Alexandre *et al* 1994). The study illustrated that the most commonly used vertical dipping method created large parallel defects on the macroscopic scale and subtle differences on the microscopic scale, although the transfer speed used of 10 mm min^{-1} was rather high, and may account for these results. In two more recent AFM studies of the same system lateral force imaging (also known as frictional force imaging, section 3.3.2.) was used along with UV-Vis and IR spectroscopy to characterise the mixed films (Alexandre *et al* 1997; Sommer *et al* 1997). The films were created by using a dilute solution (3.2 μg mL^{-1}) of glucose oxidase as the subphase in a Langmuir trough, sweeping the surface with the barriers to remove adsorbed enzyme and then spreading the fatty acid from solvent at the air-solution interface. The mixed films were allowed to equilibrate for 45 minutes before being compressed to a surface pressure of 30 mN m^{-1} and then transferred after varying time intervals onto graphite by vertical dipping. By using frictional force imaging it was possible to differentiate between the protein and fatty acid domains in the mixed films, which displayed a highly heterogeneous structure. The glucose oxidase molecules appear to adsorb to an interfacial film of behenic acid and can take up to 15 hours to reach a well ordered interfacial protein film. In essence the sample preparation mirrors the method for the production of 2D crystals of soluble proteins developed by Kornberg and Ribi (Kornberg and Ribi, 1987) and the studies therefore represent a stepwise monitoring of the events leading to crystallisation by AFM. The mixed fatty acid protein film was initially a highly heterogeneous mixture containing protein aggregates which slowly became covered by fatty acid molecules, allowing their rearrangement into ordered quasi-crystalline arrays (Alexandre *et al* 1997; Sommer *et al* 1997).

A recent study of gelatin examined the initial stages of association of the molecules, leading to the formation of gel networks, by sampling from bulk solutions and from interfacial films formed at an air-water interface (Mackie *et al* 1998). In order to achieve a high local concentration but avoid the formation of large three dimensional aggregates or gels which are difficult to image with any detail in AFM due to their softness, the process was carried out on a Langmuir trough, thereby confining the structures formed to two dimensions. At various intervals interfacial films were transferred onto mica using LB dipping and the samples imaged under butanol in dc mode. The study highlighted the importance of the thermal history of the gelatin solutions, which determined the association of the gelatin molecules and thus the rheological characteristics of the interfacial film. The results obtained supported the model of gelation which involves molecular association via triple helix formation of the gelatin followed by further association of the helical structures into bundles or fibres. Occasionally these fibres displayed a collagen-like periodicity (Fig. 5.22) consistent with the proposed reformation of a 'collagen' structure during the gelation of gelatin. Gelatin solutions which had been rapidly quenched did not form any large fibres, in contrast to slowly cooled solutions, and displayed very different interfacial rheology.

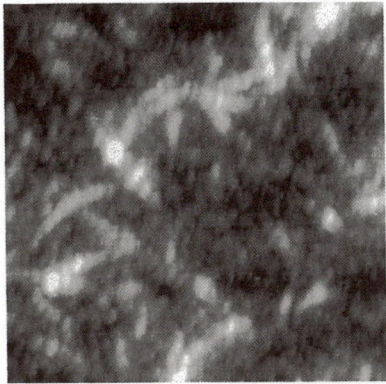

Figure 5.22. Gelatin fibres from an interfacial film occasionally display regular periodicity commonly seen in collagen fibres. Scan size: 1.6 x 1.6 µm, imaged in dc mode under butanol.

Aliquots were taken from bulk solutions at concentrations where the gelatin just gelled at room temperature (4 mg mL^{-1}). These were then diluted and examined, after drop deposition onto mica, so that the association seen in the interfacial films could be compared with the gelation process in the bulk. These bulk samples displayed the same association behaviour as the slowly cooled interfacial film samples, namely slow fibre formation over time.

An important problem in interfacial protein systems is their stability in the presence of other surface active species. In isolation both proteins and surfactants can stabilise foams or emulsions but, when the two are combined, things go wrong and stability is lost. This effect is sometimes used in a beneficial manner; for example very young babies suffer from colic (trapped wind in the intestine) caused by foaming of the proteins in milk. The foam is highly stable entrapping air which prevents the baby from passing wind (at either end!) and releasing the trapped air. Anti-colic preparations contain food grade surfactants which destroy the protein foam and so release the wind. Another less welcome example is the collapse of the foam on a glass of beer. The beer foam is composed of an interfacial protein network but surfactant molecules from a glass which hasn't been properly rinsed after washing, or lipid molecules from the lips of the drinker soon cause its destruction. Until recently the mechanisms for this effect were not understood in any detail. The competitive adsorption of proteins and surfactants to the air-water interface has been examined recently by AFM (Mackie *et al* 1999). Both spread and adsorbed interfacial protein films were created on a Langmuir trough and the experimental protocol was to progressively displace the films by the addition of surfactant to the subphase. The displacement of the protein film was monitored by periodic sampling of the film by Langmuir-Blodgett film transfer onto mica. AFM imaging of the mixed protein/surfactant film was carried out under butanol which dissolved the surfactant off the mica surface leaving just the protein network behind to be imaged. Examples of films of two different milk proteins, β-lactoglobulin and β-casein, partially displaced by the surfactant Tween-20, are shown in Fig. 5.23. The images provided qualitative information (in terms of the shapes of the surfactant domains) and quantitative information (in terms of the area and thickness of the protein domains). One of the unique advantages of AFM is that it provides real three dimensional data. Measurement of protein film area was made by simple thresholding of the AFM images to produce a binary image, and then determining the ratio of black to white pixels in the image.

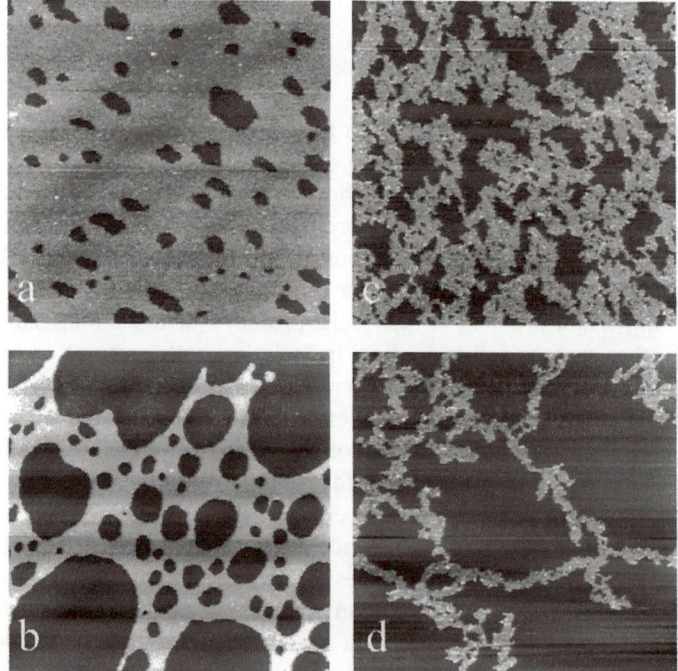

Figure 5.23. Partially displaced β-casein (a,b) and β–lactoglobulin (c,d) LB films on mica. As the surface pressure (Π) increases, due to increased adsorption of surfactant, the holes in the protein network (the areas previously occupied by surfactant) become larger. Scan sizes, surface pressure: (a) 6.4 x 6.4 μm, 16.7 mN m⁻¹ (b) 6.4 x 6.4 μm, 19.2 mN m⁻¹ (c) 3.2 x 3.2 μm, 21.8 mN m⁻¹ (d) 3.2 x 3.2 μm, 24.6 mN m⁻¹. Imaged in dc mode under butanol.

Film thickness measurement was carried out by analysing histograms of the images taking advantage of their essentially binary nature, as shown in Fig. 5.24. This provides a less subjective and therefore more accurate average film height than the more usual technique of line profile measurement at discrete points in the images. Both sets of data were combined and correlated with measurement of surface pressure from the Langmuir trough in order to reveal a detailed description of the process of surfactant displacement of proteins. By plotting protein film volume versus surface pressure of the mixed film, it became obvious that displacement of the interfacial protein film did not occur by erosion or simple exchange of protein molecules for more surface active surfactant molecules at the interface. Rather the process is heterogeneous involving nucleation of surfactant rich region at defects in the protein network, which then grow and compress the protein, forcing it to occupy less interfacial area but not displacing it completely from the interface.

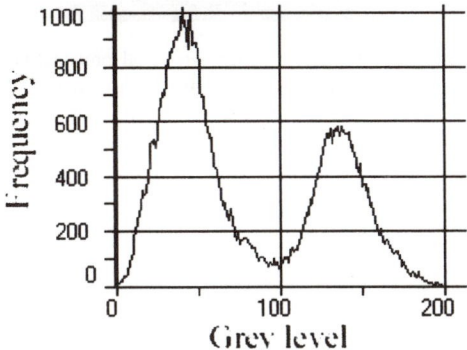

Figure 5.24. Histogram of the grey levels contained in the AFM image seen in Fig. 5.23c. The thickness of the protein film is given by the separation between the peaks, because the left hand peak represents the surfactant rich (dark) region and the right hand peak represents the protein rich (bright) region in the image.

The protein film responds to the increased surface pressure by thickening or buckling, and hence the process was termed 'orogenic', after the geological term for the creation of mountains by colliding plates in the earths' crust. At the final stages of displacement the protein network is fractured and the AFM images show islands of protein formed within a continuous surfactant phase (Fig. 5.25). Only at this final stage is the protein displaced from the interface. In this case AFM studies have produced a new model for protein displacement which could not, at present, have been deduced from other surface techniques which either lack the required resolution, or spatially average the interfacial structure. The 'orogenic' mechanism has also been demonstrated *in-situ* in a very recent 'real time' AFM study of the displacement (cleaning) of a protein film from a graphite surface under water to which surfactant solution was added during imaging (Gunning *et al* 1999).

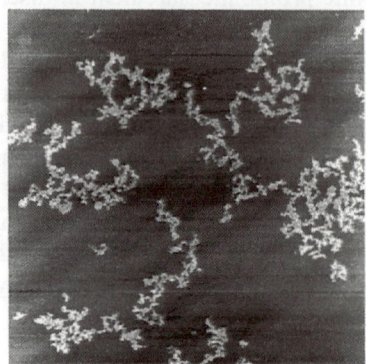

Figure 5.25. At the final stage of collapse of the interfacial protein film islands of protein are seen surrounded by a continuous **surfactant phase**. Scan size: 10 x 10 μm, surface pressure Π=27.1 mN m⁻¹.

References

Alberts, B.A, Bray, D, Lewis, J, Raff, M, Roberts, K. and Watson, J. (1994). Membrane Structure. In *Molecular biology of the cell*, 3rd ed. pp. 477-484. Garland Publishing Inc., New York & London

Alexandre, S, Dubreuil, N, Fiol, C, Malandain, J-J, Sommer, F. and Valleton, J-M. (1994). Compararison at the microscopic scale of mixed fatty acid-protein Langmuir-Blodgett films resulting from vertical or horzontal transfer. *Microsc. Microanal. Microstruct.* 5, 359-371.

Alexandre, S, Dubreuil, N, Fiol, C, Lair, D, Sommer, F, Duc, T.M. and Valleton, J-M. (1997). Analysis of the dynamic organization of mixed protein/fatty acid Langmuir films. *Thin Solid Films* 293, 295-298.

Birdi, K.S, Vu, D.T, Moesby, L, Andersen, K.B. and Kristensen, D. (1994). Structures of lipid and biopolymer monolayers investigated as Langmuir-Blodgett films by atomic force microscopy. *Surface & Coatings Technology* 67, 183-191.

Brandow, S.L, Turner, D.C, Ratna, B.R. and Gbaer, B.P. (1993). Modification of supported lipid membranes by atomic force microscopy. *Biophys. J.* 64, 898-902.

Chi, L.F, Jacobi, S. and Fuchs, H. (1996). Chemical identification of differing amphiphiles in mixed Langmuir-Blodgett films by scanning surface potential microscopy. *Thin Solid Films* 284-285, 403-407.

Czajkowski, D.M, Huang, C. and Hao, Z. (1995). Ripple phase in asymmetric unilamellar bilayers with saturated and unsaturated phospholipids. *Biochemistry* 34, 12501-12505.

Czajkowski, D.M, Sheng, S. and Shao, Z. (1998). Staphlococcal α-hemolysin can form hexamers in phospholipid bilayers. *J. Mol. Biol.* 276, 325-330.

Dufrene, Y.F, Barger, W.R, Green, J.B.D. and Lee, G.U. (1997). Nanometer-scale surface properties of mixed phospholipid monolayers and bilayers. *Langmuir* 13, 4779-4784.

Dufrene, Y.F, Barger, W.R. and Lee, G.U. (1998). Direct characterisation of mixed phospholipid/glycolipid bilayers with chemically functionalised AFM probes. *Biophys. J.* 74, A330.

Egawa, H. and Furusawa, K. (1999) Liposome adhesion on mica surface studied by atomic force microscopy. *Langmuir* 15, 1660-1666.

Fang, Y. and Yang, J. (1997). The growth of bilayer defects and the induction of interdigitated domains in the lipid loss process of supported phospholipid bilayers. *Biochim. Biophys. Acta* 1324, 309-319.

Fujihira, M. and Kawate, H. (1994). Scanning surface potential microscope for characterisation of Langmuir-Blodgett films. *Thin Solid Films*, **242**, 163-169.

Gouaux, J.E, Braha, O, Hobaugh, M.R, Song, L, Cheley, S, Shustak, C and Bayley, H. (1994). Subunit stoichiometry of staphylococcal alpha-hemolysin in crystals and on membranes: a heptameric transmembrane pore. *Proc. Natl. Acad. Sci. USA* **91**, 12828-12831.

Grandbois, M, Clausen-Schaumann, H. and Gaub, H. (1998). Atomic force microscope imaging of phospholipid bilayer degradation by phospholipase A_2. *Biophys. J.* **74**, 2398-2404.

Gunning, A.P, Wilde, P.J, Clark, D.C, Morris, V.J, Parker, M.L. and Gunning, P.A. (1996). Atomic force microscopy of interfacial protein films. *J. Colloid & Interface Science* **183**, 600-602.

Gunning, A.P, Mackie, A.R, Wilde, P.J. and Morris, V.J. (1999) *In-situ* observation of the surfactant induced displacement of protein from a graphite surface by atomic force microscopy. *Langmuir* **15**, 4636-4640.

Hui, S.K. and Sen, A. (1989). Effects of lipid packing on polymorphic phase behaviour and membrane properties. *Proc. Natl. Acad. Sci. USA* **86**, 5825-5829.

Hui, S.W, Viswanathan, R, Zasadzinski, J.A. and Israelachvili, J.N. (1995). The structure and stability of phospholipid bilayers by atomic force microscopy. *Biophys. J.* **68**, 171-178.

Inoue, T. and Yokoyama, H. (1994). Surface potential imaging of phase separated Langmuir-Blodgett monolayers by scanning Maxwell stress microscopy. *Thin Solid Films* **243**, 399-402.

Krägel, J, Grigoriev, D.O, Makievski, A.V, Miller, R, Fainerman, V.B, Wilde, P.J. and Wüstneck, R. (1999). Consistency of surface mechanical properties of spread protein layers at the liquid-air interface at different spreading conditions. *Colloids and Surfaces B: Biointerfaces* **12**, 391-397.

Kornberg, R.D. and Ribi, H.O. (1987). Formation of two dimensional crystals of proteins on lipid layers. In *Protein Structure, folding and design 2*. (ed. D.L. Oxender), pp. 175-186, Alan R. Liss: New York.

Li, B, Fuji, M, Fukakada, K, Kato, T. and Seimiya, T. (1998). In situ AFM observation of heterogeneous growth of adsorbed on cleaved mica surface. *Thin Solid Films* **312**, 20-23.

Lichtenberg, D.G, Romero, M, Menashe, M. and Biltonen, R.L. (1986). Hydrolysis of dipalmitoylphosphatidylcholine large unilamellar vesicles by porcine pancreatic phospholipase A_2. *J. Biol. Chem.* **261**, 5334-5340.

Mackie, A.R, Gunning, A.P, Ridout, M.J. and Morris, V.J. (1998). Gelation of gelatin: observation in the bulk and at the air-water interface. *Biopolymers*, **46**, 245-252.

Mackie, A.R, Gunning, A.P, Wilde, P.J. and Morris, V.J. (1999). Orogenic displacement of protein from the air/water interface by competitive adsorption. *J. Colloid & Interface Science* **210**, 157-166.

Mou, J, Yang, J. and Shao, Z. (1994). Tris(hydroxymethyl)aminomethane ($C_4H_{11}NO_3$) induced a ripple phase in supported unilamellar phospholipid bilayers. *Biochemistry* **33**, 4439-4443.

Mou, J, Yang, J. and Shao, Z. (1995). Atomic force microscopy of cholera toxin B-oligomers bound to bilayers of biologically relevant lipids. *J. Mol. Biol.* **248**, 507-512.

Racker (1985). In *Reconstitution of transporters, receptors and pathological states*, Academic Press, Orlando.

Rand, R.P, Capman, D. and Larrson, K. (1975). Tilted hydrocarbon chains of dipalmitoyl lecithin become perpendicular to the bilayer before melting. *Biophys. J.* **15**, 1117-1124.

Sackmann, E. (1983). Physical foundations of the molecular organisation and dynamics of membranes. In *Biophysics*, (eds. W. Hoppe, W. Lohmann, H. Markl & H. Zeigler), pp. 425-457. Springer-Verlag, New York.

Santos, N.C, Ter-Ovanesyan, E, Zasadzinski, J.A, Prieto, M. and Castanho, M.A.R.B. (1998). Filipin-induced lesions in planar phospholipid bilayers imaged by atomic force microscopy. *Biophys. J.* **75**, 1869-1873.

Schallamach, A. (1971). How does rubber slide? *Wear* **17**, 301-312.

Schaus, S.S. and Henderson, E.R, (1997). Cell viability and probe-cell membrane interactions of XR1 glial cells imaged by atomic force microscopy. *Biophys. J.* **73**, 1205-1214.

Seddon, J.M. (1990). Structure of the inverted hexagonal (HII) phase, and non-lamellar phase transitions of lipids. *Biochim. Biophys. Acta* **1031**, 1-69.

Shibata-Seki, T, Masai, J, Tagawa, T, Sorin, T. and Kondo, S. (1996). In-situ atomic force microscopy study of lipid vesicles adsorbed on a substrate. *Thin Solid Films* **273**, 297-303.

Shiku, H and Dunn, R. (1998). Direct observation of DPPC phase domain motion on mica surfaces under conditions of high relative humidity. *J. Phys.1 Chem. B* **102**, 3791-3797.

Singh, S. and Keller, D.J. (1991). Atomic force microscopy of supported planar membrane bilayers. *Biophys. J.* **60**, 1401-1410.

Solleti, JM, Boteau, M, Sommer, F, Duc, T.M. and Celio, M.R, (1996). Characterisation of mixed miscible and nonmiscible phospholipid Langmuir-Blodgett films by atomic force microscopy. *J. Vac. Sci. & Technol. B*, **14**, 1492-1497.

Sommer, F, Alexandre, S, Dubreuil, N, Lair, D, Duc, T.M. and Valleton, J-M. (1997). Contribution of lateral force and "Tapping Mode" microscopies to the study of mixed protein Langmuir-Blodgett films. *Langmuir* **13**, 791-795.

Song, L, Hobaugh, M, Shustak, C, Cheley, S, Bayley, H. and Gouaux, J.E. (1996). Structure of staphylococcal alpha-hemolysin, a heptameric transmembrane pore. *Science* **274**, 1859-1856.

Shwartz, D.K. (1997). Langmuir-Blodgett film structure. *Surface Science Reports* **27**, 241-334.

Tamm, L.K, Böhm, C, Yang, J, Shao, Z, Hwang, J, Edidin, M. and Betzig, E. (1996). Nanostructure of supported phospholipid monolayers and bilayers by scanning probe microscopy. *Thin Solid Films* **284-285**, 813-816.

Turner, D.C, Peek, B.M, Wertz, T.E, Archibald, D.D, Geer, R.E and Gaber, B.P. (1996). Enzymatic modification of a chemisorbed lipid monolayer. *Langmuir* **12**, 4411-4416.

Vikholm, I, Peltonen, J. and Telleman, O. (1995). Atomic force microscope images of lipid layers spread from vesicle suspensions. *Biochim. Biophys. Acta* **1233**, 111-117.

Weisenhorn, A.L, Schmitt, F-J, Knoll, W. and Hansma, P.K. (1992). Streptavidin binding observed with an atomic force microscope. *Ultramicroscopy* **42-44**, 1125-1132.

Yager, P. and Schoen, P.E. (1984). Formation of tubules by a polymerizable surfactant. *Mol. Cryst. Liq. Cryst.* **106**, 371-381.

Yang, J, Tamm, L.K, Tillack, T.W. and Shao, Z. (1993.) New approach for Atomic force microscopy of membrane proteins. *J. Mol. Biol.* **229**, 286-290.

Zasadzinski, J.A.N, Helm, C.A, Longo, M.L, Weisenhorn, A.L, Gould, S.A.C. and Hansma, P.K. (1991). Atomic force microscopy of hydrated phosphatidylethanolamine bilayers. *Biophys. J.* **59**, 755-760.

Zhai, X. and Kleijn, J.M. (1997). Molecular structure of dipalmitoyl phosphatidylcholine Langmuir-Blodgett monolayers studied by atomic force microscopy. *Thin Solid Films* **304**, 327-332.

Useful information

Gel-Pak™, Vichem Corporation, 756 North Pastoria Avenue, Sunnyvale, California 94086, USA

CHAPTER 6

ORDERED MACROMOLECULES

6.1. Three Dimensional Crystals

The study of the surfaces of 3D crystals by AFM offers the prospect of achieving the highest resolution on biological samples. The periodicity of the surface will minimise probe broadening effects and allow image reconstruction to remove noise. The structure at the surface may differ from that in the bulk and it may be possible to identify features which help in the subsequent analysis by X-ray diffraction. If the crystals are small then AFM offers an alternative to electron microscopy, with the possible advantages of acquiring images under aqueous or buffered conditions. If the crystalline structures occur naturally, then AFM allows these faces to be examined under natural conditions, with the prospect of observing biological processes such as enzymatic reactions at the surfaces. Finally the AFM permits crystal growth to be examined, and an investigation of growth kinetics and mechanisms. As a microscopic technique it is able to visualise the effects of foreign particles and the incorporation of defects into the crystal lattice. This type of information provides a basis for more systematic optimisation of crystal growth for X-ray diffraction work.

6.1.1. Crystalline cellulose

There are a number of AFM studies on isolated crystalline cellulose fibres which report high resolution images and analysis of the surface structure (Hanley *et al* 1992; Baker *et al* 1997; 1998; Kuutti *et al* 1994). The most detailed studies are on cellulose isolated from *Valonia*, which regarded as the source of the most crystalline form of cellulose I. *Valonia* cellulose contains two distinct allomorphs I_α (triclinic) and I_β (monoclinic). Recent high resolution AFM studies have been used to probe the detailed surface structure of *Valonia* cellulose. AFM images of the purified cellulose from *V. macrophysa* were enhanced by Fourier processing and compared with model Connolly surfaces generated from electron diffraction data for the I_α and I_β forms. The surface structures observed were attributed to the monoclinic phase (Kutti *et al* 1995). More recent AFM studies on *V. ventricosa* cellulose have produced images of the surface in which it has been possible to image the repeating cellobiose unit along the cellulose molecules (Fig. 6.1), though identification of the bulky hydroxymethyl group, and thus to identify a triclinic phase directly from the images (Baker *et al* 1997; 1998). The distinction between the two crystal phases required the detection of differences in the displacement of the cellulose chains along their axes by 0.26 nm. This was achieved without filtering or averaging and is believed to represent the highest resolution achieved to date on a biological sample. At present AFM has not been used to investigate possible coexistence of I_α and I_β on the crystal surfaces. The cellulose structures could be imaged under air (Kutti *et al* 1995), propanol (Baker *et al* 1997; 1998) or water (Baker *et al* 1997; 1998). The ability to image under water means that it is possible to image the surface at which biological processes occur, and that it may also be possible in the future to follow processes such as enzymatic breakdown.

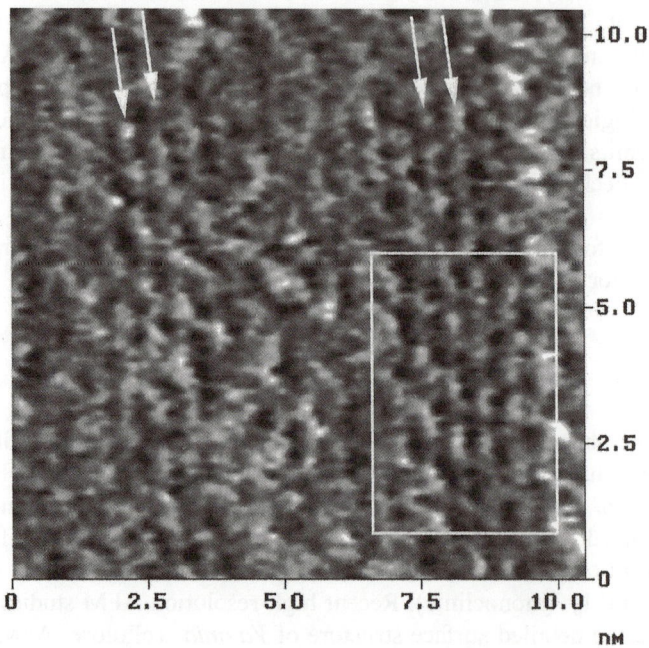

Figure 6.1. Contact error signal mode AFM images of a *Valonia* surface under water. The white arrows indicate cellulose molecules which are running almost vertically down the page. The white box highlights a region where bright spots can be seen along the length of the molecules, separated by a distance closely matching the cellobiose repeat unit. The angle of the spots within the box to the molecular axis is $64 \pm 2°$. Data provided by the authors and based on Baker *et al* 1997.

6.1.2. Protein crystals

The first AFM study of 3D protein crystals was on crystals of the membrane protein Ca-ATPase, the calcium pump from the sarcoplasmic reticulum (Lacapere *et al* 1992). For small dried crystals it was possible to measure step changes in height allowing the unit cell dimension normal to the substrate to be determined. This value was found to increase on hydration presumably due to hydration and swelling of extramembranous domains. The in-plane resolution for both dried and hydrated crystals was poor: it was not possible to resolve individual proteins or characteristic periodicities of the crystal lattice. The difficulties in imaging were attributed to displacement of layers within the crystal by the probe during scanning. A major problem identified in these early studies was the difficulty of immobilising small crystals on the substrate, particularly for studies on hydrated crystals or investigations of crystal growth in mother liquor. Approaches to resolving this problem include optimising physical adsorption to substrates (Lacapere *et al* 1992), nucleation of crystal growth directly on the substrate (Malkin *et al* 1996a,b; Kuznetsov *et al* 1997), and sticking the crystals to the substrate (Konnert *et al* 1994; Yip and Ward, 1996) with an adhesive. If the crystals are firmly attached then the second limiting factor is damage or 'etching' of the surface with the probe. Provided the crystals are resilient enough, and that the imaging force is minimised, then dc contact mode imaging can give molecular resolution (Kuznetsov *et al* 1997). It was shown for soft crystals such as insulin that if the crystal surface is damaged in contact mode by shear forces then the use of Tapping can improve the images (Yip and Ward, 1996). AFM studies have been made on a number of protein crystals including insulin (Yip and Ward, 1996), canavilin (Land *et al* 1995; 1997; Malkin *et al* 1996a), thaumatin (Malkin *et al* 1995a; 1996b), catalase (Malkin *et al* 1995a), lysozyme (Durbin and Carlson, 1992; Durbin *et al* 1993; Malkin *et al* 1995a; Konnert *et al* 1994; Kuznetsov *et al* 1997), apoferritin (Markin *et al* 1995a), bacteriorhodopsin (Kouyama *et al* 1994) and lipase (Kuznetsov *et al* 1997). In many of these studies it has been possible to obtain molecular resolution, or to observe periodicities characteristic of the crystal lattice (Konnert *et al* 1994; Land *et al* 1995; Yip and Ward, 1996; Kuznetsov *et al* 1997). The information obtained from high resolution AFM images of 3D protein crystal is unlikely to rival that obtainable from X-ray diffraction but it may complement images or image reconstruction obtainable by electron microscopy. However, AFM data can be used to obtain information useful for crystal structural analysis by X-ray diffraction: the data can be used to resolve questions on space group enantiomers, about packing of molecules within unit cells, the number of molecules per asymmetric units, or the disposition of multiple molecules within asymmetric units (Kuznetsov *et al* 1997). AFM studies of the surfaces of bacteriorhodopsin revealed a novel assemblage of bacteriorhodopsin in its 3D crystal: the crystal consists of hexagonal close-packed arrays of spherical protein clusters. Both AFM and EM have advantages for studying small crystals. AFM provides a means of observing differences in the structure of the surface when compared to the bulk; but to date no such differences

Figure 6.2. An AFM image of a canavalin crystal surface showing growth from multiple hillocks. Scan size 100 x 100 μm. Data reproduced from Land *et al* 1997 with permission.

have been reported. The major advantage of the AFM is the ability to investigate mechanisms of crystallisation and to follow crystal growth. The first studies on protein crystal growth were by Durbin and coworkers (Durbin and Carlson, 1992; Durbin *et al* 1993). Although largely non-invasive there are reports that growth rates may be modified by the scanning process (Land *et al* 1995; Yip and Ward, 1996) and the tip may displace loosely bound proteins, or protein aggregates, which might otherwise influence growth. Thus Tapping mode images of insulin crystals revealed aggregates attached to the crystal surface which were not seen in dc contact mode images (Yip and Ward, 1996). Studies on a range of protein crystals suggest that the main method of crystal growth is by the nucleation and growth of 2D islands on the surface of the crystal. This has been observed to be the dominant mechanism for lysozyme (Durbin and Carlson, 1992; Durbin *et al* 1993; Malkin *et al* 1995a; Konnert *et al* 1994), catalase (Malkin *et al* 1995a) and thaumatin (Malkin *et al* 1995a; 1996). At low supersaturations screw dislocations have been observed for lysozyme (Markin *et al* 1995a; Konnert *et al* 1994) and thaumatin Markin *et al* 1995a; 1996). At high supersaturation 3D clusters are deposited onto thaumitin surfaces and their 2D tangential growth results in multilayer stacks (Malkin *et al* 1996). The mechanisms of crystal growth, surface morphology and growth kinetics have been studied in detail for canavilin (Malkin *et al* 1995a; Land *et al* 1995; 1997). Depending on the level of supersaturation growth occurs on steps generated either by simple or complex screw dislocation sources (Fig. 6.2), 2D nucleating islands or deposited 3D clusters (Fig. 6.3). Apoferritin and occasionally trigonal catalase and tetragonal lysozyme were observed to develop very rough growing surfaces by intensive

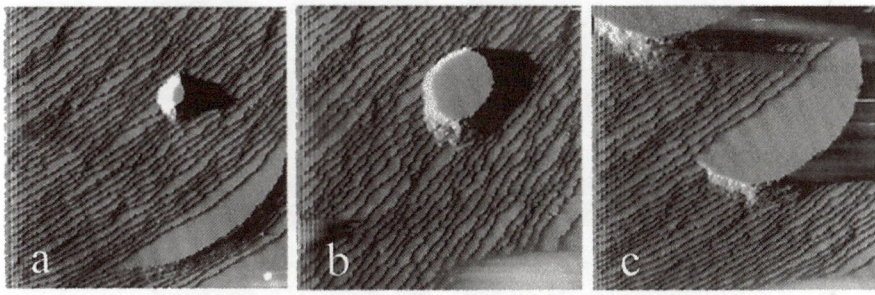

Figure 6.3. A series of AFM images (scan size 40 x 40 μm) collected at 30 secend intervals showing the growth of a macrocluster which has landed on the growing crystal surface. The nucleus grows out radially, forming macrosteps at the edge and a large plateau on top. Note that the nucleus merges flawlessly into the existing crystal. Data reproduced from Land *et al* 1997 with permission.

random nucleation; a rare process for conventional crystal growth from solution known as normal crystal growth (Malkin *et al* 1995a). For the range of proteins studied all mechanisms of crystal growth have been observed and lysozyme has been shown to exhibit all of these mechanisms under appropriate conditions. Adsorption of foreign particles results in holes which remain within the crystal upon growth of additional layers. This has been observed for canavilin (Land *et al* 1995) and lysozyme (Konnert *et al* 1994) crystals.

6.1.3. Nucleic acid crystals

When compared to proteins very few nucleic acids have been crystallised. The crystallisation process is still largely empirical because virtually nothing is known about the growth mechanisms, growth kinetics, development of surface morphology, or the effect of defects on crystallisation. Recently there has been an AFM study [Ng *et al* 1997) of tRNA crystal growth aimed at providing a basis for optimising crystallisation of RNA. Crystal growth was observed to occur in steps on steep vicinal hillocks generated by screw dislocations. Different growth mechanisms were observed in different temperature ranges which were used to vary supersaturation (Fig 6.4). At low supersaturation a 2D nucleation process occurs in which the step edges of the hillocks grow tangentially, and small islands

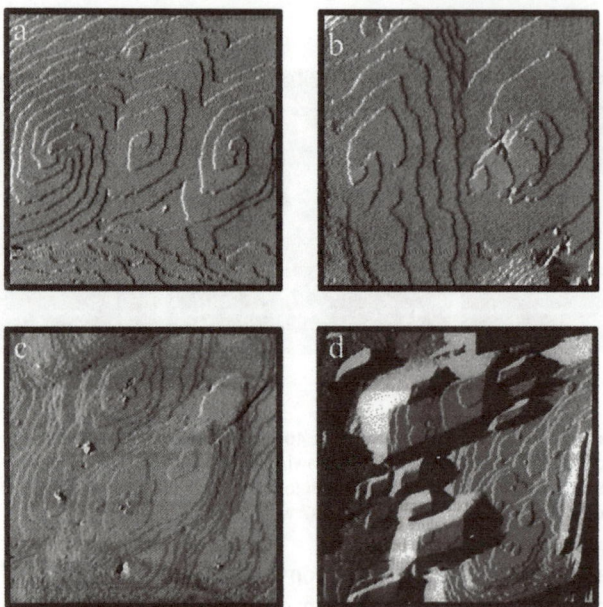

Figure 6.4. AFM images showing different surface morphologies of yeast tRNA crystals seen at different temperature ranges. (a) Dislocation hillocks formed by right handed, single left handed and double right handed screw dislocations (15° C). (b) Development of double and single screw dislocations (14°C). (c) Growth by 2D nucleation at 13°C. Also formation of a hole caused by incorporation of a foreign particle during growth. (d) Dominant mechanism of growth is by 3D nucleation with multilayer stack macrostep production at 12°C. Scan sizes (a,b) 23 x 23 μm, (c) 20 x 20 μm and (d) 34 x 34 μm. Reproduced with permission of the authors and Oxford University Press from Ng *et al* 1997.

nucleate and grow tangentially on the plateaus of the hillocks. As observed for proteins, at higher supersaturation a 3D mechanism occurs in which multilayer stacks appear and grow both normally and tangentially on the crystal surface. Adsorption of foreign particles results in holes which remain upon growth of additional layers: these types of holes have also been observed in the growth of protein (Konnert *et al* 1994; Land *et al* 1995) and virus (Malkin *et al* 1995a; 1996a) crystals. Lattice resolution images of the crystal surface were obtained but no molecular features characteristic of tRNA were observed. The study has been used to suggest ways of improving the crystallisation of RNA: it is proposed that after nucleation supersaturation is reduced to allow growth by regular and unique mechanisms (Ng *et al* 1997).

6.1.4. Viruses and virus crystals

Viruses consist of a nucleic acid packaged within a protein coat. There is interest in the detailed molecular structure, assembly and mechanisms of viral infection. Usually it is sufficient to passively adsorb the virus particles to substrates such as mica or silica wafers. If necessary substrates such as mica can be coated with poly-L-lysine to prevent the viruses particles being displaced on the surface by the probe. Images can be obtained in air, under propanol or in aqueous media. A range of types of viruses have been imaged by AFM including bacteriophages (Kolbe *et al* 1992; Thundat *et al* 1992; Imai *et al* 1993; Ikai *et al* 1994), plant viruses (Thundat *et al* 1992; Zenhausern *et al* 1992; Imai *et al* 1993; Bushell *et al* 1994; Kirby *et al* 1996; Falvo *et al* 1997) and animal viruses (Ohnesorge *et al* 1997). They have been considered as standards for height measurements or for tip deconvolution. The bacteriophage T4 polyheads were used as a standard material for appraising image processing of STM images (Engel, 1991). If the virus particles are assembled into arrays then probe broadening effects are minimised and the measured widths approach the true diameters of the virus particles (Thundat *et al* 1992; Kirby *et al* 1996). Most studies are low resolution revealing size and shape, or in the case of bacteriophages features such as heads, tails and tail fibres (Ikai *et al* 1994). There are no studies at present of the assembly of the protein coats. Submolecular resolution has been obtained by AFM on the crystalline arrays of bacteriophage T4 polyheads (Karrasch *et al* 1993). Images of pox viruses under water and buffer have revealed tubular 'protein' structures consistent with those observed previously by electron microscopy (Ohnesorge *et al* 1997). As well as intact viruses it is also possible to observe damaged virus particles from which the nucleic acid has spilled out (Fig. 6.5) onto the substrate (Kolbe *et al* 1992: Shao and Zhang, 1996).

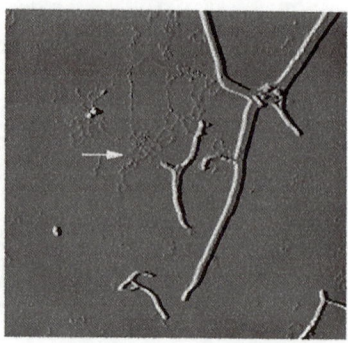

Figure 6.5. AFM image of papaya mosaic virus air dried onto mica and imaged under butanol. The arrow indicates RNA released from a damaged virus particle. Scan size 3 x 3 μm.

It has been suggested (Ikai *et al* 1994) that the heights of bacteriophage heads can be used to distinguish between normal heads and ghosts devoid of nucleic acid: the ghosts being either deformed more by the probe or collapsing down onto the substrate. Virus particles can be damaged or dissected with the probe tip (Ikai *et al* 1994; Bushell *et al* 1995; Falvo *et al* 1997): for example partial removal of the protein coat of TMV revealed the central channel (Bushell *et al* 1995). Controlled manipulation of TMV particles deposited onto graphite was carried out using a modified AFM (Fig. 6.6) and the data obtained was interpreted in terms of a mechanical model for the virus particles (Falvo *et al* 1997).

In an unique set of experiments it has been possible to study living cells infected with virus particles. Living cultured monkey kidney cells were imaged using a 'pipette-AFM' (Häberle *et al* 1991; 1992; Hörber *et al* 1992). For cells infected with pox virus it was possible to image adsorbed viruses on the cell surface (Ohnesorge *et al* 1997) and to generate a time sequence of images believed to demonstrate release of new viral particles from the cell surface (Häberle *et al* 1992; Hörber *et al* 1992; Ohnesorge *et al* 1997).

There are several reported studies on the growth of satellite tobacco mosaic virus crystals (Malkin *et al* 1995a,b; 1996a; Kuznetsov *et al* 1997). The growth mechanism differs from that seen for protein crystals (section 6.1.2), in that 2D nucleation and growth only made a major contribution to crystal growth at very low supersaturation, and no screw dislocations were observed. The dominant growth mechanism appears to be by addition of 3D clusters or microcrystals, and their subsequent expansion. If the 3D additions are misaligned with respect to the larger crystal lattice then their incorporation generates a defect structure in the crystal. Other sources of defects are adsorbed foreign particles which are also incorporated in the growing crystal. High resolution images of the crystal surfaces permit resolution of individual virus particles (Kuznetsov *et al* 1997; Malkin *et al* 1995b).

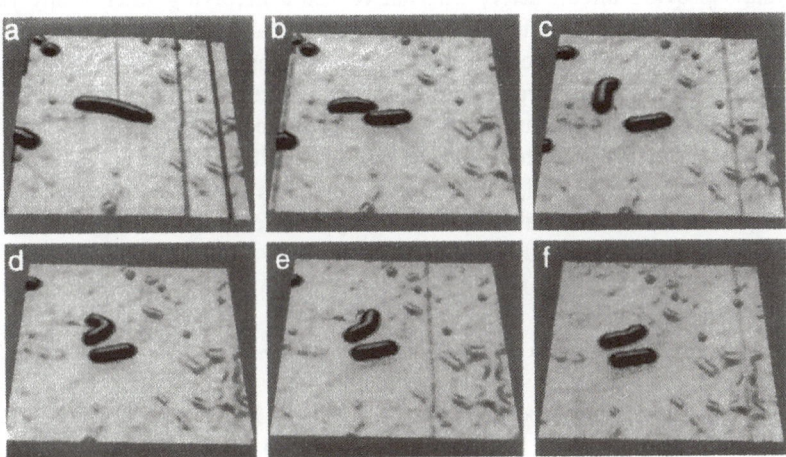

Figure 6.6. AFM images showing controlled dissection and manipulation of a TMV virus particle after deposition onto graphite. The sequence of images shows the TMV particle in its original position and orientation (a), after dissection (b), rotated (c), translated (d), and straightened so that it is parallel with the remaining segment of the virus particle (e and f). Scan size 560 x 560 nm. Reproduced with permission from Falvo *et al* 1997 and the Biophysical Society.

6.2. Two dimensional protein crystals

Two dimensional (2D) protein crystals are a ubiquitous class of proteins occurring naturally on bacterial and algal cell surfaces. The interest in these proteins from a microscopy point of view was sparked off by Henderson and Unwin (1975) who demonstrated that regular two dimensional crystalline arrays enabled elegant image processing (including electron diffraction) to be carried out *on-line* using transmission electron microscopy. In further refined studies they were able to produce a 3D map of bacteriorhodopsin (Baldwin and Henderson, 1984, Henderson *et al* 1986; 1990, Baldwin *et al* 1988) from which an atomic model of the protein has been proposed (Henderson *et al* 1990). Since then many other systems have been examined using TEM, notable examples including 2D crystals of the bacterial porins OmpF (Sass *et al* 1989) and PhoE (Jap *et al* 1988) and of the plant light harvesting complex (LHC-II) (Kühlbrandt and Downing, 1989,

Wang and Kühlbrandt, 1991). The justification for the study of 2D protein crystals is the structural determination of the proteins at atomic or near atomic resolution. Normally this would of course be done by X-ray crystallography but membrane proteins in particular are notoriously difficult to crystallise in three dimensions so 2D crystals provide an attractive alternative route involving microscopy rather than X-ray crystallography (they are typically only 5-20 nm thick which is ideal for electron diffraction but too thin for X-ray crystallography). To date most work has been carried out on naturally abundant membrane proteins which are easy to isolate and purify in large amounts. However, there are a vast array of other interesting candidates for study which are naturally rare such as receptors, channels and transporters and, with recent advances in genetics, these proteins can now be produced in reasonable quantities by over-expression of genes in cell culture (Kühlbrandt, 1992; Müller *et al* 1997a).

6.2.1.What does AFM have to offer?

AFM has, not surprisingly, been brought to bear on 2D protein crystals offering the inherent advantages of being able to work in physiological buffers, not exposing the sample to any radiation and, in terms of raw data or image contrast, achieving resolution better than TEM (Müller *et al* 1996). There are limitations to the use of AFM though; compared to TEM the image processing is more difficult since the pixel density of an AFM image cannot possibly compete with an 'on-line' electron beam which effectively has no pixel limit as such. This problem is partly obviated by the fact that AFM images are of far higher contrast than TEM images, and so the need for image processing is much less. Generally the effects of tip convolution on AFM images of specimens presents a significant problem. Image de-convolution is a non-trivial process due to factors such as compression and distortion of molecules by the AFM tip (Shao *et al* 1996b, Yang *et al* 1996) and inherently unpredictable events such as changes in the shape of the tip during scanning due to wear or, more likely with biological samples, contamination (Schabert and Engel, 1994b). However for samples which are close packed and have very low surface roughness, tip convolution becomes negligible since only the apex of the tip takes part in the imaging process as illustrated in Fig. 6.7. Incidentally this demonstrates one of the unique features of images generated by probe microscopes, namely that image resolution is enhanced as specimen roughness decreases (Bustamante *et al* 1997). This is contrary to all other forms of microscopy where such flatness leads to poor contrast. TEM with its ability to perform electron diffraction offers the advantage of being able to generate electron density maps, and from these atomic models of the protein molecules.

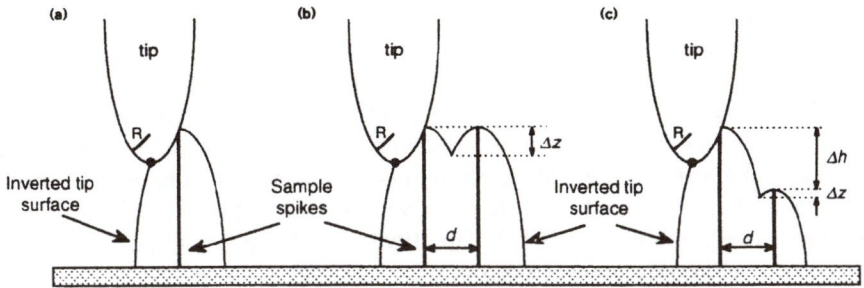

Figure 6.7 Image formation in the AFM. (a) Image of a spike obtained with a parabolic tip. (b) Image of two spikes of similar height close together. (c) Image of two spikes of different height. The ability of the AFM to resolve two spikes separated by a distance d depends on the size of the dimple ΔZ in the image and decreases with *increasing* height difference Δh between the spikes. Thus flatter samples can lead to higher resolution images. R is the radius of curvature of the tip. Reproduced from Bustamante *et al* 1997 with kind permission from Elsevier Science Ltd, London, UK.

At present AFM cannot offer this capability and it is difficult to envisage how this could ever be achieved with AFM. However, even without the capability to generate atomic models of the proteins, comparison of predicted surface topography data derived from such models with directly acquired AFM images of the surface topography which are obtained in realistic environments, can assess the reliability of the TEM data. TEM suffers particularly because the images need to be processed using averaging techniques to produce acceptable contrast, and this means that only regular repeating structures can be resolved, defects in the crystals are lost along with the noise reducing the resolution in the TEM data, and other local variations in structure such as twinning can actually produce incorrect data. AFM with its ability to generate high contrast images directly does not suffer from this drawback, and is in fact exceptionally good at visualising point defects in crystals (Devaud *et al* 1992). The conclusion from a structural determination point of view is that AFM provides a powerful complementary technique to be used alongside rather than instead of TEM. There are other areas though where AFM can provide unique information, for example by working in physiological buffers the AFM can image dynamic events *in-situ* (Müller *et al* 1996). The principle investigator involved in the transition from TEM studies of these systems to AFM studies is Engel who, along with his group, has demonstrated the undoubted contribution that AFM has to offer in this area.

6.2.2. Sample preparation: membrane proteins

Some membrane proteins such as bacteriorhodopsin occur naturally as 2D crystalline arrays and sample preparation in this case simply involves retrieving them from the cell wall and then applying various treatment steps which depend upon the system being examined. These are many and varied and beyond the scope of this book, but are described in detail in various comprehensive review articles (Boekema, 1990; Kühlbrandt, 1992; Jap, 1992) and the book, *Crystallization of membrane proteins* (Hartmut 1991).

There is, however, a more general method which, although the precise details vary for different proteins, can be applied to all membrane proteins. The best 2D crystals produced *in-vitro* have been grown from detergent solubilised and purified material. Conditions can be controlled more easily, and consequently the results are more reproducible (Kühlbrandt, 1992). The principles features of this method are as follows. Membrane proteins are insoluble in water and so the first step to forming 2D crystals is to solubilise them using detergents. They are then added in the detergent solution to a suspension of lipid-detergent micelles and finally the detergent is removed by dialysis or adsorption onto latex beads (Kühlbrandt, 1992). This leaves the proteins in a continuous lipid bilayer just as they would be in the biological membrane, resembling the situation *in-vivo*. Because they are much smaller than 3D crystals a relatively small amount of the protein is required and, furthermore, two dimensional protein crystals grow much more quickly, meaning that the protein is exposed to high levels of detergent for only a short time, favouring protein stability.

Finally reconstitution of protein membranes can lead to several different arrangements of 2D membrane protein crystals. In native membranes and in vesicle crystals all of the protein molecules face the same way in the lipid bilayer (Fig. 6.8a,b). Tubular crystals can also form which are basically vesicle crystals with the proteins arranged helically on the surface of a cylinder (Fig. 6.8c). Finally crystals with protein molecules facing alternately up and down can form when the crystals are grown from isotropic detergent solutions (Fig. 6.8d).

6.2.3. Sample preparation: soluble proteins

Although membrane proteins form 2D crystals more or less spontaneously when mixed with detergents and lipids, most soluble proteins need to be coerced into forming crystals. The general principle of 2D crystallisation for these proteins has four basic requirements. The first is that the molecules should be fixed in a plane, the second that they should have sufficient mobility within the plane to permit reordering allowing the third requirement to be satisfied, namely identical orientation of all of the molecules.

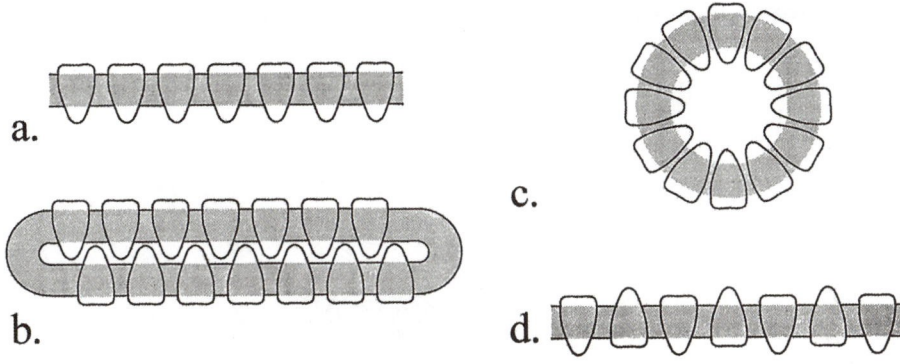

Figure 6.8. Different types of 2D protein crystals. Reproduced in part from Kühlbrandt, 1992 with permission of the author and Cambridge University Press.

The fourth and final requirement is a high concentration of the molecules in the plane so that crystallisation will be favoured over the formation of two dimensional solutions. The molecules can be fixed in two dimensions by adsorbing them onto an air-water interface but pure protein monolayers lack the necessary mobility to allow molecular rearrangement. This is overcome by adsorbing the proteins to a lipid monolayer (Uzgiris and Kornberg, 1983) either electrostatically to a charged lipid monolayer or by binding the proteins to ligands attached to the polar head groups of the lipids (Fig. 6.9). The mobility of the lipid monolayer can be controlled by adjusting packing density and/or hydrocarbon chain length of the lipid molecules, which turns out to be an important parameter for achieving 2D crystallisation of different proteins (Kornberg and Ribi, 1987). Most proteins require a high lipid density in the *fluid* phase. Lipids compete with protein for sites on the air-water interface and it is speculated that a high level of lipids prevents denaturation of the proteins, which would otherwise unfold given free access to the air-water interface, and unfolded or denatured protein will not easily crystallise (Kornberg and Ribi, 1987).

There are two options for preparing 2D crystals. The first method is to use the traditional means of constructing the lipid monolayer on a Langmuir trough and the protein is then added to the subphase (Darst *et al* 1990). Film transfer is achieved by performing an LB-dip (as described in section 5.1.4.) using a hydrophobic substrate. The lipid layer binds to the substrate on the way into the liquid and the surface should be swept with the barriers before pulling the substrate back out again in order to avoid a bilayer being transferred, in which case a second layer would be transferred with lipid molecules uppermost on the substrate surface.

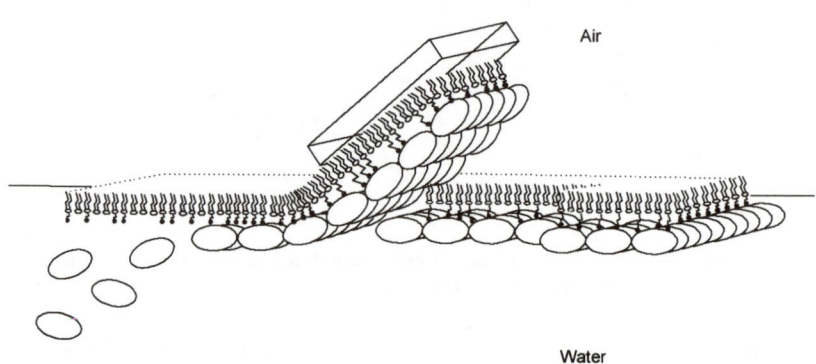

Figure 6.9. The production of 2D crystals of soluble proteins is achieved by binding the proteins to an interfacial monolayer of lipid molecules either electrostatically or chemically to allow the rearrangement and ordering needed for crystallisation. Reprinted from Kornberg and Darst, 1991 with kind permission from Elsevier Science Ltd, London, UK.

This method suffers from the drawback that most Langmuir troughs require a significant volume of liquid to fill them and, because protein concentration in the subphase needs to be relatively high (typically 1-3 mg mL^{-1}), large amounts of protein are needed, negating one of the advantages of two dimensional crystals. For small scale work the second method is to simply place the substrate directly onto the surface of protein droplets (10-20 µL) coated with lipid and then withdraw the substrate, which in this case was a carbon coated hydrophobic TEM grid (Kornberg and Darst, 1991). Some of the lipid layer and attached protein adhere to the substrate as a result of hydrophobic interactions between the hydrocarbon chains of the lipid and the carbon film. For AFM imaging HOPG could be used as an alternative, flatter hydrophobic substrate. Although this method has the advantage of requiring only very small amounts of sample it is a little more 'hit and miss', lacking the control of the Langmuir trough method. An alternative method requiring only small volumes of protein solution (~ 40 μL) is to use a Teflon cell well as a mini Langmuir trough (Czajkowsky *et al* 1998). Recently a means of monitoring, and so improving the 2D crystallisation process, has been proposed based upon ellipsometry and shear rigidity measurement of the interfacial layers on the Langmuir trough (Vénien-Bryan *et al* 1998). Ellipsometry monitors the attachment of the proteins to the lipid monolayer and shear rigidity, which measures the resistance of the interfacial film to flow, monitors the crystallisation of the protein molecules.

Another rather simpler, though less controlled method for the formation of 2D soluble protein crystals (Harris, 1992) is an adaptation of the mica spreading, negative staining carbon films procedure originally developed for forming 2D ordered arrays of viral particles for TEM (Horne and Pasquali-Ronchetti, 1974). The crucial extra step in the adapted procedure is the inclusion of polyethylene glycol (PEG). A solution of highly purified protein, around 1mg mL^{-1} in low concentration buffer, is mixed with a solution of 2% ammonium molybdate containing 0.1-0.2% PEG (molecular weight 1000-10000 D). This mixture is spread over freshly cleaved mica and excess fluid removed to leave a thin, even layer of liquid on the mica. This is then allowed to dry at room temperature, which takes approximately 5 to 10 minutes. Protein crystallisation occurs during drying, but depends upon the optimisation of conditions such as protein concentration, ionic strength of the buffer and its pH, the PEG grade and concentration, drying temperature and time. These all have to be determined by trial and error making the process potentially somewhat time consuming. Another limitation of this technique is of course the drying step which, although fine for EM, discards one of the advantages of AFM.

6.3. AFM Studies of 2D membrane protein crystals

6.3.1. Purple membrane

Probably the first 2D membrane protein crystals to be examined by AFM were the purple membranes of *Halobacterium halobium*. They are 2D crystals composed of 75% of the protein bacteriorhodopsin and 25% lipid, and function as light powered proton pumps. The bacteriorhodopsin molecules are densely packed into trigonal lattices. Structural analysis by cryo-electron microscopy has revealed a retinal unit embedded in seven closely packed α helices (Henderson *et al* 1990). Early AFM studies of purple membrane met with only limited success, which was almost certainly due to instrumental inadequacies (Worcester *et al* 1988;1990). These investigations were made in the days before optical sensing methods for AFM had been developed and, more importantly with home-made AFM tips, *and* in air where the factors of poor tip quality and high imaging forces combined to make resolution of any fine detail impossible. Things improved considerably with the advent of commercially manufactured AFM tips and AFMs with the ability to operate in liquids so that forces could be reduced, and images containing some structural information were obtained. Nevertheless, the contrast in the images was poor and the structural information was derived from Fourier analysis (Butt *et al* 1990). Much more recently however, a drastic improvement in image quality was obtained by Müller and colleagues who demonstrated the absolute importance of controlling the imaging forces in an AFM study of purple membrane, by recording a force-induced conformational change of the protein molecules on the cytoplasmic surface (Müller *et al* 1995a). The conformational change is clear and can be seen in Fig. 6.10. This represents an important step along the road to assessing the potential for sample distortion by the scanning tip of the AFM on soft biological systems at very high resolution. In addition to demonstrating the effect that the imaging force can have on such delicate systems, the conformational change itself revealed a hint about the importance of the choice of imaging environment. The conformational change was noticed to be fully reproducible in high pH buffer but less so in acidic buffer. This observation strengthened the interpretation that what was actually being seen in the AFM images was the protrusion of a loop, predicted from the sequence of the protein, which connects two helical regions in bacteriorhodopsin, since this would pop out of the cytoplasmic surface at basic pH due to electrostatic forces. The conclusions drawn from this study were twofold, firstly and perhaps not surprisingly, choice of imaging environment is particularly important if imaging is carried out in aqueous buffers, and secondly, conformational changes due to buffer conditions may be visualised with AFM. Indeed this was confirmed in a later study of HPI layers by the same group (Müller *et al* 1997).

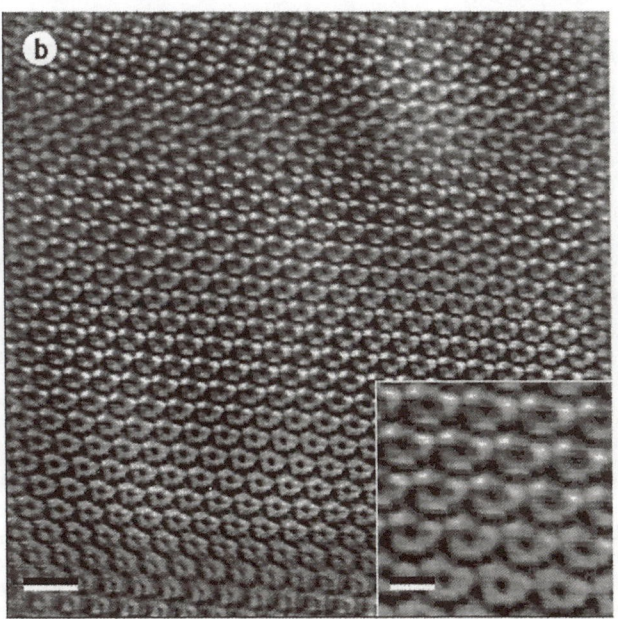

Figure 6.10. Force dependent topography of the cytoplasmic surface of purple membrane. The force was adjusted manually during the scan from 300 pN at the bottom of the image to 100 pN at the top. A distinct conformational change can be seen, doughnut shaped bacteriorhodopsin trimers transform into units with three pronounced protrusions at their edges. Scale bars (left) 10 nm (right) 4 nm. This image has been Fourier filtered to improve its clarity. Reproduced with permission from Müller *et al* 1995a.

In a related study (Müller *et al* 1995b) an attempt to determine the orientation of the purple membranes was made by adsorbing them to poly-L-lysine coated mica at different values of pH, which favour attachment by either the cytoplasmic surface (pH 7.4), or the extracellular surface (pH 3). Compared with samples adsorbed to bare mica, which produced diffraction patterns with reflections extending beyond a lateral resolution of 0.7 nm, the samples attached to poly-L-lysine treated mica were rougher, due to buckling of the membrane sheets, and generated diffraction orders to (only!) 1.18 nm, partly because the deviation of individual units in the images from the lattice symmetry was greater. Identification of the different sides of the purple membrane sheets were tentatively made but, more recently, immunolabelling has been used to confirm this and to produce 0.7 nm lateral resolution on the extracellular face (Müller *et al* 1996a); the improvement presumably being due to attachment to bare mica, which avoids the problems with poly-L-lysine coated mica. The AFM images obtained are reproduced in Fig. 6.11. below.

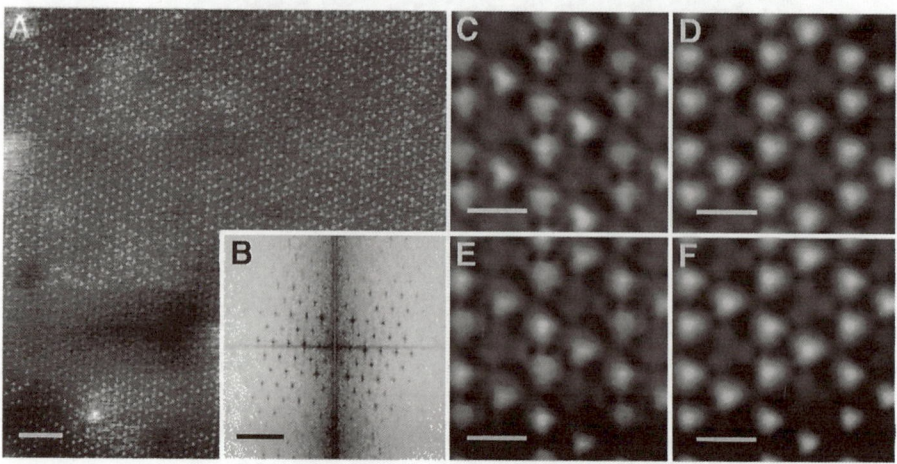

Figure 6.11. Extremely high resolution AFM images of the extracellular surface of purple membrane. (A) Image recorded in the trace direction (left to right). (B) The peaks in the power spectrum extend to a resolution of 0.7 nm. Correlation averages derived from images scanned in trace and retrace direction (right to left) calculated from 770 unit cells can be seen in C and E respectively. Symmetrized averages are shown in D and F which come from the average topography of C and E, the height in these images is 0.2 nm. Scale bars: (A) 20 nm, (B) 2 nm^{-1} and (C-F) 4 nm. Reproduced with permission from Müller *et al* 1996a.

Several interesting observations were made in the immunolabelling study. Firstly, whilst the antibody molecules could be swept off the mica surface at forces of 0.2-0.3 nN, those attached the purple membrane required the higher forces of around 1nN to remove them, suggesting greater interaction between the antibody and antigen, than between antibody and mica. Secondly, after removal of the antibody molecules from the membranes the underlying crystalline topography could then be recorded. Thirdly, areas on the membranes cleared of antibodies by the AFM tip were rapidly re-labelled indicating that the surface retained its native antigenicity, despite contact with the AFM tip. Fourthly, despite imaging with very low forces (100 pN), no high resolution images of the antibody molecules themselves were obtained - only 'blobs' of the appropriate dimensions, illustrating the advantages of forming 2D crystalline arrays of proteins if one wishes to image structural detail. A final point worth mentioning is that this study is one of only a few in which the use of antibodies was checked with a proper control sample (Müller *et al* 1996a).

6.3.2. Gap junctions

After purple membranes, the first AFM study on a protein membrane system which demonstrated good contrast of ordered arrays on a membrane surface was that carried out on gap junctions (Hoh *et al* 1991). Gap junctions are regions in the cell membranes of vertebrates which allow the free passage of small molecules (< 1 kD) and, having a low electrical resistance, provide a signalling pathway (Flagg-Newton *et al* 1979). They are made up of two plasma membranes sandwiched together with cell to cell channels bundled into an array as the 'filling' (Revel and Karnovsky, 1967). They were a good sample to pick for an early AFM study since they had already been extensively studied by X-ray diffraction (XRD) and EM, and a model for their structure proposed (Makowski *et al* 1977, Makowski 1985, Unwin and Zampighi, 1980). In the model the channels were aligned head to head across a 2-3 nm gap between the membrane sheets. In the most regular samples the channels were arranged hexagonally with lattice constants of 8 to 10 nm, but this varied according to the preparation conditions. The channel itself was composed of two connexons, one from each membrane. The connexons were proposed to be cylindrically shaped, 7.5 nm high and 7 nm wide. Each had a pore running through its centre, some 1.5-2.0 nm in diameter. Finally, the connexon was believed to consist of six identical protein subunits, since it has a six-fold symmetry. The gap junction dispersion in PBS were allowed to adsorb to glass coverslips for 10-20 minutes and then the substrate was rinsed with an excess of buffer prior to imaging. Samples revealing the hexagonal arrays were trypsinised and fixed with glutaraldehyde. AFM images, collected with forces around 1 nN, of the gap junction at low magnification under PBS buffer revealed them to be irregularly shaped flat sheets of 14 nm thickness and typically 0.5 microns in size. At higher magnification the top surface was simply seen to be undulating and featureless. Fixation of the gap junctions did not appear to change their overall morphology; they all exhibited approximately the same height, although the fixed membranes were more easily scraped off the glass surface because they are less positively charged after fixation, reducing their interaction with the negatively charged glass. When the fixed samples were imaged at higher forces (up to 15 nN) the top membrane sheet was removed. The process is shown in Fig. 6.12. and the changes in height can be seen in the line profiles beside each image (Fig. 6.12H).

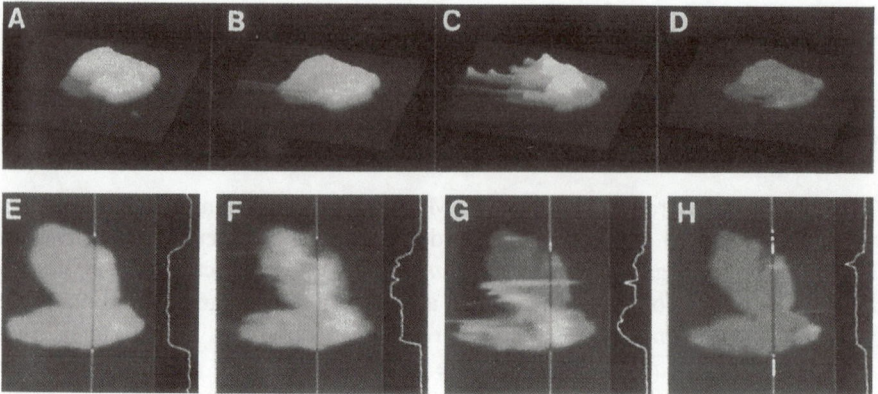

Figure 6.12. AFM tip induced dissection of gap junction membranes. Increasing the force set-point from (A) 0.8 nN (B) 3.6 nN (C) 6.1 nN and (D) 9.6 nN is shown to 'shovel' off the upper membrane from right to left on a single gap junction plaque. The bottom row of images shows a top view of another gap junction plaque which has again been subjected to increasing imaging force; (E) 0.8 nN (F) 3.1 nN (G) 10.1 nN and (H) a repeat scan at 10.1 nN. Line profiles are shown beside each image illustrating that the height of the structure is reduced in successive images to finally half the original value (from ≈ 15 nm to ≈ 7 nm). Scan sizes are 1.5 x 1.5 μm. Reprinted with permission from: Hoh, J.H, Lal, R, John, S.A, Revel, J-P. and Arnsdorf, M.F. (1991). Atomic force microscopy and dissection of gap junctions. *Science* **253**, 1405-1408. Copyright 1991 American Association for the Advancement of Science.

Once the top layer had been 'dissected' high magnification imaging clearly revealed the hexagonally packed connexons. The centre to centre spacing of the connexons was determined by Fourier analysis and yielded a value of 9.1 nm, in excellent agreement with the previous XRD and EM studies (Makowski *et al* 1977, Makowski, 1985, Unwin and Zampighi, 1980). The diameter of the connexons appeared smaller than the models at 4-6 nm and they protruded 0.4-0.5 nm from the membrane surface. Hoh and coworkers followed up this study with a more detailed re-examination of gap junctions with better control of imaging forces and other instrumental factors (Hoh *et al* 1993). The study also contained a useful discussion of potential problems associated with AFM, which at that time was a very new and underdeveloped technique (Hoh *et al* 1993). Damage induced by the AFM tip during both routine imaging and during membrane dissection was considered, as was image distortion due to instrumental factors such as tip shape and feedback-loop characteristics. The images obtained were much clearer and are reproduced in Fig. 6.13a, allowing a central pore to be discerned in the connexons (Fig. 6.13b).

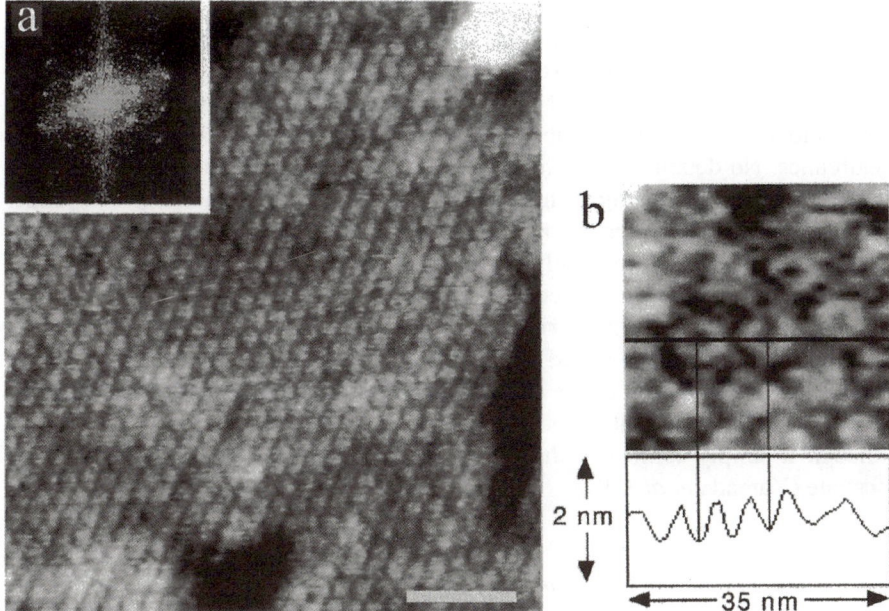

Fig. 6.13 (a) Connexon structure of the gap junction is revealed after removal of the top layer (see above Fig. 6.12). The power spectrum (inset) demonstrates the hexagonal packing. Scale bar 50 nm. (b) Cross section through selected connexons revealing pores, with a depth of approximately 0.7 nm suggesting that the AFM tip is too broad to penetrate fully, or that there is an obstruction. Reproduced with permission from Hoh *et al* 1993 and the Biophysical Society.

The AFM images were the first to reveal substructure on a membrane protein, namely details of the pore running through the connexons, but it was acknowledged that concerns regarding artifacts due to the use of AFM needed to be addressed before the full significance of the images could be assigned (Hoh *et al* 1993). In hindsight it would appear that the researchers were overcautious since many of the safeguards, such as scanning in different directions, force minimisation, feedback optimisation and thorough analysis of the images using Fourier and correlation methods have been both adopted and validated subsequently by other workers.

6.3.3. Photosynthetic protein membranes

Another early AFM study which achieved clear resolution of single molecules of a membrane protein was carried out on the photosynthetic protein membrane from the bacteria *Rhodopseudomonas viridis* (Yamada *et al* 1994). A suspension of the membranes, which had been extracted from bacterial cells was spread onto an air-water interface in a Langmuir trough and then transferred onto glass coverslips by LB dipping. In contrast to most of the other studies on membrane protein, mica was found be an unsuitable substrate causing wrinkling and aggregation of the membranes. No details are given of the imaging forces used, it must be assumed that the samples were imaged in air, since there is no mention of working under liquid. Despite this the protein molecules were imaged quite clearly as spherical features found to be hexagonally packed in most regions of the membranes. In some regions the protein molecules displayed less regular packing, and these differences were tentatively ascribed to the different faces of the membrane sheet which have different, so-called C and H subunits protruding outwards. Force modulation by oscillation of the cantilever during imaging was carried out on the samples yielding information on the local mechanical properties of the sample, although it simply confirmed that the protein membrane is softer than the glass substrate (Yamada *et al* 1994).

6.3.4. ATPase in kidney membranes

In another early example, a combined TEM/AFM study of membrane proteins examined the structures formed by sodium and potassium ATPase in purified canine kidney membranes. The AFM imaging was performed in air and the forces were kept to reasonably low levels through the use of Tapping, allowing molecular resolution to be obtained (Paul *et al* 1994). By comparing the TEM and AFM images it was concluded that the cytoplasmic face of the membranes was imaged by AFM and that the images (AFM) demonstrated that the ATPase forms channel-like structures, with a distinct pore of internal diameter 0.6-2.0 nm. Finally, an interesting observation in the study was that the dimensions of the protein molecules varied when uranyl acetate (negative stain) was applied, with some 50% shrinkage resulting from the staining procedure, demonstrating even in this early study that AFM could be used to assess the effect of common EM preparation procedures on such samples (Paul *et al* 1994).

6.3.5. OmpF porin

The outer membranes of gram-negative bacteria protect the cells from harmful agents such as proteases, bile salts, antibiotics, toxins and 'phages, and against drastic changes in osmotic pressure (Cowen *et al* 1992). The barrier contains a family of proteins known as porins which act as channels to mediate the transfer of nutrients, metabolites and waste products (Nikaido and Vaara, 1985). A major species of porin in the bacterium *Escherichia coli* is the trimeric matrix porin OmpF whose atomic structure has been resolved by X-ray crystallography (Cowen *et al* 1992). In the case of OmpF porin the channels or pores are formed by strands of β-barrel, connected by short turns on the periplasmic face, and by loops of variable length on the extracellular face. Thus the faces are smooth and rough respectively. OmpF can be reconstituted in the presence of phospholipids into 2D crystals of various forms. The first AFM study of reconstituted crystals of OmpF porin from *E. coli* was reported by Lal and coworkers in 1993, who observed a mixed pattern of rectangular and hexagonal motifs when high magnification scans were performed (Lal *et al* 1993). The researchers attributed this to the extracellular face of the porins, because of the thickness of the vesicles and the protrusion height of the features. The centre to centre spacings of the arrays were measured at 8.4 x 9.8 nm and 7.2 nm respectively, in good agreement with previously published EM data (Sass *et al* 1989). No fine detail of the actual channels was achieved, which with the benefit of hindsight, was probably due to two factors, namely, imaging in air which may induce collapse of the structure, and imaging with relatively high forces ~ 1 nN. Improvements in resolution were demonstrated the following year when AFM micrographs exhibiting 2 nm lateral resolution (Schabert *et al* 1994a) and then 1 nm lateral resolution (Schabert and Engel, 1994b) were obtained by imaging under aqueous buffers, and eliminating the drying step of the earlier study. The improved resolution allowed both extracellular and periplasmic faces to be identified in the images. The rougher extracellular face showed two distinct conformations related to the imaging force, whereas the smooth periplasmic face appeared the same over a range of forces from 200-600 pN. Furthermore, on the periplasmic face of OmpF porin a novel hexagonal crystal packing arrangement, and its transition to the more common rectangular arrangement were observed, demonstrating once again that crystal defects, (Devaud *et al* 1992) and different crystalline packing arrangements are within the grasp of AFM (Schabert and Engel, 1994b). The paper by Schabert and Engel also details the range of image analysis techniques which are possible on such highly ordered systems, many of which have been 'borrowed' from electron microscopy. Techniques such as correlation averaging (Saxton and Baumeister, 1982) and Fourier analysis were used to allow accurate measurement of unit cell parameters, and to assess the accuracy of the measurements respectively (Schabert and Engel, 1994b). Before applying such treatment to the images scans were performed in forward and reverse direction, and the images added to compensate for frictional effects. Reconstitution of the membranes generally produced a double layer where the extracellular sides were sandwiched between the periplasmic sides and so not accessible to the AFM tip. This was overcome by imaging regions where the overlap was not perfect, and also by 'teasing away' edges of the upper layers with the AFM tip to expose sections of lower extracellular layer, although this was only possible on bilayers which lacked the two loosely bound lipopolysaccharide molecules per OmpF porin trimer (Schabert and Engel, 1994b). This 'nanodissection' was repeated in a later study, and an example of how this technique reveals both faces of the 2D OmpF porin crystal is shown in Fig. 6.14 (Schabert *et al* 1995).

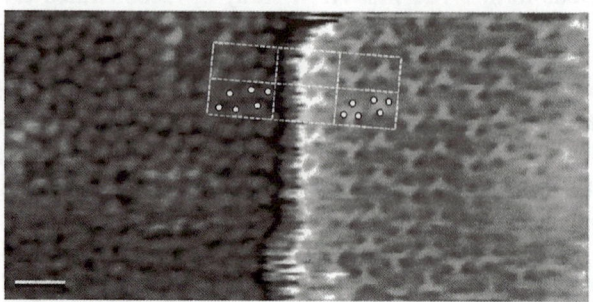

Fig 6.14. Visualisation of both faces of OmpF porin was possible after removal of the upper layer. The corrugated extracellular surface is visible on the left side of the AFM image, and the smoother periplasmic face to the right. Rectangular unit cells (a=13.5 nm, b=8.2 nm) which comprise two trimers are framed, the dots mark the positions of the trimers on both faces. Scale bar 10 nm. Reprinted with permission from: Schabert, F.A, Henn, C. and Engel, A. (1995). Native *Eschericia coli* OmpF porin surfaces probed by atomic force microscopy. *Science*, **268**, 92-94. Copyright 1995 of the American Association for the Advancement of Science.

Although AFM cannot produce atomic models of 2D protein crystals it can be used to validate such models. The protein-protein and protein-lipid interactions in OmpF porin were modelled and AFM topographs used to verify which model was the most accurate. AFM images also revealed details about the flexibly of protein loops in the structure which were 'hidden' in X-ray crystallography studies (Schabert *et al* 1995).

6.3.6. Bacterial S layers

Virtually all archaebacteria have cell envelopes which incorporate a regular surface layer (S layer) of proteins or glycoproteins (König, 1988). S layers make up some 7-12% of the total protein content of the cells, and provide an important interface between the cell and its surroundings (Beveridge, 1981). The first S layer to be examined by AFM was from the bacterium *Sulfolobus acidocaldarius* (Devaud *et al* 1992). Indeed it was probably the first AFM study on 2D protein crystals to attain lattice resolution without resorting to Fourier processing of the images. The reason for this success was almost certainly due to the coating of the sample with titanium by electron beam evaporation, making it able to withstand the forces encountered when imaging in air. The resolution, low by present standards at around 10 nm, was nevertheless sufficient to distinguish between the extracellular and cytoplasmic faces of the crystals. Perhaps the most important finding in this study was that twin boundaries in 2D crystals which are not easily visible in TEM images, are highly structured areas, the two S-layer crystal domains merging in such a way as to preserve the overall pattern of pores. Thus even in this early study the role that AFM can to play as a rather powerful complementary technique to TEM was demonstrated (Devaud *et al* 1992).

Another bacterial S-layer, and one which has been extensively studied, is the hexagonally packed intermediate (HPI) layer of *Deinococcus radiodurans* (Baumeister *et al* 1982). It is extracted from the outer membrane of whole cells with detergent (Baumeister *et al* 1982). Assembled from a 107 kD protein it forms hexamers of 655 kD, producing a lattice of unit cell size of 18 nm. Each hexamer is composed of a massive core from which spokes that connect adjacent hexamers emanate. Modelling of electron microscopy data indicates that the core encloses a pore which is surrounded by six relatively large openings, centred around the three fold line of symmetry (Baumeister *et al* 1886). In what was an important step for the application of AFM to biological molecules at high magnifications, an early study of HPI layers quantitatively compared AFM images, obtained under buffers, with TEM data, obtained *in vacuo*, in order to try to verify the reliability of the AFM in the study of protein structure (Karrasch *et al* 1994). A resolution of 1 nm laterally and 0.1 nm vertically was achieved and, furthermore, measurements taken from the AFM images were in excellent agreement with previous electron microscopy data. It was also clear that regions of differing rigidity in the structure could be discriminated in the AFM by evaluating the variations between images of many molecules. The authors (Karrasch *et al* 1994) tentatively concluded that this might allow observations of function related structural changes in such systems using AFM. Within two years, such structural changes were actually observed on HPI layers; the opening and closing of pores (Müller *et al* 1996b). An important instrumental prerequisite to such delicate work was the use of very low imaging forces of around 100 pN in order to eliminate probe forces as the dominant mechanism of structural alteration. In a further effort to eliminate tip-induced effects both hydrophilic oxide-sharpened Si_3N_4 tips and electron beam deposited carbon hydrophobic tips were used, and produced the same results. Fig. 6.15a,b shows the unprocessed AFM images that were obtained (Müller *et al* 1996b). It can be seen that the hydrophobic inner surface of the HPI layer exhibits two conformations in the AFM images. These have been attributed to the opening and closing of the central pore of the hexameric core because, in some images, this region is unobstructed (round circles) and in others it is plugged (square boxes). Most importantly upon re-scanning of the same region, some five minutes later, different hexamers were unobstructed (square boxes-Fig. 6.15b) eliminating the possibility that the original image simply represented a natural heterogeneity in the structure of the hexamers.

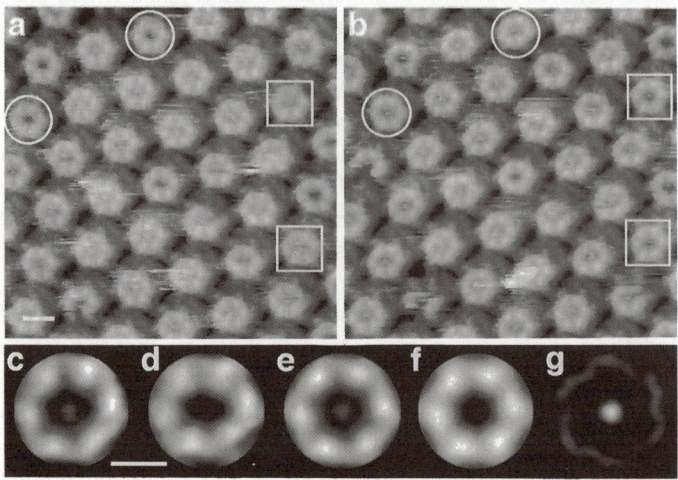

Figure 6.15. Conformation changes of the inner surface of HPI layer captured in successive AFM images of the identical area taken some 5 minutes apart. (a) Open pores in the image are marked with circles and closed or plugged pores are marked with squares. (b) Pores which were open have now closed (circles) whilst some that were closed have opened (squares). (c-f) Images resulting from various mathematical operations (details in text) on the raw images demonstrate that there are real and measurable differences between the images of the two states of the pores. Scale bars / grey level range (a-b) 10 nm / 6 nm (c-g) 6 nm / 3 nm. Reproduced with permission from Müller *et al* 1996b.

To quantify the observed structural changes, 330 units from ten different images were aligned by adjusting their translational and rotational position (or put another way, jiggling them around until the best fit was found) and then performing multivariate statistical analysis. The results of this can be seen in Fig. 6.15c-g. The difference in terms of height between the obstructed and empty pore was 0.8 ± 0.5 nm, only just above the uncertainty level, but probably significant since this figure is derived from the average of many images. When one considers the small size of the pore, electron microscopy data suggests it is about 2.2 nm in diameter, it is easy to see that this will limit the penetration of the AFM tip, which is not infinitely sharp, meaning that the depth measured by AFM is bound to be underestimated. The only "fly in the ointment" is that it is not known what induces the observed conformational change, since no changes to the imaging or buffer conditions occurred between scans, leaving the conclusion that random switching between open and closed states is a specific property of the HPI layer somewhat tentative (Müller *et al* 1996b; 1997b). Nevertheless, the very high resolution achieved in this study (approximately 0.8 nm laterally and 0.1 nm vertically) proved that dynamic events can be followed with AFM, even at the sub-molecular level, and this surely represents a very significant milestone in its application to biology.

6.3.7. Bacteriophage φ29 head-tail connector

The assembly of bacteriophage particles is a complex process involving interaction between unassembled components into various intermediates of the virus. Typical 'phages consist of a head and a tail which are connected by proteins, known as connector or portal proteins. These are implicated in the translocation of DNA inside the viral head. Necks from the bacteriophage φ29, made up of connector proteins p10 and p11, have a 12 fold rotational symmetry (Carrascosa *et al* 1982). Electron microscopy data from 2D crystals of the φ29 connector revealed an open channel in the system (Carazo *et al* 1986). However this channel was found to be closed in necks extracted from native viral particles (Carazo *et al* 1985). This suggested that the different conformational states may play a role in the packing of DNA within the viral particles (Carrascosa *et al* 1990). Although this system had been studied fairly extensively there was still relatively little information on the actual topography of the inner and outer surfaces of the connector and so AFM was applied to tackle this problem (Müller *et al* 1997a). For the purposes of 2D crystal formation and AFM imaging, relatively large amounts of φ29 connector protein were obtained by the over-expression of the gene encoding it in *E. coli*. In contrast to the usual methods, 2D crystals were formed by very slowing raising the ionic strength of a 3-4 mg mL^{-1} solution of the connector proteins up to 2M NaCl. In all other respects the sample treatment was as detailed in the section 6.4.1. As well as 2D crystals, 3D crystals were also prepared and imaged, and interestingly these failed to produce sub-molecular resolution, this being attributed to their relative lack of stability to shear forces of the AFM tip. It seems that the mica substrate plays an important role in supporting the sample, and indeed this conclusion is probably applicable to other interfacial systems if high resolution AFM images are required, justifying Langmuir-Blodgett film transfer. As for bacteriorhodopsin a force induced conformational change was observed in the AFM images of the φ29 connectors and this is shown in Fig. 6.16.

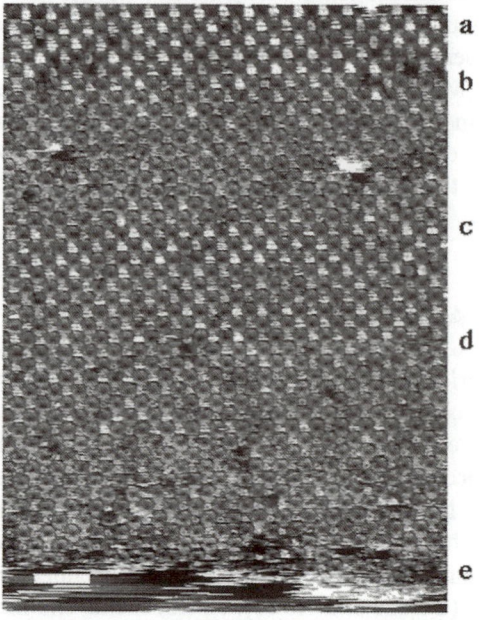

Fig 6.16. Force induced conformational change of φ29 connectors. Extended protrusions from the narrow connector end which are visible at low force (50-100 pN, region a-b) were pushed down by the AFM tip with only a moderate increase in the force (150 pN, region b-c). This process was fully reversible as can be seen when the force was lowered again (region c-d) and raised (d-e). Finally if the force was increased to too high a value (300 pN) the sample was damaged by the tip (region e). Scale bar 5 nm, Grey range 5 nm. Reproduced with permission from Müller *et al* 1997a.

At the top of the image in Fig. 6.16, which was scanned at low force ~50-100 pN, the narrow domains of the connectors dominate the image contrast (regions a-b), but, as the force between AFM tip and sample is increased (between regions b-c, force ~150-200 pN), the narrow part was squashed down onto the surface of the crystal, allowing the wider central channel beneath to be more clearly resolved. When the force was reduced again the extended conformation of the narrow, and apparently flexible, end of the connectors once more became visible (regions c-d). To prove the reproducibility of the conformational changes the force was again raised and the wide connector domain became clear (region d-e). Finally, at the bottom of the image the probe force was increased to 250-300 pN, resulting in the AFM tip pushing right through the sample to the mica underneath (region e), demonstrating the delicate nature of the sample, and the need to control the imaging force accurately. The process of disruption is probably actually due to the shearing force of the tip in contact mode, rather than simple piercing of the crystal sheet by the loading (normal) force of the tip. When the force was increased to image the wider connector domains even the raw data showed, with reasonable clarity, the presence of twelve subunits in good agreement with previous EM studies (Carazo *et al* 1986). Additionally, however, the AFM data revealed that the wide end of the connector had a right handed orientation vorticity. This may have some bearing on the DNA packing mechanism of the connectors. The AFM images clearly show that there is a difference between the channel on each side of the connector; one wide with a diameter of 3.7 nm, and one narrow with a diameter of 1.7 nm. Furthermore, the AFM data indicated that the shape of the channel is not cylindrical but conical, an important observation, since knowledge of the precise nature of the channel enables verification of models for the packing of DNA (Carazo *et al* 1985; Carrascosa *et al* 1990). The data obtained by AFM is consistent with the model for channel opening and closing involving small concerted movements of the subunits, and closing of the channel has been linked to the final step of DNA packaging (Carrascosa *et al* 1990) by the ϕ29 connectors.

6.3.8. Gas vesicle protein

Gas vesicles are hollow tubular structures composed entirely of proteins which provide buoyancy for aquatic micro-organisms. The gas vesicle from the cynaobacterium *Anabaena flos-aquae* was examined recently under propanol after spray deposition onto mica by AFM (McMaster *et al* 1996). The study found that the major protein of the vesicles, GvpA, was packed into ordered arrays of rib-like structures with a periodicity of 4.6 nm. In contrast to all previous studies reported in this section, AFM imaging was carried out in the error-signal mode (sometimes also called deflection mode, section 3.2.3) at high scan speeds. It was speculated that the high scan speed reduced instrumental drift during the scan resulting in less image distortion. Furthermore, working in error signal mode rendered the instrument more sensitive to the small deviations of the AFM tip on this flat specimen, since the data collection bandwidth is not limited by the control loop as it is in constant force mode. In error signal mode the only limiting factor on bandwidth is the inertia of the tip-cantilever assembly, hence a very low spring constant lever was used ($k = 0.06$ N m^{-1}). Resolution of the rib-structure formed by the proteins was clearly observed and, in some areas, an even finer level of detail, namely the β-sheet secondary structure of the protein molecules which had a periodicity of 0.57 nm, was seen. An advantage of using AFM was that by imaging the surface of the vesicles directly the packing sense of the molecules was determined, something which is impossible with X-ray diffraction, since the diffraction pattern is a projection map which has contributions from both sides of the vesicle superimposed. The study achieved a level of resolution believed to be amongst the highest to be obtained to date on a biological system (0.57 nm laterally). The downside of the error signal mode images was that only lateral dimensions could be determined from the images, since no height information is available in this imaging mode. The study nevertheless suggests an interesting alternative technique for achieving very high resolution on ordered arrays of proteins.

6.4. AFM studies of 2D crystals of soluble proteins

The soluble proteins ferritin and catalase have been imaged by AFM as 2D crystalline arrays (Ohnishi *et al* 1993;1996; Furono *et al* 1998). Ferritin concentrates insoluble Fe(III) ions into a soluble protein-mineral complex. The iron is transferred through the protein into an internal cavity in the molecule, achieving concentrations of iron some one thousand times greater than for the free ion. Ferritin is found in all animals, plants and many bacteria, and provides iron for the proteins involved in respiration, nitrogen fixation, cell division and biosynthesis. Catalase is one of the most abundant enzymes found in nature, and protects cells from the damaging effects of hydrogen peroxide by catalysing its breakdown into water and free oxygen. Beef liver catalase is commonly used as a calibration standard in electron microscopy, forming a tetrameric structure composed of four identical subunits.

Both proteins were crystallised in two dimensions by binding them to a charged lipid monolayer of poly (1-benzyl-L-histidine) (PBLH) which had been spread at an air-water interface, using the method described above for soluble proteins (Kornberg *et al* 1987;1991). The charged lipid, PBLH, played a double role in the studies. Firstly, it bound the opposite charged protein molecules allowing crystallisation to occur, and secondly, once the crystals were transferred onto the substrate for AFM imaging in pure water, it acted to screen the charges on the protein itself, as demonstrated by the absence of hysteresis in force versus distance curves upon tip approach and withdrawal from the sample. This allowed low force imaging to be performed (Ohnishi *et al* 1993). In common with the studies on 2D membrane protein crystals described in the preceding sections, low forces (less than 100 pN) were a prerequisite to successful imaging. In the earliest study, which was carried out on ferritin, regular hexagonally packed arrays of the proteins were clearly seen, and there was also a hint of sub-molecular detail, paradoxically in areas where the crystalline packing was imperfect (Ohnishi *et al* 1993).

These studies were extended to catalase where a strong dependence of the image quality on buffer pH during the crystallisation step was demonstrated (Ohnishi *et al* 1996). Catalase is negatively charged in buffers above its isoelectric point at pH 5.7. This meant that in order to cancel the charge on the catalase the PBLH layer needed to be positively charged. In the range pH 6-7 PBLH bears only a slight positive charge, which is not enough to counteract the negative charge contribution from the catalase. The AFM tip is also negatively charged in pure water under which the 2D catalase sheets were imaged, and the net result was repulsion between the AFM tip and sample, preventing proper tracking of the tip, and hence poor images were obtained. This effect can be likened to a record-player stylus skipping or jumping when trying to play a record which has an excessive amount of static charge on its surface. When the 2D crystals were prepared at pH 6.0 however, the force-distance curves obtained suggested that the positive charge on the PBLH was slightly greater than the reduced negative charge on the catalase at this pH, being as it is close to the isoelectric point of the protein (pH 5.7), since there was a small amount of adhesion present between tip and sample. This led to optimum imaging by the AFM since now the tip could track the sample surface accurately. This optimum pH value sits within the range of isoelectric point of the protein and pKa of the histidyl residues in the PBLH. At lower values of pH during crystal formation (pH <5.0), both the PBLH and catalase are positively charged, so of course the net charge on the 2D crystals was also positive. This produced excessive adhesion between the negatively charged AFM tip and the 2D crystals, rendering AFM imaging impossible due to destruction of the sample. The results obtained by the AFM of the catalase molecules were in general agreement, in terms of unit cell dimensions, with electron microscopy data, but the tetrameric subunits of the protein were not resolved (Ohnishi *et al* 1996). Nevertheless, the study clearly demonstrated the need to control the electrostatic interactions between the AFM tip and sample which arise when imaging is carried out in aqueous solutions (Ohnishi *et al* 1996), and this point is discussed further in the next section.

Not surprisingly, the electrostatics of the catalase-PBLH system also effected the actual formation of the 2D protein crystals, since they determine the extent of interaction between the two molecules, as confirmed by TEM studies (Sato *et al* 1993). It can now be appreciated that the formation and subsequent AFM imaging of 2D crystals of *soluble* proteins is a little more complicated than for membrane protein crystals since the interaction between the protein and lipid layer, needed to induce crystallisation, can effect AFM image quality. In the most recent study using a specially prepared electron beam-deposited 'supertip', 2D crystals of ferritin and catalase were re-examined and higher resolution was achieved (Furuno *et al* 1998). Images were obtained both in water and in air at molecular resolution. Because of the capillary forces present when working in air (2-8 nN) the samples required fixation with negative stain (methylamine tungstate) to prevent probe damage. Under water much lower forces were achieved and better images were obtained with no need for negative staining. The AFM images are reproduced in Fig.6.17.

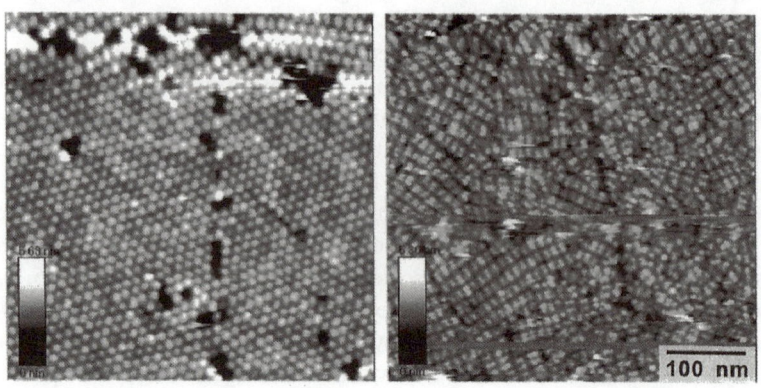

Figure 6.17. AFM images of 2D crystals of the soluble proteins (a) ferritin and (b) catalase in water. Reproduced with permission from Furuno *et al* 1998.

It was noted that the crystal sheets remained flatter in water than when imaged in air, perhaps due to buckling of the film during the drying step. In both environments the catalase molecules appeared to pack into a rectangular unit in some regions of the images and into a different unit form in others. The 'supertip' used to image the proteins was grown by focusing the electron beam from a field-emission scanning electron microscope (FESEM) onto the apex of a standard Si_3N_4 tip. FESEMs have a much smaller spot size compared to conventional SEMs enabling sharper tips to be grown, typically with radii of curvature of around 2.5-3.0 nm.

6.4.1. Imaging conditions

It is worth mentioning at this point some details of the imaging conditions under which such high resolution studies were accomplished. For these, and indeed virtually all of the highest resolution studies of 2D protein crystals, the AFM was operated in constant force (dc) mode and under aqueous buffers, but there are two notable exceptions. In one case high resolution images were obtained in air, but on a metal coated sample (Devaud *et al* 1992) and in the other high resolution was obtained by operating the AFM in propanol in error-signal mode at high scan speeds (McMaster *et al* 1996). Because dc mode is prone to thermal drift, the samples contained in the liquid cell were often left for several hours to reach stability before imaging (Müller *et al* 1995b). In principle the problem of thermal drift could be overcome by using an ac mode of operation, such as Tapping in liquids, since the amplitude of oscillation upon which the set-point is controlled is unaffected by thermal fluctuation. However, in a study of specimen heights using Tapping mode AFM, HPI layers were imaged by both contact mode and Tapping mode in liquid, and the following pertinent observations were made (Schabert and Rabe, 1996). The edges of the membrane sheets appeared to be damaged in successive scans using contact mode, whereas Tapping mode eliminated this effect, indicating that Tapping mode in liquid may be less damaging than contact mode. However, when small areas on the surface of the membrane patches were examined at high resolution, contact mode produced no obvious damage, and revealed much greater detail than Tapping mode images of the same area using the same AFM tip, proving that for ultimate resolution contact mode imaging appears to be better. Another point to note is that generally short cantilevers (~100 μm long) were used, these having greater angular sensitivity than their longer counterparts and greater stability against flexure. When it comes to the actual AFM tip itself, it is reported that for 2D crystals oxide sharpened silicon nitride probes are the best, although they did suffer more frequently from multiple tip artifacts and tip astigmatism (Schabert and Engel, 1994). The quality of the AFM tip was checked before imaging 2D crystals in two ways. First, contamination of the tip was checked for by the presence of hysteresis in the force-distance curve upon retraction of the tip from bare mica surface. Only tips with no hysteresis were used. Secondly, the stick and slip effect which occurs at the end of each scan line, as the tip changes its fast scan direction, seemed to be correlated with tip quality, and so only tips which displayed the minimum amount of this behaviour were selected (Schabert and Engel 1994).

Scan speed, or tip velocity, was also found to be an important factor. There is a critical limit of 2 μm s^{-1} which cannot be exceeded for high resolution imaging (Butt *et al* 1993) and speeds of around 1 μm s^{-1} were typical for most of the 2D crystals cited here. A useful trick when engaging the AFM tip was to set the scan size to zero in order to avoid sample damage and tip contamination (Müller *et al* 1995b). This last step may not be necessary depending upon how the microscopes' approach mechanism works. Deformation of the sample was monitored by comparing height profiles obtained in forward and reverse scan directions and at differing scan angles. A very important point noted in the work cited here (Müller *et al* 1995b) is that the applied force was adjusted manually to compensate for thermal drift of the cantilever. This advice is good for any sample, one should never assume that the force will not vary as a scan progresses, despite the term 'constant force' imaging, since the instrument only controls the deflection of the cantilever relative to a predetermined null point, where the tip was not in contact with the surface. This null point can vary with time and temperature, and the instrument has no way of detecting such variation during image acquisition, other than de-coupling the tip from the sample which can, of course, ruin the image. Choice of substrates is another important factor to successful imaging at very high resolution. Generally mica is the best. The charged nature of its surface in aqueous buffers allows control over the binding of the 2D crystals through choice of ions and ionic strength (Schabert and Engel, 1994b). A general feature of these studies was that membrane sheets which were not tightly bound to the substrate, after an incubation of the protein solutions on the mica, made imaging conditions unstable and so were best removed by gentle rinsing of the substrate with pure buffer solution. Functionalised glass coverslips were also tried as a substrate but resulted in buckling of the sheets, which increased their apparent roughness to the detriment of image resolution (Schabert and Engel, 1994b), although glass coverslips were used successfully to deposit photosynthetic membranes (Yamada *et al* 1994) and gap junctions (Hoh *et al* 1991). The studies on soluble proteins successfully employed silicon wafers as the substrate (Ohnishi *et al* 1993; 1996; Furono *et al* 1998).

Another interesting point to note is that in many of the studies, low magnification images of entire sheets were obtained with forces of ~ 500 pN with no apparent damage to the specimen. This illustrates an important point in AFM studies of soft samples, which is that as the magnification increases, the density of scan lines increases and so the damage induced by the AFM tip increases (and of course the converse is also true accounting for the higher forces allowed at low magnifications). So it is that, very often, optimum image resolution is not obtained by simply increasing the magnification endlessly, but rather scanning at slightly lower magnification, where sample damage is minimised.

6.4.2. Electrostatic considerations

When AFM imaging is performed under aqueous buffers electrostatic forces play a pivotal role in determining how the tip interacts with the sample. Since the tip must accurately track the surface of the sample to produce a faithful image of its topography this means that particular attention needs to be paid to this interaction.

Some background on the origin of electrostatic forces is discussed in chapter 3 (sections 3.1.2 & 3.1.4). In practice the two most important force contributions which need to be considered when working in aqueous liquids are the double layer forces between tip and sample, which are effected by the ionic strength and pH of the buffer solution, and the van der Waals force between the tip and sample surface. The van der Waals force is a long range force which is always attractive, but becomes significant only over very short distances (<1 nm), when working in buffered aqueous liquids where it is often opposed by repulsive double layer forces. Biological molecules and AFM tips both exhibit a surface charge in aqueous environments, since acidic and functional groups at their surfaces dissociate according to their pK values. The magnitude and sign of the charge depends on the pH and temperature of the surrounding buffer solution. As explained in chapter 3, the result of this surface charge is to create a form of ionic 'atmosphere' of opposite charged ions surrounding the surfaces and this is known as a *double layer*. The double layer *force* comes about because an electrostatic interaction occurs when the electrical double layers of the tip and the sample overlap. The distance over which the double layer force acts is known as the Debye length and this is affected by the ionic strength of the solution, higher ionic strength reducing the Debye length. The trick to successful imaging at high resolution is to try to balance the attractive van der Waals force with a repulsive double layer force of very short Debye length so that the AFM tip can track the sample surface intimately. Since the van der Waals forces are more or less fixed, this can best be achieved by manipulation of the double layer force by variation of ionic strength and, if need be, the pH of the buffer solution in order to make the double layer force repulsive.

In a recent paper, Müller and coworkers set out experimental protocols for achieving this situation (Müller *et al* 1999). They demonstrate that what may intuitively seem the best situation, where perfect balance between attractive van der Waals forces and repulsive double layer forces is achieved, isn't actually ideal. This is because the AFM controls the force interaction between the sample and tip in a proper sense only in repulsive situations, where there is sufficient gradient in the force versus distance characteristics of the system to allow it to detect positive deflections of the cantilever. Remember that the AFM cannot determine the force directly, it can only estimate it by measuring cantilever deflection, and because AFM control loops cannot work with negative force set-points this limits stable control to positive deflections of the lever. When perfect balance is achieved between the repulsive double layer force and the attractive van der Waals forces the repulsive force versus distance gradient is very small, making it impossible for the AFM control loop to work properly. When such conditions prevail one can appreciate that the result will be that the attractive van der Waals force will pull the AFM tip onto the sample and may deform it, even if the applied external force due to bending of the cantilever is low (in fact a *negative* bend on the lever would be required to counteract this adhesion rendering the instrument unstable, since any small reduction in the adhesive force would cause the tip to detach from the surface). Indeed this attractive force is shown to be large enough to generate frictional effects in AFM images at very high resolution, such that forward and reverse scans of the same surface of purple membranes produce different topographies (Müller *et al* 1999). The best situation, particularly when a realistic tip shape is considered - this being one where a blunt hemispherical tip has a small sharp asperity on the end, is to have a small but measurable repulsive double layer force of magnitude approximately 0.1 nN counteracting the attractive van der Waals forces. In this way the tip can image in a stable manner in constant force mode, with the bulk of the applied force being distributed by the longer range double layer force acting over the blunter portion of the tip, but the sharp asperity on the apex will nevertheless track the surface intimately and with very low net force (Müller *et al* 1999). For the samples studied, HPI layers, purple membranes and OmpF porin, the optimum buffer conditions were between 100-200 mM for monovalent cations and around 50 mM for divalent cations at neutral pH (pH 7.6). These figures vary according to the charge density of the sample and the AFM tips themselves, which are not always identical even from the same wafer (Müller *et al* 1999). The take-home message is that happily there is no need to go through long-winded calculations to achieve electrostatic balance, rather the electrolyte concentrations under which the AFM works can simply be tuned empirically, with reference to force versus distance curves for each sample and tip, in order to achieve optimum imaging conditions. A series of force versus distance curves taken in buffer of varying ionic strength is shown in Fig. 6.18 to illustrate this principle.

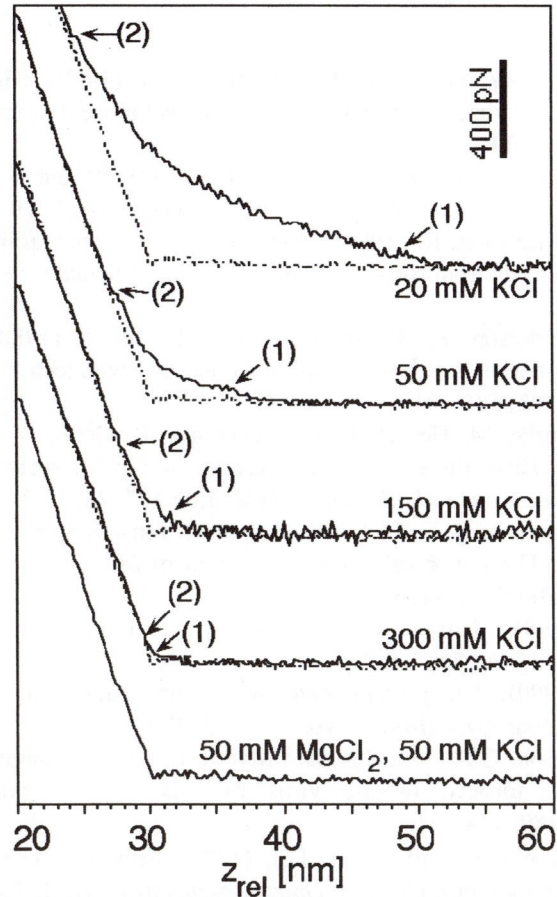

Figure 6.18. Force-distance curves recorded on the extracellular surface of purple membrane. The ionic strength of the imaging liquid was varied at constant pH, in order to alter the thickness of the electrostatic double layer. In order to show the effect of the distance dependence of the electrostatic double layer and its effect on the repulsive force between sample and tip, the dotted lines show an approach curve for a sample where there is no repulsion (i.e. after electrostatic balancing). The sharp turning point on the dotted lines marks the position of tip-sample contact. Taking this into consideration the electrostatic repulsion in the top curve for 20 mM KCl is seen to extend some 50 nm from the sample surface (arrow 1) and prevents the tip from making contact with the sample surface until point 2, where the force due to cantilever flexure is high enough to overcome the repulsion (>400 pN). Such a large force will damage the sample and prevent high resolution imaging. Optimum imaging conditions require most of the electrostatic repulsion to be removed, and this is achieved by raising the ionic strength which reduces the thickness of the double layer. Perfect balance between the attractive van der Waals force and repulsive electrostatic double layer force is achieved in 300 mM KCl (and in 50 mM $MgCl_2$, 50 mM KCl). However the optimum images were obtained when there was a small repulsive force (100 pN) between sample and tip as seen in the curve for 150 mM KCl (arrow 1). Reproduced with permission from Müller *et al* 1999 and the Biophysical Society.

References

Baker, A.A, Helbert, W, Sugiyama, J. and Miles, M.J. (1997). High-resolution atomic force microscopy of native *Valonia* cellulose I microcrystals. *J. Structural. Biol.* **119**, 129-138.

Baker, A.A, Helbert, W, Sugiyama, J. and Miles, M.J. (1998). Surface structure of native cellulose microcrystals by AFM. *Appl. Phys. A* **66**, S559-S563.

Baldwin, J. and Henderson, R. (1984). Measurement and evaluation of electron diffraction patterns from two-dimensional crystals. *Ultramicroscopy* **14**, 319-336.

Baldwin, J.M, Henderson, R, Beckmann, R. and Zemlin, F. (1988). Images of purple membrane at 2.5 Å resolution obtained by cryo-electron microscopy. *J. Mol. Biol.* **202**, 585-591.

Baumeister, W, Barth, M, Hegerl, R, Guckenberger, R, Hahn, M. and Saxton, W.O. (1986). Three dimensional structure of the regular surface layer (HPI layer) of *Deinococcus radiodurans*. *J. Mol. Biol.* **187**, 241-253.

Baumeister, W, Karrenberg, F, Rachel, R, Engel, A, Ten-Heggler, B. and Saxton, W.O. (1982). The major cell envelope protein of *Micrococcus radiodurans* (R1). *Eur. J. Biochem.* **125**, 535-544.

Beveridge, T.J. (1981). Ultrastructure, chemistry and function of the bacterial wall. *Int. Rev. Cytol.* **72**, 229-317.

Boekema, E.J. (1990). The present state of two-dimensional crystallisation of membrane proteins. *Electron Microsc. Rev.* **3**, 87-96.

Bushell, G.R, Watson, G.S, Holt, S.A. and Myhra, S. (1995). Imaging and nano-dissection of tobacco mosaic virus by atomic force microscopy. *J. Microscopy* **180**, 174-181.

Bustamante, C, Rivetti, C. and Keller, D.J. (1997). Scanning force microscopy under aqueous solutions. *Current Opinion Structural Biol.* **7**, 709-716.

Butt, H-J, Downing, K.H. and Hansma, P.K. (1990). Imaging the membrane protein bacteriorhodopsin with the atomic force microscope. *Biophys. J.* **58**, 1473-1480.

Butt, H-J, Siedle, P, Seifert, K, Fendler, K, Seeger, E, Bamberg, E, Weisenhorn, A.L, Goldie, K. and Engel, A. (1993). Scan speed limit in atomic force microscopy. *J. Microscopy* **169**, 75-84.

Carazo, J.M, Santisteban, A. and Corrascosa, J.L. (1985). Three dimensional reconstruction of the bacteriophage ϕ29 neck particles at 2.2nm resolution. *J. Mol. Biol.* **183**, 79-88.

Carazo, J.M, Donate, L.E, Herranz, L, Secilla, J.P. and Corrascosa, J.L. (1986). Three dimensional reconstruction of the connector of bacteriophage ϕ29 at 1.8nm resolution. *J. Mol. Biol.* **192**, 853-867.

Carrascosa, J.L, Vinuela, E, Garcia, N. and Santisteban, A. (1982). Structure of the head-tail connector of bacteriophage ϕ29. *J. Mol. Biol.* **154**, 311-324.

Carrascosa, J.L, Carazo, J.M, Herranz, L, Donate, L.E. and Secilla, J.P. (1990). Study of two related configurations of the neck of bacteriophage ϕ29. *Computers Math. Appl.* **20**, 57-65.

Cowan, S.W, Schirmer, T, Rummel, G, Steiert, M, Ghosh, R, Pauptit, R.A, Jansonius, J.N. and Rosenbusch (1992). Crystal structures explain functional properties of two *E. coli* porins. *Nature* **358**, 727-733.

Czajkowsky, D.M, Sheng, S. and Shao, Z. (1998). Staphylococcal α-Hemolysin can form hexamers in phospholipid bilayers. *J. Mol. Biol.* **276**, 325-330.

Darst, S.A, Ahlers, M, Kubalek, E.W, Meller, P, Blankenburg, R, Ribi, H.O, Ringsdorf, H. and Kornberg, R.D. (1990). Two dimensional crystals of streptavidin on biotinylated lipid layers and their interactions with biotinylated macromolecules. *Biophys. J.* **59**, 387-396.

Devaud, G, Furcinitti, P.S, Fleming, J.C, Lyon, M.K. and Douglas, K. (1992). Direct observation of defect structure in protein crystals by atomic force and transmission electron microscopy. Biophys. *J.* **63**, 630-638.

Durbin, S.D. and Carlson, W.E. (1992). Lysozyme crystal growth studied by atomic force microscopy. *J. Cryst. Growth* **122**, 71-79.

Durbin, S.D, Carlson, W.E. and Saros, M.T. (1993). *In situ* studies of protein crystal growth by atomic force microscopy. *J. Phys.* **D50**, B128-B132.

Engel, A. (1991). Biological applications of scanning probe microscopes. *Ann. Rev. Biophys. Biophys. Chem.* **20**, 79-108.

Falvo, M.R, Washburn, S, Superfine, R, Finch, M, Brooks Jnr, F.P, Chi, V. and Taylor, R.M. (1997). Manipulation of individual viruses: friction and mechanical properties. *Biophys. J.* **72**, 1396-1403.

Flagg-Newton, J, Simpson, I. And Loewenstein, W.R. (1979). Permeability of the cell to cell membrane channels in mammalian cell junctions. *Science* **205**, 404-407.

Furuno, T, Sasabe, H. and Ikegami, A. (1998). Imaging two dimensional arrays of soluble proteins by atomic force microscopy in contact mode using a sharp supertip. *Ultramicroscopy* **70**, 125-131.

Häberle, W, Hörber, J.K.H. and Binnig, G. (1991). Force microscopy on living cells. *J. Vac. Sci. & Technol. B* **9**, 1210-1213.

Häberle, W, Hörber, J.K.H, Ohnesorge, F, Smith, D.P.E. and Binnig, G. (1992). In situ investigations of single living cells infected by viruses. *Ultramicroscopy* **42-44**, 1161-1167.

Hanley, S.J, Giasson, J, Revol, J-F. and Gray, D (1992). Atomic force microscopy of cellulose microfibrils; comparison with transmission electron microscopy. *Polymer* **33**, 4639-4642.

Harris, J.R. (1992). 2D crystallisation of soluble protein molecules for TEM: the negative staining carbon film procedure. *Microscopy & Analysis* **30**, 13-16.

Hartmut, M. (1991). General and practical aspects of membrane protein crystallisation. In *Crystallization of membrane proteins.* (ed. Hartmut Michel), pp. 73-87. CRC press: London.

Henderson, R, Baldwin, J.M, Downing, K.H, Lepault, J. and Zemlin, F. (1986). Structure of purple membrane from *Halobacterium halobium:* recording, measurement and evaluation of electron micrographs at 3.5Å resolution. *Ultramicroscopy* **19**, 147-178.

Henderson, R, Baldwin, J.M, Ceska, T.A, Zemlin, F, Beckmann, E. and Downing, K.H, (1990). Model for the structure of bacteriorhodopsin based on high resolution electron cryo-microscopy. *J. Mol. Biol.* **213**, 899-929.

Hoh, J.H, Lal, R, John, S.A, Revel, J-P. and Arnsdorf, M.F. (1991). Atomic force microscopy and dissection of gap junctions. *Science* **253**, 1405-1408.

Hoh, J.H, Sosinsky, G.E, Revel, J-P. and Hansma, P.K. (1993). Structure of the extracellular surface of the gap junction by atomic force microscopy. *Biophys. J.* **65**, 149-163.

Hörber, J.K.H, Häberle, W, Ohnesorge, F, Binnig, G, Liebich, H.G, Czerny, C.P, Mahnel, H. and Mayr, A. (1992). Investigation of living cells in the nanometer regime with the atomic force microscope. *Scanning Microscopy* **6**, 919-930.

Horne, R.W. and Pasquali-Ronchetti, I. (1974). A negative staining-carbon film technique for studying viruses in the electron microscope. I. Preparative procedure for examining icosahedral and filamentous viruses. *J. Ultrstructural Res.* **47**, 361-383.

Ikai, A, Imai, K, Yoshimura, K, Tomitori, M, Nishikawa, O, Kokawa, R, Kobayashi, M. and Yamamoto, M. (1994). Scanning tunneling microscopy/atomic force microscopy studies of bacteriophage T4 and its tail fibres. *J. Vac. Sci. & Technol. B* **12**, 1478-1481.

Imai, K, Yoshimura, K, Tomitori, M, Nishikawa, O, Kokawa, R, Yamamoto, M, Kobayashi, M. and Ikai, A. (1993). Scanning tunnelling and atomic force microscopy of T4 bacteriophage and tobacco mosaic virus. *Jpn. J. Appl. Phys.* **32**, 2962-2964.

Jap, B.K, (1988). High resolution electron diffraction of reconstituted PhoE porin. *J. Mol. Biol.* **199**, 229-231.

Jap, B.K, Zulauf, M, Scheybani, T, Hefti, A, Baumeister, W, Aebi, U. and Engel, A. (1992). 2D crystallisation: from art to science. *Ultramicroscopy*, **46**, 45-84.

Karrasch, S, Dolder, M, Schabert, F, Ramsden, J. and Engel, A. (1993). Covalent binding of biological samples to solid supports for scanning probe microscopy in buffer solution. *Biophys. J.* **65**, 2437-2446.

Karrasch, S, Hegerl, R, Hoh, J.H, Baumeister, W. and Engel, A. (1994). Atomic force microscopy produces faithful high resolution images of protein surfaces in an aqueous environment. *Proc. Natl. Acad. Sci.* **91**, 836-838.

Kirby, A.R, Gunning, A.P. and Morris, V.J. (1996a). Imaging polysaccharides by atomic force microscopy. *Biopolymers* **38**, 355-366.

Kolbe, W.F, Ogletree, D.F. and Salmeron, M.B. (1992). Atomic force microscopy imaging of T4 bacteriophages on silicon substrates. *Ultramicroscopy* **42**, 1113-1117.

König, H. (1988). Archaebacterial cell envelopes. *Can J. Microbiol.* **34**, 395-406.

Konnert, J.H, D'Antonio, P. and Ward, K.B. (1994). Observation of growth steps, spiral dislocations and molecular packing on the surface of lysozyme crystals with the atomic force microscope. *Acta Cryst.* **D50**, 603-613.

Kornberg, R.D. and Ribi, H.O. (1987). Formation of two dimensional crystals of proteins on lipid layers. In *Protein Structure, folding and design 2.* (ed. D.L. Oxender), pp. 175-186, Alan R. Liss: New York.

Kornberg, R.D. and Darst, S.A. (1991). Two dimensional crystals of proteins on lipid layers. *Current Opinion in Structural Biol.* **1**, 642-646.

Kouyama, T, Yamamoto, M, Kamiya, Iwasaki, H, Ueki, T. and Sakurai, I. (1994). Polyhedral assembly of a membrane protein in its three-dimensional crystal. *J. Mol. Biol.* **236**, 990-994.

Kühlbrandt, W. and Downing, K.H. (1989). Two-dimensional structure of plant light harvesting complex at 3.7 Å resolution by electron crystallography. *J. Mol. Biol.* **207**, 823-828.

Kühlbrandt, W. and Wang, D.N. (1991). Three-dimensional structure of plant light harvesting complex determined by electron crystallography. *Nature* **350**, 130-134.

Kühlbrandt, W. (1992). Two-dimensional crystallisation of membrane proteins. *Quart. Rev. Biophys.* **25**, 1-49.

Kuutti, L, Peltonen, J, Pere, J. and Teleman, O. (1995). Identification and surface structure of crystalline cellulose studied by atomic force microscopy. *J. Microscopy* **178**, 1-6.

Kuznetsov, Yu.G, Malkin, A.J, Land, T.A, DeYoreo, J.J, Barba, A.P, Konnert, J. and McPherson, A. (1997). Molecular resolution imaging of macromolecular crystals by atomic force microscopy. *Biophys. J.* **72**, 2357-2364.

Lacapere, J-J, Stokes, D.L. and Chateny, D. (1992). Atomic force microscopy of three-dimensional membrane protein crystals. *Biophys. J.* **63**, 303-308.

Lal, R, Kim, H, Garavito, M. and Arnsdorf, M.F. (1993). Imaging of reconstituted biological channels at molecular resolution by atomic force microscopy. *American J. Physiology* **265** (*Cell Physiol. 34*), C851-C856.

Land, T.A, DeYoreo, J.J. and Lee, J.D. (1995). An in-situ AFM investigation of canavalin crystallisation kinetics. *Surface Sci.* **384**, 136-155.

Land, T.A, Malkin, A.J, Kuznetsov, Yu.G, McPherson, A. and DeYoreo, J.J. (1997). Mechanisms of protein crystal growth: an atomic force microscopic study of canavalin crystallisation. *Phys. Rev. Letts.* **75**, 2774-2777.

Makowski, L. (1985). In *Gap Junctions*, (ed. M.V.L. Bennett, D.C. Spray) pp. 5-12. Cold Spring Harbor: New York.

Malkin, A.J, Kuznetsov, Yu.G, DeYoreo, J.J. and McPherson, A. (1995a). Mechanisms of growth for protein and virus crystals. *Nature Structural Biol.* **2**, 956-959.

Malkin, A.J, Land, T.A, Kuznetsov, Yu.G, McPherson, A. and DeYoreo, J.J. (1995b). Investigation of virus crystal growth mechanisms by *in situ* atomic force microscopy. *Phys. Rev. Letts.* **75**, 2778-2781.

Malkin, A.J, Kuznetsov, Yu.G. and McPherson, A. (1996a). Incorporation of microcrystals by growing protein and virus crystals. *Proteins: Structure, Function, Genetics* **24**, 247-252.

Malkin, A.J, Kuznetsov, Yu.G, Glantz, W. and McPherson, A. (1996b). Atomic force microscopy studies of surface morphology and growth kinetics in thaumatin crystallisation. *J. Phys. Chem.* **100**, 11736-11743.

Makowski, L, Casper, D.L.D, Philips, D.A. and Goodenough, J. (1977). Gap junction structures II. Analysis of the X-ray diffraction data. *J. Cell Biol.* **74**, 629-645.

McMaster, T.J, Miles, M.J. and Walsby, A.E. (1996). Direct observation of protein secondary structure in gas vesicles by atomic force microscopy. *Biophys. J.* **70**, 2432-2436.

Müller, D.J, Büldt, G. and Engel, A. (1995a). Force induced conformational change of bacteriorhodopsin. *J. Mol. Biol.* **249**, 239-243.

Müller, D.J, Schabert, F.A, Büldt, G. and Engel, A. (1995b). Imaging purple membranes in aqueous solutions at sub-nanometer resolution by atomic force microscopy. *Biophys. J.* **68**, 1681-1686.

Müller, D.J, Schoenenberger, C, Büldt, G. and Engel, A. (1996a). Immuno-atomic force microscopy of purple membrane. *Biophys. J.* **70**, 1796-1802.

Müller, D.J, Baumeister, W. and Engel, A. (1996b). Conformational change of the hexagonally packed intermediate layer of *Deinococcus radiodurans* imaged by atomic force microscopy. *J. Bacteriol.* **178**, 3025-3030.

Müller, D.J, Engel, A, Carrascosa, J.L. and Velez, M. (1997a). The bacteriophage φ29 head-tail connector imaged at high resolution with the atomic force microscope in buffer solution. *EMBO J.* **16**, 2547-2553.

Müller, D.J, Schoenenberger, C-A, Schabert, F. and Engel, A. (1997b). Structural changes in native membrane proteins monitored at subnanometer resolution with the atomic force microscope. *J. Structural Biol.* **119**, 149-157.

Müller, D.J, Fotiadis, D, Scheuring, S, Müller, S. A. and Engel, A. (1999). Electrostatically balanced subnanometer imaging of biological specimens by atomic force microscope. *Biophys. J.* **76**, 1101-1111.

Ng, J.D, Kuznetsov, Yu.G, Malkin, A.J, Keith, G, Giege, R. and McPherson, A. (1997). Visualisation of RNA crystal growth by atomic force microscopy. *Nucleic Acids Res.* **25**, 2582-2588.

Nikaido, H. and Vaara, M. (1985). Molecular basis of bacterial outer membrane permeability. *Microbiol. Rev.* **49**, 1-32.

Ohnesorge, F.M, Hörber, J.K.H, Häberle, W, Czerny, C-P, Smith, D.P.E. and Binnig, G. (1997). AFM review study on pox viruses and living cells. *Biophys. J.* **73**, 2183-2194.

Ohnishi, S, Hara, M, Furuno, T. and Sasabe, H. (1996). Imaging two dimensional crystals of catalase by atomic force microscopy. *Jpn. J. Appl. Phys.* **35**, 6233-6238.

Paul, J.K, Nettikadan, S.J, Ganjeizadeh, M, Yamaguchi, M. and Takeyasu, K. (1994). Molecular imaging of Na^+, K^+-ATPase in purified kidney membranes. *FEBS Letters*, **346**, 289-294.

Revel, J.P. and Karnowski, M. (1967). Hexagonal array of subunits in intercellular junctions of the mouse heart and liver. *J. Cell Biol.* **33**, C7-C12.

Sass, H.J, Beckmann, R, Zemlin, F, van Heel, M, Zeitler, E, Rosenbusch, J.P, Dorset, D.L. and Massalski, A. (1989). Densely packed ß-structure at the protein lipid interface of porin is revealed by high-resolution cryo-electron microscopy. *J. Mol. Biol.* **209**, 171-175.

Sato, A, Furuno, T, Toyoshima, C. and Sasabe, H. (1993). 2-dimensional crystallisation of catalase on a monolayer film of poly (1-benzyl-L-histidine) spread at the air-water interface. *Biochim. Biophys. Acta* **1162**, 54-60.

Schabert, F.A, Hoh, J.H, Karrasch, S, Hefti, A. and Engel, A. (1994a). Scanning force micrscopy of *E coli* OmpF porin in buffer solution. *J. Vac. Sci. & Technol. B*, **12**, 1504-1507.

Schabert, F.A. and Engel, A. (1994b). Reproducible acquisition of *Eschericia coli* porin surface topographs by atomic force microscopy. *Biophys. J.* **67**, 2394-2403.

Schabert, F.A, Henn, C. and Engel, A. (1995). Native *Eschericia coli* OmpF porin surfaces probed by atomic force microscopy. *Science*, **268**, 92-94.

Schabert, F.A. and Rabe, J.P. (1996). Vertical dimension of hydrated biological samples in tapping mode scanning force microscopy. *Biophys. J.* **70**, 1514-1520.

Shao, Z. and Zhang, Y. (1996a). Biological cryo atomic force microscopy: a brief review. *Ultramicroscopy* **66**, 141-152.

Shao, Z, Mou, J, Czajkowsky, D.M, Yang, J. and Yuan,J-Y. (1996b). Biological atomic force microscopy: what is achieved and what is needed. *Adv. Phys.* **45**, 1-86.

Thundat, T, Zheng, X-Y, Sharp, S.L, Allison, D.P, Warmack, R.J, Joy, D.C. and Ferrel, T.L. (1992). Calibration of atomic force microscope tips using biomolecules. *Scanning Microscopy* **6**, 903-910.

Unwin, P.N.T. and Henderson, R. (1975). Molecular structure determination by electron microscopy of unstained crystalline specimens. *J. Mol. Biol.* **94**, 425-440.

Uzgiris, E.E. and Kornberg, R.D. (1983). Two dimensional crystallisation technique for imaging macromolecules, with an application to antigen-antibody-complement complexes. *Nature* **301**, 125-129.

Vénien-Bryan, C, Lenne, P-F, Zakri, C, Renault, A, Brisson, A, Legrand, J-F and Berge, B, (1998). Characterisation of the growth of 2D protein crystals on a lipid monolayer by ellipsometry and rigidity measurements coupled to electron microscopy. *Biophys. J.* **74**, 2649-2657.

Worcester,D.L, Miller, R.G. and Bryant, P.J. (1988). Atomic force microscopy of purple membranes. *J. Microscopy*, **152**, 817-821.

Worcester,D.L, Kim, H.S, Miller, R.G. and Bryant, P.J. (1990). Imaging bacteriorhodopsin lattices in purple membranes with atomic force microscopy. *J. Vac. Sci. & Technol.* **A8**, 403-405.

Yamada, H, Hirata, Y, Hara, M. and Miyake, J. (1994). Atomic force microscopy studies of photosynthetic protein membrane Langmuir-Blodgett films. *Thin Solid Films* **243**, 455-458.

Yang, J, Mou, J, Yuan, J-Y and Shao, Z. (1996). The effect of deformation on the lateral resolution of atomic force microscopy. *J. Microscopy* **182**, 106-113.

Yip, C.M. and Ward, M.D. (1996). Atomic force microscopy of insulin single crystals: direct visualisation of molecules and crystal growth. *Biophys. J.* **71**, 1071-1078.

Zenhausern, F, Adrian, M, Emch, R, Taborelli, M, Jobin, M. and Descout, P. (1992). Scanning force microscopy and cryo-electron microscopy of tobacco mosaic virus as a test specimen. *Ultramicroscopy* **42-44**, 1168-1172.

CHAPTER 7

CELLS, TISSUE AND BIOMINERALS

7.1. Imaging methods

The AFM is designed to produce high resolution images on hard, flat surfaces. Very few biological systems approximate to that ideal. Biominerals such as teeth, bone or shells are hard, and often partially crystalline. However, the major problem lies in the production of sufficiently flat surfaces. As will be seen later in section 7.12 this can often be achieved by cutting and polishing surfaces. Cells are normally large with respect to the size of the tip and highly deformable. This has led to novel methods of sample presentation (Fig. 7.1) and the analysis of the deformation and indentation of the cells.

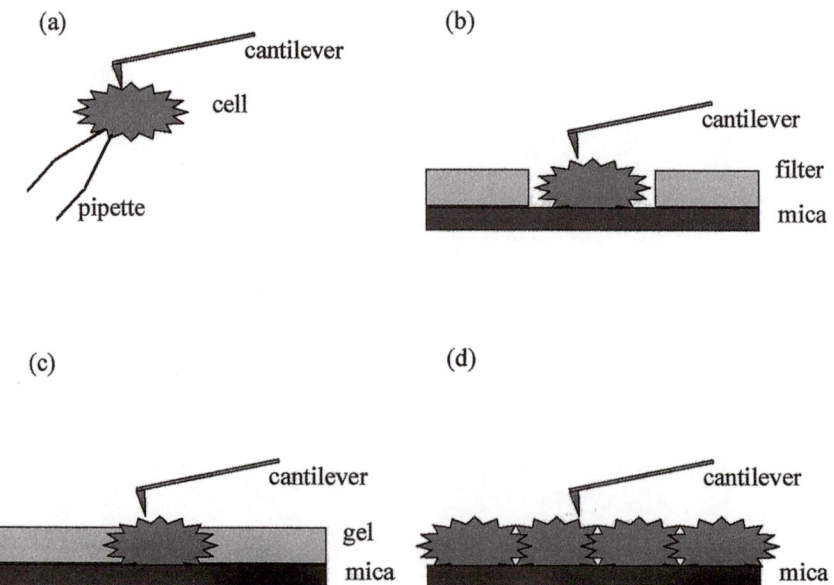

Figure 7.1. Schematic diagram showing different methods for immobilising cells. (a) Pipette method. (b) Use of porous media. (c) Trapping in a gel. (d) Confluent monolayer.

Tissues again present unique problems. In general, tissues can be examined using modified procedures developed for light and electron microscopy, with a major problem being the need to produce surfaces which are flat enough to allow imaging with the AFM.

7.1.1. Sample preparation

A range of methods have been developed for immobilising cells. Ideally such methods should allow imaging of living cells under physiological conditions. An elegant solution to this problem is to pinion individual cells on the end of a micropipette (Fig. 7.1a). The cells retain their natural shape, they can be immersed in physiological media, and the pipette can be used to scan the surface of the cell beneath the AFM probe tip (Haberle *et al* 1991; 1992; Horber *et al* 1992; Ohnesorge *et al* 1997). If individual cells are deposited onto flat substrates several difficulties occur. If the cells are fairly rigid and retain their shape then they effectively result in a rough surface, which is difficult to image without damaging or displacing the cells, results in distorted images heavily convoluted by the shape of the probe, or an inability to image the entire cell because the height of the cell exceeds the available z displacement of the piezoelectric scanner. The solution to this problem is to smooth the surface by effectively embedding the cells. Two approaches have been tried: entrapping the cells in porous media (Fig. 7.1b) (Holstein *et al* 1994; Kasa and Ikai, 1996) or completely enclosing the cells in a medium such as agar (Fig. 7.1c) (Gad and Akai, 1996). In the former case the pores restrict cell growth. In media such as agar the cells can grow but, once they grow out onto the surface of the gel, the imaging difficulties return. Growth of confluent monolayers provides a smoother sample surface allowing the surfaces of the cells to be imaged (Fig. 7.1d). For individual cells it is possible to immobilise them by air drying onto the substrate, the use of non-specific surface coatings such as poly-L-lysine (Butt *et al* 1990) or specific coatings such as bound antibodies (Prater *et al* 1990). Individual cells grown on flat substrates tend to be flattened and spread out on the surface. These structures are easier to image although in the vicinity of large organelles such as nuclei the structure may be too high to image. Cells are soft materials which will deform during scanning. As will be described later in section 7.1.2 such deformation increases the contact area between the tip and the sample reducing the achievable resolution. Resolution can be enhanced by stiffening the structure by fixation. Deformation of the outer surface of the cell results in an unexpected bonus: the cell moulds itself over the stiffer internal organelles or cytoskeleton allowing visualisation of cellular substructure even in living cells (Fig. 7.2).

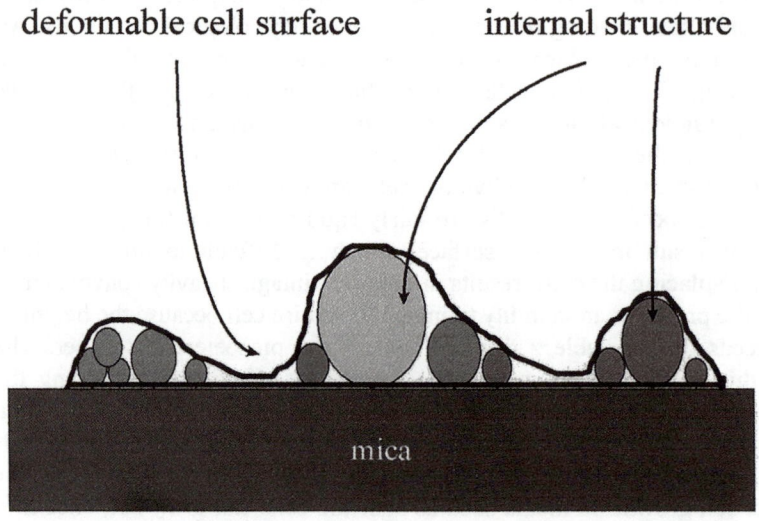

Figure 7.2. Schematic diagram showing a pliable outer cell surface moulding itself down onto more rigid intracellular components.

For tissue samples it is usually necessary to employ methodology developed for light or electron microscopy. Samples normally need to be fixed, embedded and sectioned or fractured. The final surface needs to be sufficiently smooth to allow imaging. Generally steps such as dehydration or metal coating can be avoided, although dried, metal coated samples, or metal replicas can be imaged with the AFM. Biomineral surfaces can generally be cut and polished. Powdered materials can be dried down onto substrates, or imaged after embedding in media such as KBr discs.

7.1.2. Force mapping and mechanical measurements

Force-distance curves

In the common mode of operation the feedback loop of the AFM operates to maintain a constant cantilever deflection, and images are nominally acquired at constant force. The assumption is that locally the force-distance curve is the same and hence, at constant cantilever deflection (assumed constant force), the image is determined solely by the topography of the sample surface. If the force-distance curves varies between the sample and substrate, or locally across the sample, then the image contrast is not simply a reflection of the sample topography, but also depends on material properties. Fig. 7.3 illustrates a variety of types of force-distance curves which may be observed for biological systems.

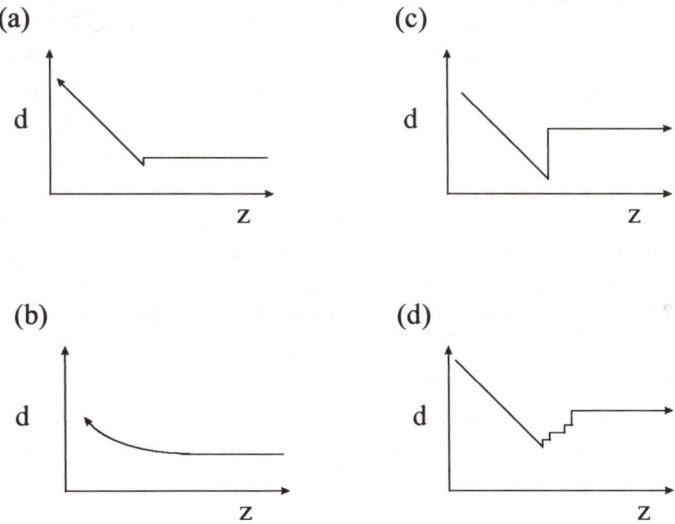

Figure 7.3. Schematic diagram illustrating the types of curves of cantilever deflection 'd' versus sample displacement 'z' observed for biological samples. Curves (a,b) are approach curves and curves (c,d) are retraction curve, as indicated by the arrows. (a) Hard, homogeneous sample and van der Waals forces. (b) Soft, and/or charged sample surface, or surface containing attached polymer which leads to an entropic repulsive force. (c) Adhesive interactions or capillary forces. (d) Binding of the tip to molecules on the surface.

Different types of interaction can all lead to differences in contrast. Important factors for cellular systems will be surface charge, elasticity, mobile surface layers, adhesion and localised molecular binding. Several, or all of these factors, can contribute to image contrast, or can be emphasised to select surface features or properties. At present the most important factor influencing contrast in studies on cells is sample deformation.

Force mapping

Because the image contrast is sensitive to the details of the force-distance curve there has been increasing interest in its measurement and interpretation for soft biological systems. In a new method called force mapping force-distance curves are recorded at each sample point of an image. An increasingly accepted way of displaying such data is as force-volume plots, and an excellent account of the acquisition of force-volume data has been written by Hoh and coworkers (Hoh *et al* 1997) who have also reviewed the general interpretation of such data for mapping biological surfaces (Heinz and Hoh, 1999a). The force volume plot consists of a topographical image of an area of the sample together with an array of force-distance curves recorded over the same area. The data can be displayed as slices of the volume plot showing force at constant height, slices corresponding to constant force images, or as individual force-distance curves. Data is recorded for both approach and retraction from the surface. In studies of soft biological systems the force-distance data can contain information on surface charge, sample viscoelasticity, adhesion or other factors modifying the force-distance curve such as specific binding. Thus the force volume data can be used to generate different types of force maps which can be used to generate different types of contrast for comparison with normal topography images (Heinz and Hoh, 1999a). A major application of this approach has been the analysis of the mechanical properties of cellular surfaces.

Elasticity and elasticity mapping

As mentioned earlier, the major factor which has been considered in depth is the effect of sample deformation during imaging. When the AFM is used to image hard samples the force-distance curve is of the form shown in Fig. 7.3a. As the tip approaches the sample there is flat region where the tip is not in contact with the surface. Once contact is achieved then the force (or cantilever deflection) increases linearly with decreasing tip-sample separation, and a plot of cantilever deflection against probe tip-sample separation would have a slope of 1. For soft samples the increase in cantilever deflection following contact is more gradual reflecting deformation of the sample (Fig. 7.3b). In this case the cantilever deflection will be dependent on the viscoelastic properties of the sample and this will contribute to contrast in the image. On a positive note this effect permits the mapping of the mechanical properties of the sample (Fig. 7.4f).

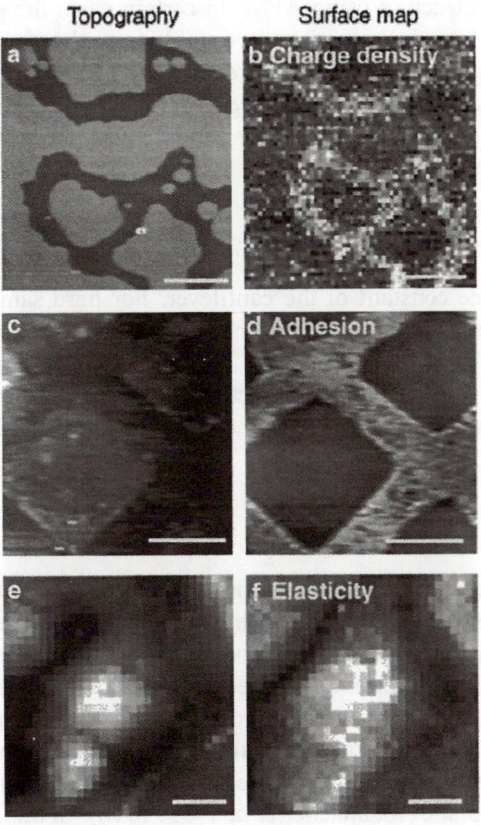

Figure 7.4. Surface maps. Topographical (a) and charge (b) maps (scale bars 1 µm) of a phospholipid bilayer on mica. The more highly charged mica surface appears bright in the charge map whereas the higher bilayer region appears bright in the topographic image. Topographical (c) and adhesion (d) maps (scale bars 500 nm) of a patterned streptavidin surface. The streptavidin is bound on the rows between the squares and appears bright in the adhesion map made with a biotinylated tip. Topographical (e) and elasticity (f) maps (scale bars 10 µm) of mitotic epithelial cells produced by FIEL mapping. This figure is reprinted from: Heinz, W.F. and Hoh, J.H. (1999a). Spatially resolved force spectroscopy of biological surfaces using the atomic force microscope. *Trends Biotechnol.* **17**, 143-150. Copyright (1999), with permission from Elsevier Science. The original experimental data maps are (a,b) from Heinz and Hoh (1999b), (b,c) from Ludwig *et al* 1997 and (e,f) from A-Hassan *et al* 1998.

However, a consequence of the sample deformation is that the contact area between the tip and the sample increases and this reduces the practical resolution which can be achieved. In effect the sample deformation on scanning blurs the image by effectively increasing the size of the tip. Another, and perhaps originally unexpected, consequence of the deformation of the sample surface is the ability to image cellular substructure using AFM. In certain cases the pliable outer coating of the cell can be folded down onto more rigid internal organelles or cytoskeleton revealing these structures even in living cells (Fig. 7.2).

The contact region of the force-distance curve contains information on the viscoelasticity of the sample surface. The loading force (F) is proportional to the deflection of the cantilever (d):-

$$F = -k \, d. \tag{7.1}$$

where k is the force constant of the cantilever. For hard samples (Fig. 7.3a) the magnitude of the cantilever deflection (d) is directly proportional to the z displacement (z) of the sample: $d = z$. For soft samples (Fig. 7.3b) the same z deflection (z) will cause a smaller deflection (d) of the cantilever as a result of an elastic deformation (δ): $z = d+\delta$. Thus the force-distance curve for the soft sample will be of the form:-

$$F = -k \, d = -k \, (z-\delta). \tag{7.2}$$

By subtracting the data for the soft sample and for the hard reference material (substrate) it is possible to generate an indentation-force curve which can then be modelled to determine the elastic modulus of the sample. Such curves are modelled using the Hertz model (Hertz, 1881) which describes the elastic deformation of two touching surfaces under load. Different geometries can be used to model the tip-sample interaction. Thus the Youngs' modulus of platelets was determined by treating the tip as a cone with an opening angle of about 30^0 and the platelet as flat plane (Radmacher *et al* 1995; 1996). In this case:-

$$F_{cp} = 0.5 \, \pi\delta^2 \, E \, (1-v)^{-1}. \, \tan(\alpha) \tag{7.3}$$

where α is the opening angle of the cone, E Youngs' modulus and v the Poisson ratio of the soft sample. For studies on vesicles where the radius of the sample is similar to that of the probe tip a more appropriate model is a spherical tip and a spherical particle (Laney *et al* 1997) which gives the relationship:-

$$F_{ss} = 1.335\delta^{1.5} E (1-v)^{-1}.[R_T R_V/(R_T + R_V)]^{0.5} \qquad (7.4)$$

where R_T and R_V are respectively the radii of the tip and the sample. Different expressions of this kind are obtained for different tip and sample geometries. The best model applicable can be found by plotting the F-δ data as a log-log plot. Estimates of tip shape and size can be made by direct measurement (e.g. scanning electron microscopy) or by deconvolution of images of biological or non-biological standards. The model assumes that the sample is homogeneous. Under these limiting conditions the analysis yields reasonable values for Youngs' modulus. In practice any limitations imposed by the use of the Hertz model are not too severe because, in general, the main interest is in the variation of mechanical properties, rather than measurement of absolute modulus values. Examples of the use of force-distance curves to measure elastic properties of biological material are studies on cartilage (Jurvelin *et al* 1996), glial cells (Haydon *et al* 1996) and epithelial cells (Putman *et al* 1994; Hoh and Schoenenberger, 1994). Good examples of the application of mechanical mapping of biological systems are work on chromosomes (Fritzsche and Henderson, 1997), cholinergic synaptic vesicles (Laney *et al* 1997), MDCK cells (Hoh and Scoenenberger, 1994), platelets (Radmacher *et al* 1996), bone (Tao *et al* 1992) and cardiocytes (Hofmann *et al* 1997; Domke *et al* 1999).

A-Hassan and coworkers (A-Hassan *et al* 1999) describe a refined mapping method called force integration to equal limits (FIEL) mapping which eliminates difficulties involved in the measurement of the tip-sample contact point and the spring constant of the cantilever. In this approach mechanical properties are determined from measurements of the work done by the cantilever, which is calculated from the area under the force-distance curve. In this approach it is necessary to isolate the purely elastic contributions to elasticity and this is done by separating time dependent (viscous) components from time independent (elastic) components of the deformation. The FIEL mapping method is designed to monitor and display spatial or time dependent variations in mechanical properties (A-Hassan *et al* 1999). In practice the force-distance curve may contain elastic (reversible) and inelastic or plastic (irreversible) components. The elastic components may also be time dependent. Thus the force volume data will be sensitive to imaging conditions such scan rates or whether contact or non-contact modes are used. For example, ac modes, such as Tapping, minimise or eliminate frictional effects, but also indent the sample at high frequency.

The indentation of the surface increases the contact angle between the probe tip and the sample. This in turn reduces the obtainable resolution and it is possible to estimate the expected resolution if the tip size and geometry and the elastic properties of the sample are known. This type of analysis has been made for soft gelatin films where the modulus was varied by imaging the sample in different propanol/water mixtures (Radmacher *et al* 1995).

An alternative way for measuring elastic behaviour is to use the AFM as a microrheometer. This approach has been used to study the elastic properties of bacterial sheaths (Xu *et al* 1996). The sheaths were suspended between bars on a GaAs grating and compressed with the AFM tip (Fig. 7.5). Equations developed to describe the force-indentation curves were first tested on model plastic films and then used to evaluate the elasticity of the bacterial sheaths.

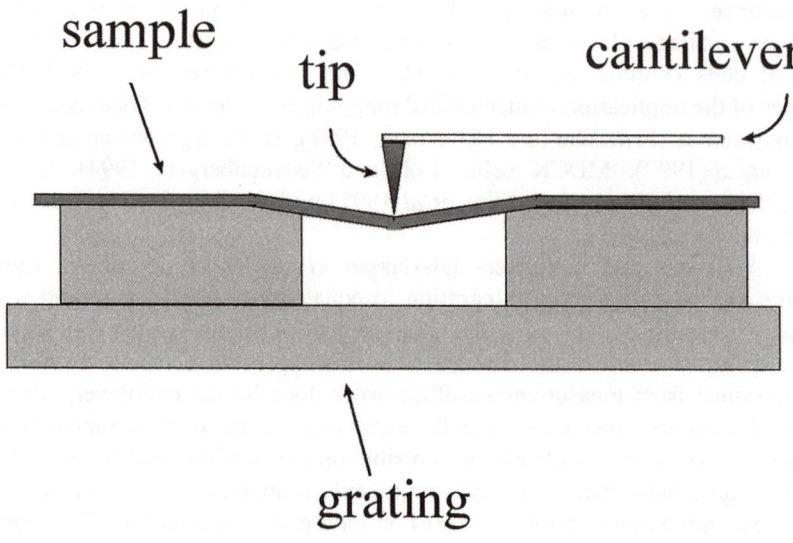

Figure 7.5. Schematic diagram illustrating the use of an AFM as a microrheometer to study the elasticity of bacterial sheaths. Based on a diagram in Xu *et al* 1996.

Charge, adhesion and other maps

Most cellular systems will be charged and surface charge will influence image contrast. If the charge varies over very short distances on the surface then surface deformation may smooth the charge distribution eliminating this effect. However, if charge differences occur in large patches, or are localised at specific surface structures, then this may provide additional contrast (Fig. 7.4b). At present the only charge maps for biological samples appear to be for bacteriorhodopsin membrane patches (Butt, 1992; Rotsch and Radmacher, 1997; Heinz and Hoh, 1999b) and phospholipid bilayer patches (Heinz and Hoh, 1999b) on hard substrates. From the known surface charge density of the substrate it was possible to calculate a reasonable value for the surface charge density of the membrane (Butt, 1992). By producing relative maps of surface charge density (or surface potential) it is possible to avoid the need to measure tip charge and radius, the cantilever spring constant and the tip-sample contact position, necessary to evaluate absolute tip-sample separations. This is achieved by subtracting iso-force surfaces collected at different ionic strengths; a method known as D-D mapping (Heinz and Hoh, 1999b). Charge mapping has been used to illustrate phase separation in mixed surfactant films (Yuan and Lenhoff, 1999).

Adhesion maps can be used to map tip-sample binding and hence, if functionalised tips are used, specific structures can be identified on sample surfaces. Adhesion maps can be generated by selecting the most negative force on the retraction curve and plotting iso-force maps as a function of position on the surface (Fig. 7.4d). This approach overcomes one of the difficulties with AFM imaging which is identification of surface features. The feasibility of this approach was shown by using biotinylated tips to map distributions of streptavidin on a patterned surface (Ludwig *et al* 1997). The method was termed affinity imaging and was used to compare topographical, elasticity and adhesion maps of the sample. Antibody coated tips can be used for selective imaging and have been used to image monolayers of intracellular adhesion molecules (Willemson *et al* 1998). The only attempt to map structures on the surfaces of living cells appears to be the work of Gad and coworkers (Gad *et al* 1997). These authors used concanavillin A coated tips to detect mannan polysaccharides on the surface of living yeast cells. The mapping shows a non-uniform distribution of the polysaccharide on the cell surface.

As described in section 4.5.3, neurofilaments are branched and the branches give rise to long range repulsive (entropic) interactions with the tip. Brown and Hoh (Brown and Hoh, 1997) have used iso-force difference mapping to reveal such interactions.

The use of derivatised tips to map surface features is an area that is likely to become of increasing importance and use in the future.

7.2. Microbial cells: bacteria, spores and yeasts

7.2.1. Bacteria

Bacterial surface layers

A number of bacterial surface layers have been studied extensively by AFM. These structures are generally rigid, well ordered and can be isolated as large sheets. Examples include the HPI layer of *Deinococcus radiodurans* (Schabert *et al* 1992), *Halobacterium* purple membrane (Muller *et al* 1995), S-layers from *Bacillus coagulans* and *Bacillus sphaericus* (Ohnesorge *et al* 1992) and *Escherichia coli* porin surfaces (Schabert and Engel, 1994; Schabert *et al* 1995). Other more complex structures include the sheath, hoops and plugs of the surface structures of *Methanospirillum hungatei* (Southam *et al* 1993) and the gas vesicles of the cyanobacterium *Anabaena flos-aquae* (McMaster *et al* 1996a). The highest resolution images reported are observations of β-sheet protein secondary structure in *A. flos-aquae* gas vesicles (McMaster *et al* 1996a). Atomic force microscopy studies on these ordered naturally occurring structures have already been discussed in section 6.3. In addition to obtaining high resolution images of cell surface structures it is also possible to measure and model their elastic properties using the AFM. This type of study is well illustrated by investigations on the sheath of the methanogen *Methanospirillum hungatei* (Xu *et al* 1996). The sheaths were air dried onto grooved GaAs plates and the AFM tip used to compress the sheaths into the grooves (Fig. 7.5). The measured Youngs' modulus suggested that these surface features were more than adequate to maintain structural integrity of the cells, and may therefore play an additional role as a pressure regulator controlling release of methane generated by these bacteria (Xu *et al* 1996).

Bacterial cells

Isolated bacteria deposited on substrates can be imaged in air (Fig. 7.6a) or under aqueous conditions. The cell wall structures are rigid and it is possible to reveal the roughness of the surface (Gunning *et al* 1996; Braga and Ricci, 1998), although the origin of these surface features (Fig. 7.6b) are still obscure. Bacterial flagella can be imaged in either the dc (Jaschke *et al* 1994; Beech *et al* 1996) or ac (Gunning *et al* 1996) modes (Fig. 7.7d), and substructure and flagella motors have been observed (Jaschke *et al* 1994). AFM has been used to 'read' photoresists generated by X-ray microscopy of *E. coli*. The images revealed the outer Gram-negative envelope and internal structure attributed to chromosomal DNA (Rajyaguru *et al* 1997). The action of antibiotics on bacteria has been investigated by AFM. Studies include observation of the action of penicillin on *Bacillus subtilis* (Kasas *et al* 1994) and the use of AFM to examine changes in surface structure of *E. coli* due to exposure to the β-lactam antibiotic cefodizime (Braga and Ricci, 1998). The major advantages of AFM in such studies are the ability to image under natural conditions, and the simpler sample preparation, when compared to electron microscopy methods. To the authors knowledge there are no reported studies on bacterial spores, the surface structures of spores or of spore germination.

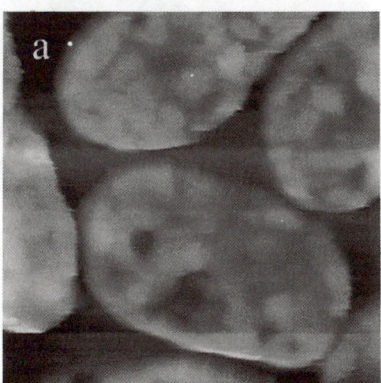 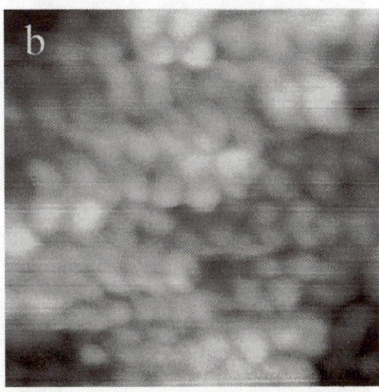

Figure 7.6. Low and high resolution AFM images of *Pseudomonas putida* bacteria. Images obtained by contact mode imaging in air, after deposition onto mica. (a) Individual bacteria, scan size 3 x 3 μm. (b) High resolution image of the bacterial surface, scan size 500 x 500 nm.

Bacterial biofilms

In addition to imaging bacteria deposited onto solid supports it is also possible to use AFM to investigate the formation and structure of biofilms formed at interfaces or on solid surfaces. There are a range of microscopic methods which can be applied to study biofilms (Surman *et al* 1996; Beech *et al* 1996) and AFM complements the use of optical and electron microscopy methods. The major advantages of the use of AFM are the ability to cover the magnification range of both optical and electron microscopy, but under natural imaging conditions with minimal sample preparation, and the production of quantifiable 3D images of the surfaces. The major limitation in the use of AFM is the curvature, or roughness of the substrates, or the biofilms formed on them. As the curvature or roughness of the sample increases the AFM images become more distorted and, eventually, when the surface curvature or roughness becomes comparable to the height of the probe, it is no longer possible to obtain images by AFM.

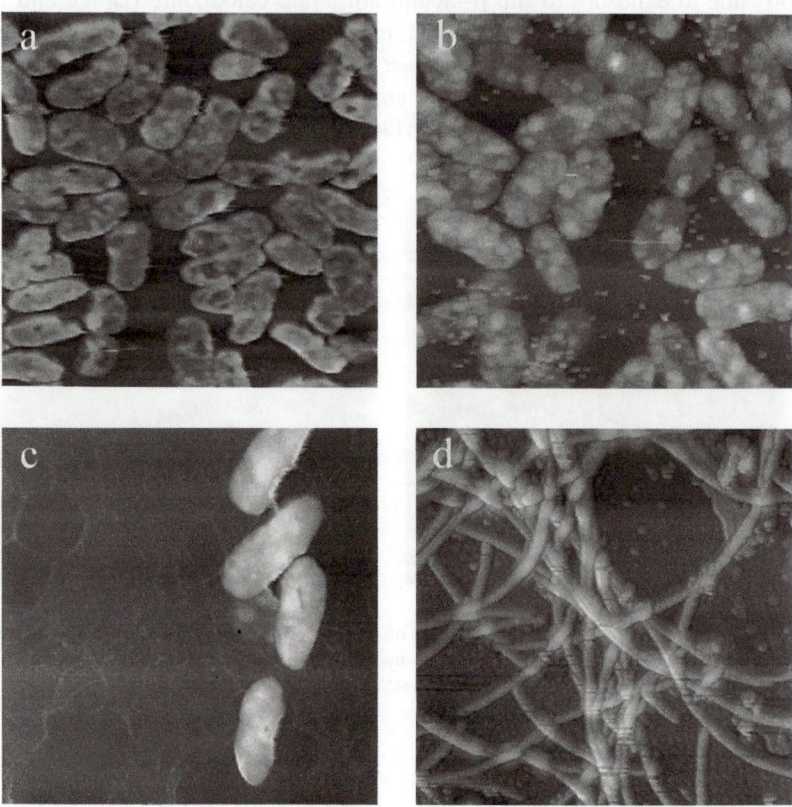

Figure 7.7. AFM images of a bacterial *P. putida* biofilm prepared at an oil-water interface, transferred onto mica and imaged in air. Images (a-c) are contact mode images. Topographic (a) and antibody labelled topographic (b) images, both having scan sizes of 10 x 10 µm. The antibody labels have been enhanced by gold labelled secondary antibodies and silver enhancement. The bright labels are seen attached to the bacterial cells and regions between the cells. The image (c) shows bacterial flagella trapped in an extracellular polysaccharide layer secreted by the bacteria. Scan size 10 x 10 µm. The antibodies are labelling this polysaccharide film. (d) Non-contact ac images of the bacterial flagella. Scan size 2 x 2 µm.

At the present time AFM has been used to study biofilms formed on glass (Surman *et al* 1996), metal surfaces (Bremer *et al* 1992; Steele *et al* 1994; Beech, 1996;

Beech *et al* 1996), hydrous Fe(III)-oxides (Maurice *et al* 1996) and at oil-water interfaces (Gunning *et al* 1996). Studies of hydrated bacterial biofilms formed on copper surfaces have revealed that the bacteria are observed bound adjacent to pits with extracellular polymer extending into the pits (Bremer *et al* 1992). This is important because metal ion binding by extracellular polysaccharides has been suggested as a basis for pit corrosion of copper surfaces (Geesey *et al* 1986; Jolley *et al* 1988). More recent studies of the formation of *Pseudomonas* species biofilms on copper also illustrated the importance of extracellular polysaccharide in film formation (Beech *et al* 1996). Related studies of bacterial biofilms on steel surfaces have highlighted the presence of extracellular polysaccharide, and have been used to examine pitting of the surface caused by biofilm formation (Steele *et al* 1994; Beech, 1996; Beech *et al* 1996). AFM has also demonstrated the importance of extracellular polysaccharide in the *Pseudomonas putida* bacterial biofilms formed at oil-water interfaces in emulsions (Gunning *et al* 1996). These biofilms are stiff, can be isolated from emulsions, and then imaged by light and electron microscopy (Parker *et al* 1995). Such isolates are rough and the resolution of the AFM images is poor, revealing only the packing of the bacterial cells (Gunning *et al* 1996). If, however, biofilm formation is modelled by growing such biofilms at flat oil-water interfaces, then high resolution AFM images can be obtained for samples of these flat biofilms pulled from the interface onto mica substrates by Langmuir-Blodgett techniques (Gunning *et al* 1996). In addition to visualising the packing of the bacteria it is also possible to identify remnants of bacterial flagella trapped within the extracellular polysaccharide matrix (Fig. 7.7d). AFM images of flat films labelled with polyclonal antibodies, gold labelled secondary antibodies and silver enhanced suggest that the antibodies bind to the bacterial surface, the extracellular polysaccharide and the flagella. There is probably further scope for the use of AFM to study the early stages of biofilm formation and/or the efficiency of present, or novel cleaning methods for removing such biofilms.

7.2.2. Yeasts

A few AFM studies have been made on yeasts, all of which are on *Saccharomyces cerviceae* (Henderson, 1994; Pereira *et al* 1996; Kasas and Ikai, 1996; Gad and Ikai, 1996). Yeast cells are large and malleable making dynamic studies of living cells in natural environments difficult: the cells are easily deformed or displaced and large rough samples are often impossible to image due to the limited vertical motion of the tip. Air drying yeasts onto glass substrates immobilises them and allows surface morphology to be imaged in air. Characteristic features such as bud scars can be recognised and it is possible to demonstrate differences in surface morphology for different strains (Henderson, 1994; Pereira *et al* 1996). Methods have been developed to allow imaging of live yeasts under natural conditions. The first approach involved immobilising the yeasts in Millipore filters (section 7.1) with pore sizes similar to the dimensions of the yeast cells (Kasa and Ikai, 1996). This is similar to the approach used by Holstein and coworkers (Holstein *et al* 1994) to immobilise and dissect fixed *Hydra vulgaris* polyps. This reduced the roughness of the sample allowing imaging in a natural environment; namely liquid culture medium. Although morphological features, such as bud scars, could be identified no growth activity was observed, possibly due to the restrictive effects of the filter paper pores. A refinement of the above procedure is to immobilise yeasts in agar gel (section 7.1 & Fig. 7.7) again permitting imaging in liquid culture medium (Gad and Akai, 1996).

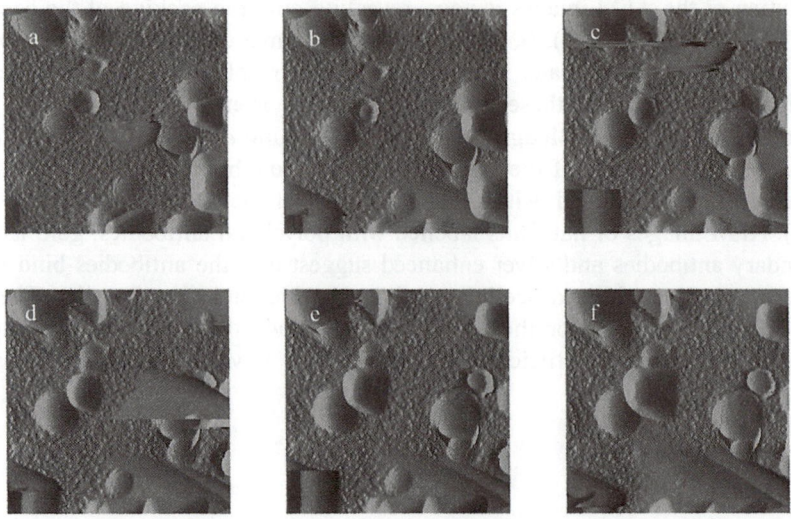

Figure 7.7. AFM deflection mode real-time images of yeast cell growth and budding. The yeast cells have been embedded in and on an agar gel. The scan size is 20 x 20 μm and the time lapse between successive frames is 6 minutes. Data described by Gad and Ikai, 1996 and reproduced with permission of the authors and the Biophysical Society.

High resolution images revealed birth and bud scars and, in addition, it was possible dynamically to follow growth and budding of the yeast cells (Fig. 7.7). As the cells grow on the gel surface the images gradually lose their spherical shape adopting a pyramidal character; an artifact caused by the finite size and shape of the probe tip. Eventually the cell size makes the surface impossible to image. The use of Con A labelled tips has allowed the mapping of the distribution of mannan polysaccharide on the surface of living yeast cells. This is at present the only example of affinity mapping (section 7.1.2) to identify and highlight surface structures on living cells (Gad *et al* 1997).

7.3. Blood cells

AFMs were originally designed for high resolution studies on surfaces. Thus the earliest AFMs had sub micron scan ranges which did not permit studies of intact cells. Once the scan range had been enhanced then cellular systems began to be studied, and blood cells were amongst the earliest samples investigated.

7.3.1. Erythrocytes

Red blood cells are readily available, easy to recognise, and were ideal for the early assessment of applications of AFM to study cellular systems. The early studies were concerned with imaging intact cells and in determining the level of resolution at which surface morphology could be imaged. The shape of red cells is characteristic of particular diseases and there is a growing interest in the identification of ultrastructural features associated with diseases, or infection of erythrocytes with parasites. The earliest images of red blood cells were on fixed cells either in air (Gould *et al* 1990) or in buffer (Butt *et al* 1990). Fixation prevented deformation of the cells by the probe and the images revealed the characteristic doughnut shape of the cells. Higher resolution scans revealed surface detail but the origin of these features remained obscure. Spectrin is the most abundant protein of the red blood cell membrane and isolated spectrin preparations have been imaged by AFM (Almqvist *et al* 1994; Zhang *et al* 1996) revealing structures of various oligomeric forms whose shape and dimensions are consistent with structures observed by TEM. Using Tapping mode in air Zhang and coworkers (Zhang *et al* 1995) have mapped the surface structure of fixed red blood cell surfaces down to nanometre resolution. They observed close packed arrays of particles of varying shape and in size from nanometres to several hundred nanometres. Cryo-AFM studies (Zhang *et al* 1996) of glutaraldehyde fixed erythrocytes revealed the presence of domains with closed boundaries, having lateral dimensions in the range of several hundred nanometres in size, similar to those revealed by cryo-AFM on red blood cell ghosts (Han *et al* 1995). Takeuchi and coworkers (Takeuchi *et al* 1998) describe the optimisation of methods for imaging the skeletal network of red blood cell ghosts by conventional AFM. These authors compared various fixation, drying and freezing methods and recommend rapid freezing in a liquid cryogen, followed by freeze drying, as the best method for preparing ghost specimens on glass cover slips for imaging by AFM (Takeuchi *et al* 1998). Their studies clearly showed that air drying is not suitable for preserving the intact membrane skeletal structure even after fixation in glutaraldehyde. Labelling with gold particles coated with anti-spectrin antibodies, and the effects on images of partial extraction of spectrin molecules, were used to confirm that the networks observed were spectrin networks. Images of the membrane structure on the cytoplasmic and extracellular surfaces have been used to discuss the 3D folded structure of the spectrin network and its possible influence on cell deformation during circulation, or on how abnormalities in the spectrin network can lead to loss of mechanical strength and/or deformability of the red blood cell membrane. Specific labels can be used to locate and identify specific sites on the surface of red blood cells. Neagu and coworkers (Neagu *et al* 1994) used superparamagnetic beads coupled to antibodies to locate the transferrin receptors on the surface of erythrocytes.

AFM has been used to compare the structures of normal and pathological red blood cells (Zachee *et al* 1992; 1994; 1996) and differences in the cytoskeleton networks of normal and *Plasmodium falciparum* infected erythrocyte ghosts (Garcia *et al* 1997). AFM has been used to image erythrocytes from patients with hereditary sperocytosis revealing the abnormal surface pseudopodia (Zachee *et al* 1992). These cells become normal on removal of the spleen. The peripheral blood plasma of uremic patients contain echinocytic erythrocytes. AFM has been used to image type 3 echinocytes showing the smaller more rounded ovoid shape with an even distribution of spicules (needle-like projections) across the cell surface (Fig. 7.8). Unlike the normal discoid shaped cells the type 3 echinocytes show only a small central crater. The echinocytic erythrocytes revert to a normal discoid shape when the uremic plasma is washed away and replaced by normal blood plasma. There is clearly scope for both low and high resolution studies of the interconversion of pathogenic and normal erythrocytes due to the influence of pharmocological or pathogenic factors. The malarial parasite *P. falciparum* alters the morphology of the surfaces of the erythrocytes. AFM studies of normal and infected red blood cell ghosts show that the surface of the infected cells is smoother, contains identifiable parasites and exhibits large (0.2 - 0.7 µm) particulate protrusions. Higher resolution images are claimed to show differences in the density of the spectrin networks for normal and infected cells (Garcia *et al* 1997).

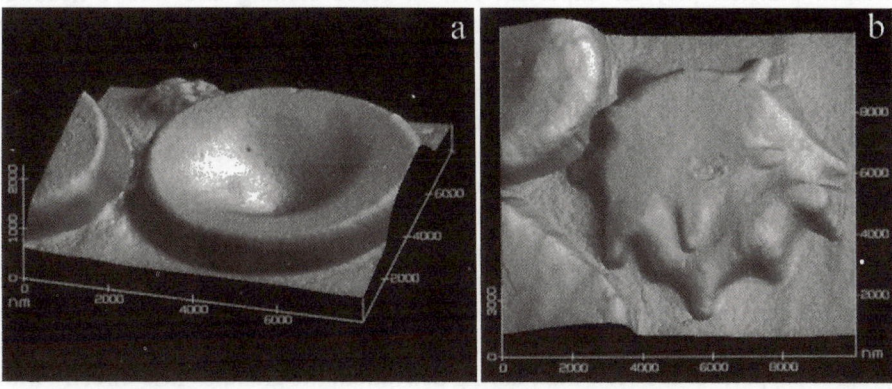

Figure 7.8. Comparative AFM images of (a) a normal erythrocyte and (b) an uremic echinocyte. The uremic echinocyte shows spicules and a smaller central cavity. Data reproduced from Zachee *et al* 1994.

7.3.2. Leukocytes and lymphocytes

Fixed white blood cells can also be imaged in phosphate buffered saline. If diluted blood is applied to a silanised glass cover slip and then washed, the red blood cells are removed leaving the white cells adhered to the derivatised glass surface (Butt *et al* 1990). The white cells have a more irregular spread-eagled shape and a rougher surface. The use of combination optical and AFM microscopes (Putman *et al* 1992) and/or immunological labelling methods (Putman *et al* 1993) are of obvious importance in identifying cell types and in the subsequent use of AFM for high resolution studies under natural conditions. Gold labelled antibodies, with or without silver enhancement, can be observed bound to the surface of air dried lymphocytes and offer the prospect of marking specific surface features of the cells (Putman *et al* 1993; Neagu *et al* 1994). Comparative imaging of cells before and after labelling provides a means of identifying specific surface features and imaging them at high resolution. Specific labels can be used to isolate and bind particular types of cells. Thus antibody coated glass slides have been used to preferentially bind B-lymphocytes for imaging by AFM (Prater *et al* 1990).

7.3.3. Platelets

Human platelets play an important role in blood clotting and wound healing. Normally they exist in the bloodstream in a 'resting state' in which the cells are discoid in shape. Injury to a blood vessel promotes activation of the cells: this involves a drastic change in the cytoskeletal structure leading to a marked change in cell shape. Activation can be induced by contact with wettable surfaces and it has been possible to obtain real time AFM images of the activation of platelets adhered to glass surfaces under physiological conditions (Fritz *et al* 1993; 1994). The resting platelets adhere poorly to the glass substrate and are difficult to image. However, the time dependent changes in the cells following activation can be visualised: initially thin filopodia (spike-like protrusions) are seen to protrude from the interior of the cell which become filled with granules transported from the centre of the cell eventually forming a flat lamelliporadium. The AFM studies provide support for the view that, during activation, the granules fuse directly with the plasma membrane. Details of the cytoskeleton and changes during activation could be resolved in unstained living cells. In unfixed platelets the tip does deform the cells during scanning to different extents depending on the height of the cell, structural differences within different regions within the cell, and on the magnitude of the applied force. Platelets which have bound to the surface but have not been activated can be scanned at low forces without any noticeable time dependent changes in shape. However, scanning at higher forces has been reported to promote activation (Fritz *et al* 1993). Deformation of the cells permits information to be obtained on the elasticity of cell structure (Fritz *et al* 1993; 1994; Radmacher *et al* 1996). Using force mapping techniques (section 7.1.2) it has been possible to generate simultaneous topographic and elasticity maps (Fig. 7.9) of the living activated cells (Radmacher *et al* 1996). The variation of elastic properties across the cell can be correlated with standard features of the cell such as the pseudonucleus, the inner and outer filamentous zones and the cortex, and then interpreted in terms of the ultrastructure of these regions. Derivatised tips can also be used for selected mapping of cell surfaces and one such study has been reported on platelets (Siedlecki and Marchant, 1998). Identification of specific features on platelet surfaces can be achieved through the use of specific labels. 14 nm gold particles conjugated to fibrinogen receptors have been used to map the distribution of these sites on the cell surface. The AFM studies were validated through complementary low voltage, high resolution scanning electron microscopy (LVHRSEM) (Eppell *et al* 1995).

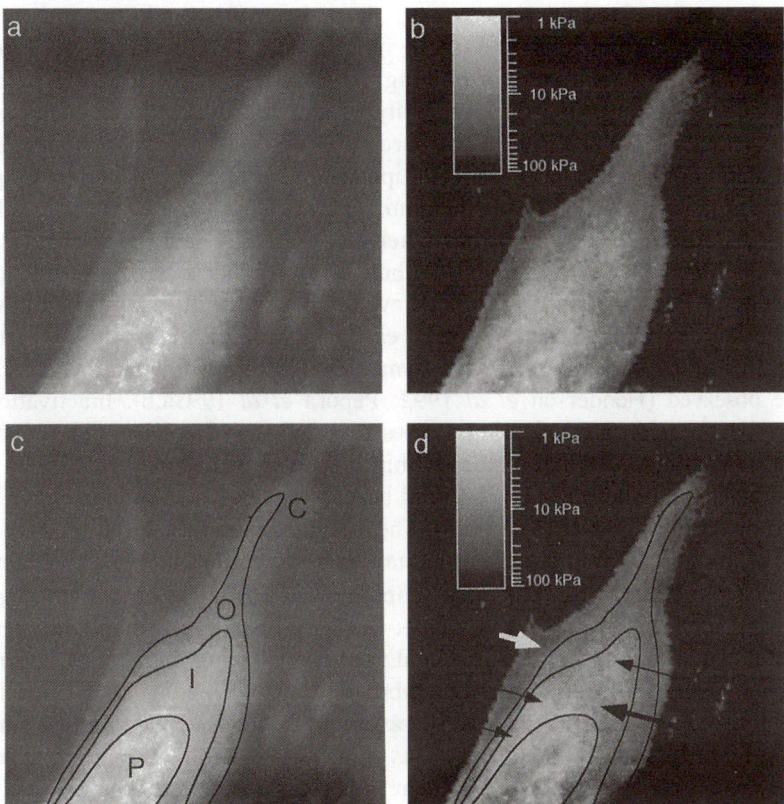

Figure 7.9. Topography (a) and elasticity (b-d) maps of human platelets. Scan size 4.3 x 4.3 μm. The range of grey scales in image 'a' is 2 μm. In image 'b' the elastic modulus is encoded logarithmically so that black corresponds to 100 kPa and white to 1 kPa (as shown in the grey scale bar in the diagram). Different parts of the platelet are identified in 'c': P-pseudonucleus, I-inner web, O-outer web, and C-cortex. In 'd' it can be seen that the pseudonucleus is softest (1.5- 4 kPa), the inner web has a stiffness about 4 kPa, and the outer web ranges about 10-40 kPa. Some areas in the cortex (white arrow) are stiffest (50 kPa). There are some areas that deviate from the general pattern; the area indicated by the thick black arrows is 10 kPa and the area indicated by the thin, black arrows is 4 kPa . Data reproduced from Radmacher *et al* 1996 with permission of the authors and the Biophysical Society.

7.4. Neurons and Glial cells

Glial cells provide a system in which it has been possible to probe the structure and dynamics of intercellular components in living cells using the AFM. Biologically glial cells play an important role in the development, maintenance and regeneration of the nervous system. There is an interest in the growth and development of these cells and their interactions with surfaces or neurons. Although AFM is a technique for probing the structures of surfaces, early studies on glial cells demonstrated the possibility of imaging intracellular components. The cells used (XR1) are from a standard cell line derived from *Xenopus laevis* retinal neuroepithelium, or from rat hippocampi, and were plated onto glass coverslips and imaged in growth medium. The nucleus of the cells can easily be identified in the images because of its height, suggesting that the nucleus is less deformable than the rest of the cell. Although mitochondria have been identified in fixed cells (Papura *et al* 1993a) they have not been seen in living cells by AFM, possibly because these structures may easily deform during scanning. At the extremities of the spreading cells the dynamics of internal filamentous structures can be observed (Henderson *et al* 1992; Papura *et al* 1993a,b). Inactivation of actin polymerisation and labelling studies revealed that the AFM was observing actin filaments but not microtubules within the cells (Henderson *et al* 1992). The microtubules are believed to be obscured because isolated microtubules have been imaged by AFM (Vinckier *et al* 1995). These studies raised the question as to how the AFM imaged internal structures. It has been shown that imaging of filaments requires a minimum threshold force (Papura *et al* 1993a). Two suggestions for visualising internal structure were made: firstly, a pliable membrane structure is pressed down onto the more rigid skeletal framework during imaging or, secondly, the probe penetrates the 'fluid' membrane directly imaging the underlying filamentous network. In this context it was found that holes created in the surface by imaging at high force would reseal as shown by subsequent imaging at lower forces (Henderson *et al* 1992). At present the evidence suggests that, at least with the use of standard tips, the membrane deforms around the underlying rigid filaments: this has been deduced from force-distance studies, evidence that AFM imaging does not release intracellular trapped fluorescent dyes or impair physiological functions such as signal transduction mechanisms (Haydon *et al* 1996). Recent investigations have shown that repeated AFM imaging of live cells does not reduce cell viability or increase cell death rates. However, there is clear evidence for the accumulation of cell membrane components on the tip, indicating substantial probe-membrane interactions during imaging. These effects were found to be more pronounced in contact rather than Tapping mode imaging (Schaus and Henderson, 1997).

   In studies of mixed cultures of neurons and glial cells it was possible to observe neuron-glial cell interactions in both fixed and live cells (Papura *et al* 1993a). In fixed cells high resolution images could be obtained of the growth cones of the neurons, and the actin-based filopodia extending from the body of the growth cones. By controlling the imaging force it was possible to manipulate the biological system; e.g. cut neurites, remove growth cones or selectively displace the weaker bound neurons (Papura *et al* 1993a). For glial cells alone, if the imaging force is gradually increased, then it is possible to observe selective removable of different features of the cells, providing relative indications of the levels of adhesion to the surface (Papura *et al* 1993b). A very recent paper compares AFM images of glial cells with images acquired using a surface plasmon resonance microscope (SPRM) (Giebel *et al* 1999). Contrast or height in SPRM images are related to cell-substrate contact distance and the microscope provides a new method for probing cell adhesion.

AFM has been used to investigate the localisation of individual calcium channels on the release face of a presynaptic nerve terminal. The channels were located by binding of biotinylated neurotoxin ω-conotoxin, fixation and then enhancement by tagging with avidin-gold. The resolution obtained was a 10-fold improvement over the use of light microscopy, and the detection of a 40 nm interchannel spacing, which was found to be independent of channel density, has been taken to imply anchoring of the channels on the release face of the transmitter (Haydon *et al* 1994). Synaptic vesicles play an important role in the transmission of nerve signals. The vesicles release the neurotransmitter acetylcholine into the synaptic cleft, where it triggers an action potential in postsynaptic cells. AFM has been used to visualise synaptic vesicles in fluid environments and to characterise changes in size and shape (Papura *et al* 1995). Force mapping methods (section 7.1.2) have been used to show that the vesicles possess a substructure consisting a stiffer central core region, the elasticity of which varies in different buffers, becoming harder in the presence of calcium ions (Laney *et al* 1997), surrounded by a more elastic, less rigid outer region. It is suggested that the core, which may correspond to electron dense particles seen by TEM, may in fact be proteoglycan.

7.5. Epithelial cells

Cell monolayers provide useful model systems for studying cell-cell interactions and cell differentiation. They are also attractive systems for AFM investigations of soft cells. The formation of a monolayer effectively reduces sample roughness: the scanning probe no longer has to climb over the entire height of individual cells but rather scans along the surface of the monolayer, without the need to penetrate down onto the substrate. A number of investigations have been made on the Martin Darby canine kidney (MDCK) cell line. AFM has been used to examine the morphology of the membrane surface of polarised renal epithelial cells (MDCK cells) (Grimellec *et al* 1994). The most detail was observed on air-dried cells. At the lowest magnification entire cells and their junction zones within the monolayer were visible. With increasing magnification it was possible to visualise microvilli and surface pits, and at the highest magnification, globular structures possibly attributable to proteins or protein aggregates. Such fine structure is also observable in fixed cells imaged under phosphate saline buffer. However, treatment with 'Pronase' smoothes the surface suggesting that the protruding particles are proteins. Images of the surfaces of living cells under buffer were fuzzy or blurred (Grimellec *et al*1994; Hoh and Schoenenberger, 1994) although in some cases the nucleus of cells could be identified (Hoh and Schoenenberger, 1994). Treatment with enzymes, such as neuraminidase which partially degrade the glycocalix, improved imaging allowing visualisation of globular protrusions which were then susceptible to treatment with 'Pronase', suggesting that they are surface glycoproteins (Grimellec *et al* 1994). Surface microvilli observed in SEM images are not seen on living cells by AFM, although progressive fixation does reveal surface corrugation, possibly due to mechanical stabilisation of microvilli, allowing them to be seen by AFM (Hoh and Schoenenberger, 1994). Fixation also allowed visualisation of microvilli on the surface of rat carcinoma cells RCMD cells (Pietrasanta *et al* 1994) and CC531 cells (Braet *et al* 1998a) in confluent monolayers. An alternative model for studying cell polarisation and differentiation is the human adenocarcinoma cell line HT29. As with MDCK monolayers the AFM images of HT29 monolyers are fuzzy and blurred. However, if the cells are air-dried and then imaged under butanol, then the microvilli are clearly visible (Fig. 7.10) and globular features on the surface of the microvilli, attributed to glycoproteins can be seen (Kirby *et al* 1998). In the normal hydrated state where the microvilli extend out from the cell the probe can permeate between the microvilli deforming the structure on scanning. Air-drying flattens the microvilli down onto the surface and butanol appears to maintain this state allowing imaging (Kirby *et al* 1998). If the probe deforms the microvilli during scanning then, at low applied forces, it may be possible to reveal these structures. Such studies have been made by Lesniewska and coworkers (Lesniewska *et al* 1998) on living MDCK cell monolayers. At normal imaging forces the cell surfaces are fuzzy whereas for forces < 300 pN a close packed array of the tips of the extended microvilli could be seen. On living MDCK cells it was possible to follow dynamic events such as the formation of bulges and spikes (Hoh and Schoenenberger, 1994; Schoenenberger and Hoh, 1994), and these events appear to be intrinsic properties of the cells and not effects triggered, or stimulated by the action of the probe. Measurement of force-distance curves allowed the elastic properties of the cells to be monitored and it was possible to follow the mechanical stiffening of the cells during fixation (Hoh and Schoenenberger, 1994). A more complete force mapping of living MDCK cells provided unexpected contrast variations within the

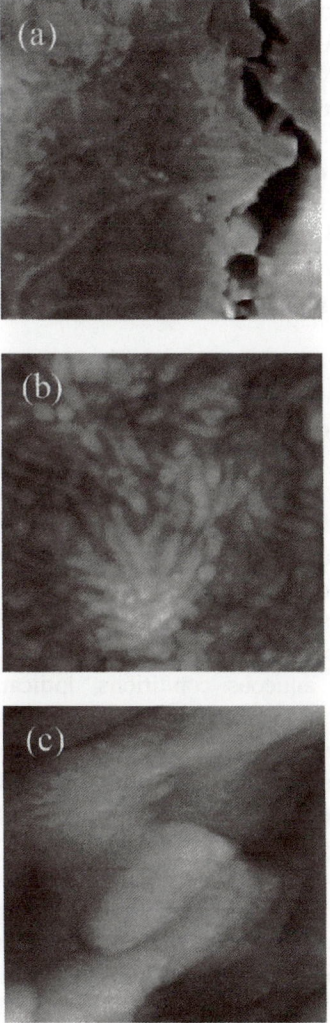

Figure 7.10. AFM contact mode images of an HT29 confluent monolayer imaged under butanol. (a) Boundary between two cells showing microvilli and possibly filopdia, scan size 30 x 30 μm. (b) Magnified image of a cell surface clearly showing microvilli, scan size 4.8 x 4.8 μm. (c) High resolution image of the surface of a microvillus showing globular 'glycoproteins', scan size 924 x 924 nm. Data reproduced from Kirby *et al* 1998.

elastic profiles revealing structural features of the cells. The inter-cell contact boundaries were found to be stiffer than the centres of the cells. Intercellular features highlighted included a boundary between the nucleus and the cytoplasm (A-Hassan *et al* 1998). An intriguing use of AFM to measure changes in the mechanical properties of MDCK cells is an attempt to develop a biosensor based on micromechanical interrogation of living cells (Antonik *et al* 1997). The biosensor consists of a tip-less cantilever coated on one side with a cultured layer of MDCK cells. Exposure of the biosensor to the lytic bee venom melittin and the respiratory inhibitor sodium azide caused a deflection of the cantilever, detected in an AFM, and attributed to changes in the rigidity of the cells. It is suggested that such sensors could be developed for studying cellular responses to chemicals in areas such as the screening of pharmaceuticals, or toxicity testing (Antonik *et al* 1997).

7.6. Non-confluent renal cells

AFM images of the inner cytoplasmic leaflet of the plasma membrane of MDCK cells in air or aqueous media reveal a distribution of globular structures believed to be surface proteins. In air this layer is partially decorated with cytoskeleton features such as actin fibres but these decorative features disappear when imaged in aqueous media (Le Grimellic *et al* 1995). Treatment with enzymes such as neuraminidase is unnecessary in order to obtain high resolution images of the 'protein' structure under aqueous conditions, indicating that glycosylation is confined to the outer layer of the membrane.

Certain derivatives of MDCK cells do not form homogeneous polarised monolayers. R5 cells spread out on glass substrates and, in their highly flattened extremities, it is possible to image features such as cytoskeletol fibres and vesicles, which cannot be imaged by SEM (Hoh and Schoenenberger, 1994; Schoenenberger and Hoh, 1994). Time resolved events can be observed by AFM: these include dynamics of stress fibres, wave-like rearrangements of the cytoplasm and even motion of vesicles along fibres (Schoenenberger and Hoh, 1994). Dynamic processes have also been observed by AFM in migrating MDCK-F cells (Oberleithner *et al* 1993; 1994; 1995; 1996; 1997a). The AFM has been used to excise membrane patches from the surface of MDCK cells (Oberleithner *et al* 1996) and to measure changes in the height of cloned and isolated ROMK 1 potassium channels, arising from addition of ATP (Henderson *et al* 1996a,b; Oberleithner *et al* 1997b). The latter studies used the 'molecular sandwich' technique: the protein is bound to the tip and then touched down onto, and interacts with the mica substrate. Changes in shape of the sandwiched protein can be detected by variations in the deflection of the cantilever.

Studies on monkey kidney cells by both contact and Tapping mode AFM suggest that it may be possible to simultaneously acquire images of the surface structure of the cell membrane and the underlying cytoskeleton network. This has been achieved either by comparing topographic images (Putman *et al* 1994) or deflection (error signal mode) images (Le Grimellec *et al* 1997) in contact and Tapping mode. Both static (Le Grimellec *et al* 1997) and dynamic (Putman *et al* 1994) studies have been made of the morphology of the cells. Phase imaging provides an alternative method for probing the viscoelasticity of cell surfaces: phase images of cultured CV-1 kidney cells have recently been reported by Lesniewska and coworkers (Lesniewska *et al* 1998).

Finally, time dependent AFM studies on individual monkey kidney cells have been used to follow the release of virus particles from cells infected with pox virus (Haberle *et al* 1992; Horber *et al* 1992; Ohnesorge *et al* 1997). Individual cells were held on a pipette and scanned beneath an AFM tip in physiological medium. By imaging the surface of the cell as a function of time it was possible to observe what is believed to be the exocytosis of a virus through the cell wall. The images can be combined to produce a video of this biological process. Measurements of the elastic properties of the cell surface have been used in an attempt to identify changes in the structure of the cell surfaces during infection or expulsion of virus particles.

7.7. Endothelial cells

The endothelium acts as a mechanical transducer between blood and the vascular walls of blood vessels, and the effects of flow can induce a range of cellular responses. The earliest AFM studies on living endothelium cells investigated the effects of shear stress on cellular structure. Comparative studies were made on unsheared and sheared confluent monolayers (Barbee *et al* 1994) of cultured bovine aortic endothelial cells in phosphate buffered saline. Whereas the unsheared cells were polygonal with well defined cell boundaries, the sheared cells were elongated and aligned in the flow direction, their boundaries were less distinct, and new aligned fibrous surface ridges were seen. Comparative fluorescent studies on fixed cells suggested that the changes observed were due to shear-induced rearrangement of the cytoskeleton structure, with the surface ridges corresponding to aligned bundles of actin fibres (Barbee *et al* 1994). AFM has also been used to probe changes in surface structure with time after confluence (Barbee, 1995) in order to assess structural factors which may influence the mechanical response of the cells to flow.

Sinusoidal liver endothelial cells (LEC) possess fenestrae which act as sieves controlling the exchange of fluids, solutes and particles between blood and the microvilli coats of parenchyma cells. AFM has provided information on the structure and dynamic response of fenestrae to different stimulants (Braet *et al* 1995; 1996a,b; 1997ba,b; 1998b).

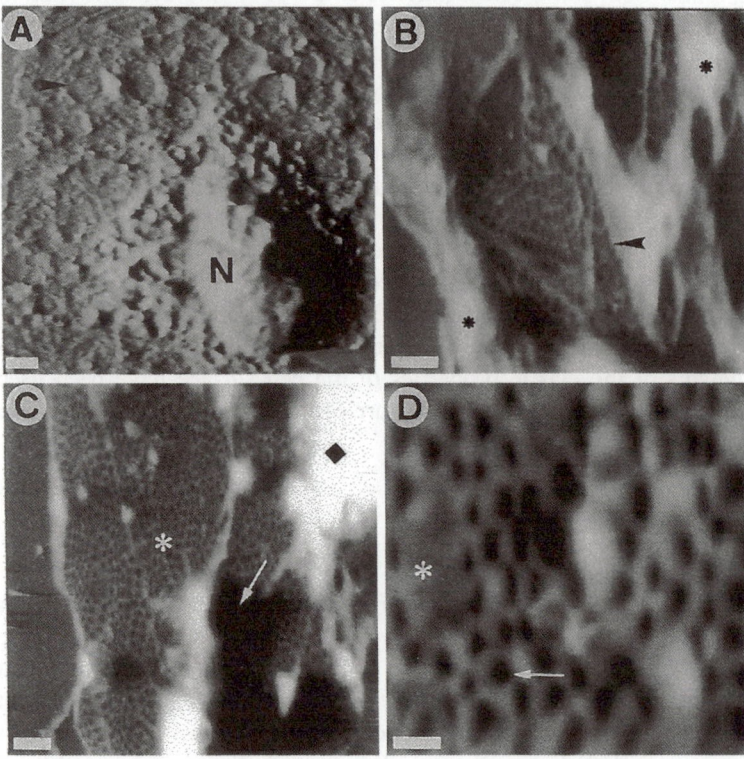

Figure 7.11. AFM images of wet fixed endothelial cells. (A) Low magnification (scale bar 2.5 µm) image showing the central nucleus (N) and sieve plates (arrowhead). (B) Higher magnification (scale bar 1 µm) image showing a sieve plate (arrowhead) in surrounding cytoplasm. The asterisks show white bumps. (C) Liver endothelial cells treated with 10 µg mL^{-1} of cytochalasin B for two hours, inducing a highly fenestrated cytoplasm (asterisk). White bumps (black diamond) and shadowing of structures (arrow) are also present; scale bar 1 µm. (D) Higher magnification (scale bar 500 nm) of the fenestrated cytoplasm after cytochalasin B treatment, which show fenestrae (arrow). Notice also the presence of typical small unfenestrated dots (asterisk) that occur after treatment with the microfilament-inhibiting drug. Data reproduced from Braet *et al* 1996a with permission. Copyright of the Royal Microscopical Society.

Comparative SEM and AFM images on dried, gold coated cells revealed similar features: fenestrae arranged in sieve plates surrounded by an elevated border corresponding to an underlying tubular structure. Height measurements suggest that the sieve plates are about 200 nm below the surface of the nearby cytoplasm (Braet *et al* 1996a,b). AFM images could also be obtained on dried, uncoated cells and wet, glutaraldehyde-fixed cells, with the latter yielding the best images (Braet *et al* 1996a; 1997b). Comparative studies of the elasticity of fixed and unfixed cells has confirmed that fixation improves image quality by stiffening the cells and inhibiting distortion on scanning (Braet *et al* 1998b). Use of non-contact imaging modes improved the quality of the images of wet cells by further eliminating deformation artifacts, such as distortion of the fenestrae in the scan direction (Braet *et al* 1997b). Imaging wet cells showed that dehydration, critical point drying or drying by evaporation of hexamethyldisilazane resulted in a considerable shrinkage of fenestrae (Braet *et al* 1996a,b; 1997a). This is a clear demonstration of the advantages of the AFM in allowing imaging without drying the samples. It was also confirmed that treatment with ethanol and serotonin led respectively to enlargement and shrinkage of the fenestrae diameter, and that treatment with cytochalasin increased the number of fenestrae (Braet *et al* 1996a) (Fig. 7.11).

7.8. Cardiocytes

The ability of the AFM to image structure in living cells at molecular resolution, plus the ability to map mechanical properties of cells at high spatial and temporal resolution, provides a means of studying mechanisms of beating in contractile cells. AFM images of living rat atrial cardiomyocytes revealed the centrally located nucleus plus the submembraneous fibrous cytoskeleton structure concentrated at the extremities of the cells (Shroff *et al* 1995). Fixation enhanced the images of the cytoskeleton. Light microscopy of cells stained for F-actin showed that the fibrous structures seen by AFM are actin bundles. Force-distance measurements on quiescent cells showed that the cells are softer above the nuclear region, and become stiffer towards the periphery, where the fibrous structures are visible. Increases in cell stiffness were observed on the addition of calcium, or on fixation with formalin. By following changes in cantilever deflection, due to expansion or contraction of the cells, it was possible to monitor the localised activity of beating cells at high spatial (1-3 nm) and temporal (60-100 μs) resolution, and to probe changes in stiffness during single contractions (Shroff *et al* 1995). Force mapping methods (section 7.1.2) have been used to map the elastic properties of living chicken cardiocytes (Hofmann *et al* 1997). Using a simple geometrical model (Hertz, 1882) for the tip indenting the cell the force-distance curves have been interpreted to map the Youngs' modulus across the cell. The fibres appear as stiffer structures embedded in the softer cell. The theoretical model matches the force-distance curves well in the softer regions, but fails in the vicinity of the fibres, probably because the model assumes an homogeneous structure for the cell surface. Imaging at increasing force enhanced the submembraneous cytoskeleton. Topographic and elastic maps have been used to follow the breakdown of the actin network due to the addition of the drug cytochalasin B (Hofmann *et al* 1997). Most recently AFM has been used to analyse the mechanical pulsing of individual active single chicken cardiocytes and the synchronised beating of cells in a confluent layer (Domke *et al* 1999). With the AFM it was possible to map the localised beating across complete cells (Fig. 7.12).

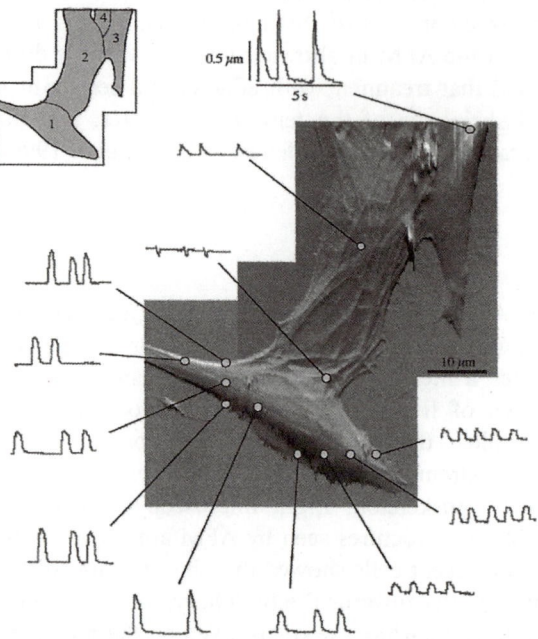

Figure 7.12. Use of the AFM to pulse map a group of active cardiomyocytes. Although the cells were beating it was still possible to image them. In the figure the deflection images of two scans were superimposed. The inset shows a sketch of the margins of the cells. Several time sequences were recorded at different locations on the cells. The presented sequences are are scaled identically; their locations on top of the cell are marked with white spots. No pulses were found that showed negative amplitudes. Only the series between cells '1' & '2' showed a biphasic pulse shape. Data reproduced from Domke *et al* 1999 with permission of the authors and Springer-Verlag GmbH & Co., KG who own the copyright.

7.9. Other mammalian cells

There are a number of other studies on cells which do not fit easily into the above classifications. In some cases these studies represent new methodology or new types of investigation which are worth recording. A few such studies are collected in this section. Novel methods for improving the imaging and identification of subcellular structures are described by Pietrasanta and coworkers (Pietrasanka *et al* 1994). In fixed and dried rat mammary carcinoma cells (RMCD cells) it was possible to recognise subcellular structures such as the cell nucleus and nucleoli, plus microfilaments of the cytoskeleton, and also to visualise surface features such as microvilli and microspikes. Treatment of the fixed cells with Triton X-100 to remove lipids and soluble protein allowed visualisation of mitochondria, the identification of which was confirmed by rhodamine 123 staining and fluorescence microscopy. The detergent resistant sample surface was then dry-cleaved with cellotape to remove the dorsal part of the cell, and presumably the membrane associated microfilament layer, in order to reveal the stress fibres and an underlying network of finer filamentous structures, some associated with granular structures. Fluorescence labelling revealed the presence of both actin and microtubules although individual components could only be tentatively assigned, and not specifically recognised from the AFM data alone. This type of dissection approach, coupled with labelling methods may provide a route to assigning features observed in intact living cells.

The visualisation of submembraneous cytoskeleton structure makes it possible to image time dependent structural changes and/or the associated changes in mechanical properties of the cell. A number of such studies have been reported including work on skin fibroblasts (Braet *et al* 1998a), cardiocytes (Hofmann *et al* 1997), and MDCK epithelial cells (A-Hassan *et al* 1998). Time lapse images have been obtained showing the disruption of the cytoskeleton network due to the action of drugs such as latrunculin (Fig. 7.13) on skin fibroblasts (Braet *et al* 1998a), colchicine on RBL-2H3 rat leukemic mast cells (Chang *et al* 1993), and cytochalasin B on cardiocytes (Hofmann *et al* 1997). Of perhaps more interest is the observation of events resulting from specific stimulation of cells. AFM images of living unstimulated RBL-2H3 mast cells are featureless. However, if IgE antibodies are bound to the cells and they are then stimulated by the addition of a multivalent antigen, then the cells show a well defined cytoskeleton structure (Chang *et al* 1993). Time dependent changes in quiescent and activated cells can be imaged by AFM (Braunstein and Spudich, 1994). In quiescent cells granular motion along cytoskeleton filaments and cytoplasmic surface waves were imaged. Granular motion has also been observed in R5 cells (Schoenenberger and Hoh, 1994), and cytoplasmic surface waves have been seen in R5 (Schoenenberger and

Hoh, 1994) and lung cells (Kasas *et al* 1993). Following activation it was possible to observe an increase in the number of cell associated granules, granular motion and changes in cytoskeleton at the cell borders (Kasas *et al* 1993).

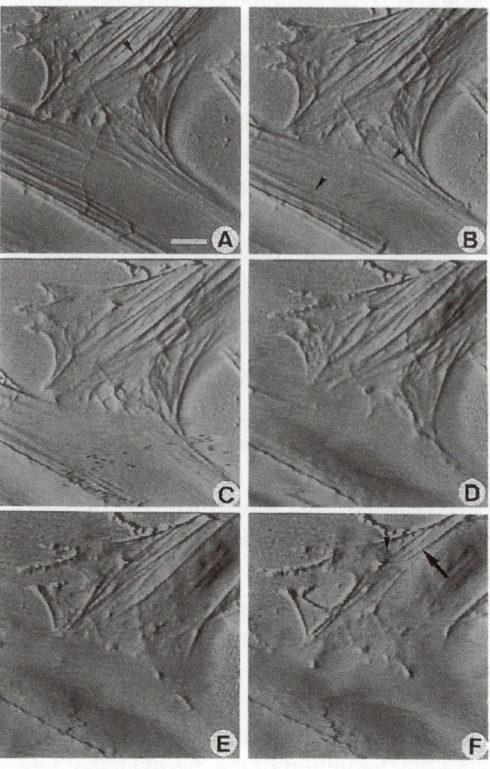

Figure 7.13. Time lapse series of AFM images of latrunculin A induced disruption of microfilaments in skin fibroblasts. (A) Untreated fibroblasts showing a parallel fibre orientation indicated by the black arrowheads. Latrunculin was added and the images B-F were recorded sequentially afterwards. Image acquistion time is about 10 minutes. (B) First signs of microfilament disruption indicated by black arrowheads. (C-E) General loss of latrunculin A sensitive fibres can be seen; peripheral filaments disappear first, followed later by the central filaments. (F) After 45-50 minutes treatment, only a few latrunculin A insensitive fibres remain (indicated by black arrow and arrowhead). Scan bar 10 μm. Data reproduced from Braet *et al* 1998a with permission. Copyright of the Royal Microscopical Society.

The detergent extraction method (Pietrasanka *et al* 1994) was used on fixed dried cells to reveal details of the changes in submembraneous structures between unactivated and activated cells (Kasas *et al* 1993). Such images demonstrated the spreading of the filamentous networks after activation, the fact that granules and/or vesicles sit on top of the cytoskeleton network, and also provided very detailed images of nuclear pore complexes on the nuclear surface of the activated cells (Kasas *et al* 1993). The role played by the softness of the outer cell surface in allowing imaging of intracellular structures is dramatically illustrated in images of ingested latex beads (Beckman *et al* 1994; Braet *et al* 1998a) and zymosan particles (Beckman *et al* 1994) in macrophages after phagocytosis. Once again use of the detergent extraction method enhanced imaging of intracellular components allowing, for example, investigation of particle-filament interactions and the elimination of the role of filaments in phagocytosis (Beckman *et al* 1994).

AFM has been used to image sperm and to investigate the structure and hydration of sperm chromatin. Combined AFM and X-ray microscopy of rat sperm suggested that a significant fraction of the volume of air-dried nucleus must be hydrated (Da Silva *et al* 1992). AFM has been used to compare the volume of bull and mouse sperm heads and amembraneous nuclei in both a fully hydrated and dehydrated state, and to follow the effect on single cells of the addition of increasing concentrations of propanol (Allen *et al* 1996). Previous estimates of volume, and their implications for the packing of sperm DNA, were based on EM data. The AFM studies showed that the EM preparation resulted in substantial dehydration: they suggested that the chromatin is extensively hydrated with water which comprises at least half the volume occupied by the chromatin in the nuclei. AFM has been used to investigate the structural organisation of chromatin and some of this work has already been discussed in section 4.2.3. Studies of marsupial spermatozoa by AFM revealed differences in the organisation of chromatin packaged by histones or protamines (Soon *et al* 1997): the nucleoprotamine particles appearing as tighter bundles than the nucleohistones particles. Fig. 7.14 shows AFM images of ram sperm showing how different imaging methods and analysis can be used to reveal structural detail on living sperm.

7.10. Plant cells

Plant cells were one of the first living systems to be imaged using AFM (Butt *et al* 1990). Small sections from plant leaves were cut out and stuck to a stainless steel disc and then imaged by AFM under water. Images of the underside of leaves of *Lagerstroemia subcostata*, a small Indian tree, showed cellular features. High resolution images failed to reveal structural features below 200 nm in size and this was attributed to the presence of a thick cuticle. The leaves of the water lily *Nymphaea odorata* are believed to possess thinner cuticles and AFM images revealed more detail. In addition to features resembling cells it was possible to identify fibrous structures. At large applied force the sample surface was scraped and damaged but, provided the imaging force was kept below 2 nN, then features down to 12 nm could be resolved (Butt *et al* 1990).

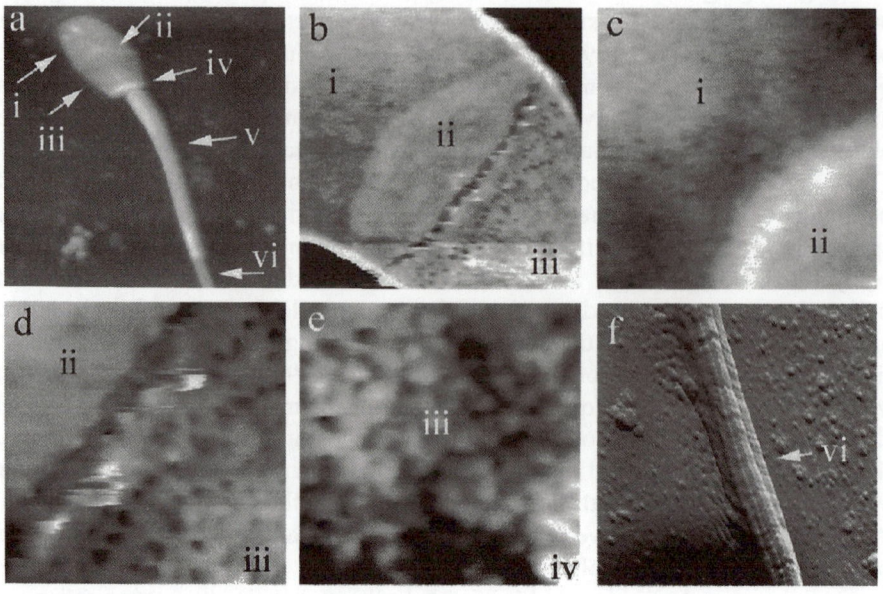

Figure 7.14. AFM images of live ram sperm. The sperm have been treated with sodium azide to reduce their motility. (a) Tapping mode image in air of an intact sperm showing different structural features: (i) acrosome, (ii) equatorial region, (iii) post acrosome, (iv) neck region, (v) mid-piece, and (vi) principal piece. Scan size 20 x 20 μm. (b) Topographical contact mode image of a sperm head in buffer using an elongated tip of height 7 μm. Structural detail has been enhanced by 'unsharp' masking (Adobe Photoshop). Scan size 5 x 5 μm. Contact mode imaging in buffer of regions (c) 'i' & 'ii', scan size 3 x 3 μm, (d) 'ii' & 'iii', scan size 2 x 2 μm, and (e) 'iii' & 'iv', scan size 1.5 μm. (f) Error signal mode image in air of the principal piece of the tail, scan size 20 x 20 μm. Data obtained in collaboration with P James, R Jones (IAH, Babraham UK) and C Wolf (IFR, Norwich UK).

More recently AFM studies have been reported on isolated ivy leaf cuticles (Canet *et al* 1996). The cuticles were extracted enzymatically from plant leaves. Transverse sections were cut with a microtome after embedding in Epon. Images of the inner and outer faces of the isolated cuticles were obtained after binding them down with double sided cellotape. In AFM images of sections the stacked lamella could be seen in the outer lamella zone. The inner reticulate region was largely amorphous with evidence for fibrous inclusions at regions which would have been close to the epidermal cell wall. The outer surface of the cuticle appeared featureless and difficult to image because the probe tip tended to stick to the surface. Low resolution images of the internal surface of the cuticles showed imprints of the epidermal cells surrounded by high walls. Higher resolution images of the imprints revealed a helicoidal stacking of fibrous structures. These fibrous structures were removed by acid hydrolysis suggesting that they are polysaccharide fibres emanating out from the epidermal cell wall into the wax cuticle. The fibrous structures seen on the inner face by AFM are consistent with the fibres seen in transverse sections near the cell wall by AFM and TEM (Canet *et al* 1996).

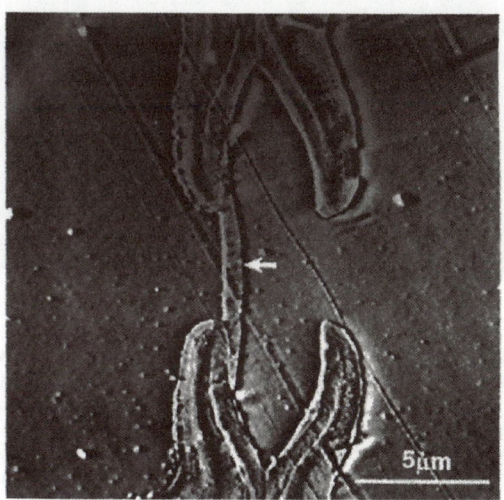

Figure 7.15. AFM deflection mode image showing the surface of a transverse section of black spruce (*Picea mariana*), showing a bordered pit joining a pair of tracheids. The white arrow indicates the torus. It is possible to distinguish the middle lamella, primary cell walls and secondary cell walls. The diagonal lines arise from imperfections in the diamond knife and indicate the knife direction. Data reproduced from Hanley and Gray, 1994 with permission of the authors.

The surfaces of mechanically pulped fibres, and transverse and radial sections of black spruce (*Picea marianna)* wood, have been examined by AFM (Hanley and Gray, 1994). Thin sections were microtomed from Epon embedded pieces of wood. In sections AFM revealed details of the cell wall and of characteristic features such as bordered pits (Fig. 7.15). Within images of sections it is possible to resolve the middle lamella and different regions of the secondary cell wall (Fig. 7.16). These 'textural' differences revealed by AFM are thought to arise on cutting the sections. The orientation of microfibrils within the cell wall, relative to the cut direction, will influence the roughness of the cell wall, and possibly the extent to which it is deformed during scanning, thus influencing the contrast as imaged by AFM. Woody tissues are lignified and phase images of wood pulp fibres reveal bright patches, attributed to residual lignin, which are invisible in the normal topographical images (Hansma *et al* 1997). The lignin is considered to be more hydrophobic than the cellulose thus giving rise to the difference in contrast.

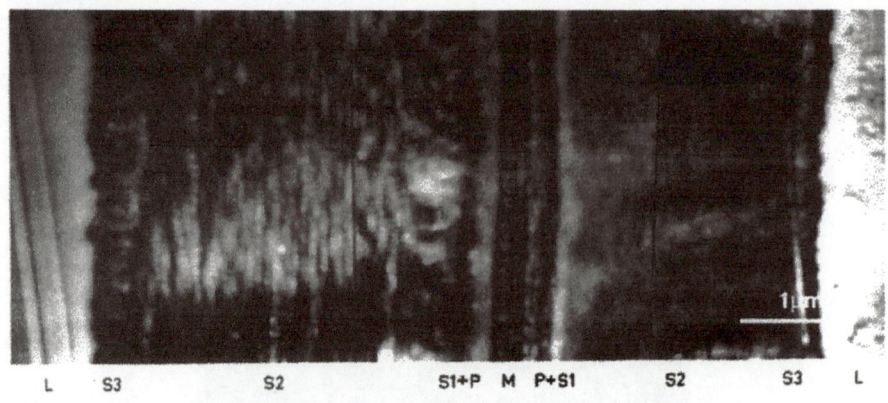

Figure 7.16. A composite of three AFM images of adjacent fibre walls of black spruce. The sections have been taken radially through the tangential wall. The middle lamella (M), primary cell wall (P), and the secondary cell wall (S1, S2 & S3) can be easily distinguished. The resin filled lumen (L) is also shown in the image. The knife marks in the lumen on the left-hand side of the image indicate the knife direction. The differences in contrast between adjacent cell walls arises due to differences in surface roughness on cutting and is attributed to different orientations of microfibrils in the cell wall (e.g. the transition regions S1- S3). Data reproduced from Hanley and Gray, 1994 with permission of the authors.

There are a few studies of pollen grains (Rowley *et al* 1995; van der Wel *et al* 1996; Demanet and Sankar, 1996). In general the images are comparable to SEM data although the sample preparation is easier, and AFM can reveal higher resolution data on exine surface substructure. Rowley and coworkers studied sections cut from grains embedded in resin (Rowley *et al* 1995). Pollen grains from *Kalanchoe blossfeldiana* and *Zea mays* were held in place on the substrate with double-sided cellotape and then imaged in the contact mode. The surface detail seen in error signal mode images (Fig. 7.17) is similar to that seen with field emission scanning electron microscopy (FESEM) (van der Wel *et al* 1996) but higher resolution images could be obtained by AFM: for example, both AFM and FESEM images of the exine surface of *Z mays* pollen grains reveal globular projections interspersed with small holes, but AFM shows substructure of the protrusions not seen by FSEM. Demanet and Sakar used the simplest sample preparation: the pollen grains were deposited directly onto stainless steel discs and then imaged in the non-contact mode (Demanet and Sakar, 1996).

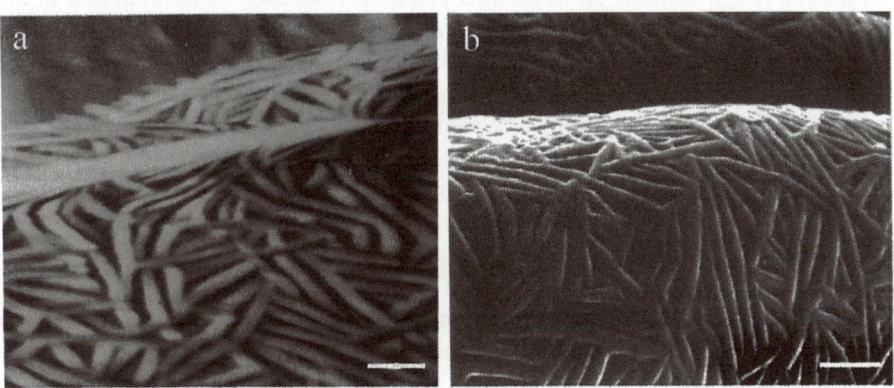

Figure 7.17. Surface structures of fresh *Kalanchoe blossfeldiana* pollen grain. (a) AFM error signal mode image in air, bar marker is 1 μm. (b) Comparative FESEM image of *K blossfeldiana* pollen after coating with Au/Pt, bar marker is 1 μm. Data reproduced from van der Wel *et al* 1996 with permission.

The detailed imaging of isolated plant cell walls and the extracted cellulose microfibres is discussed in detail in section 4.4.4. There is a least one reported studies on plant cell protoplasts (van der Wel *et al* 1996). Images were obtained for fixed samples in air after drying onto glass slides, in water for cells attached to poly-L-lysine coated slides, or for unfixed protoplasts in air, buffer or incubation medium. The best images were obtained for fixed, air-dried protoplasts. No attempt was made to identify features revealed in the images (van der Wel *et al* 1996).

7.11. Tissue

Intact tissue is too large and rough to be directly imaged by AFM. In addition the AFM would only probe the outermost structure of the sample. Standard preparative methods in histology and cytology convert soft biological material into rigid specimens and replicas. Furthermore there are standard techniques for fracturing or sectioning specimens to reveal internal structures. Are there advantages in combining such methods with the use of AFM imaging? AFM should be easier to carry out on such specimens because they are more rigid, and are thus less deformable than untreated biological material. Because the AFM 'feels' the surface structure then, if fracture or sectioning results in excessively rough surfaces, it will not be possible to obtain images at all. Provided the surfaces are not too rough then the AFM offers a potential resolution better than that of the light microscope and the SEM, and comparable with that of the TEM. At the highest magnifications it is possible that the AFM may, in some circumstances, be limited with respect to TEM by surface roughness. However, if sample roughness is small and results from specimen related properties then this may provide unique image contrast in the AFM.

7.11.1. Embedded sections

Early AFM studies of thin sections of embedded biological material revealed the sensitivity of AFM to the roughness of the surface of the cut section (Amako *et al* 1993). In TEM images of thin sections of Epon embedded *Vibrio cholerae*, cut either with a diamond or glass knife, the bacterial cells are clearly visible. Whereas bacterial cells are identifiable in AFM images of diamond cut sections, the additional roughness of the glass cut sections obscures the bacterial cells in the AFM image. This clearly illustrates the different contrast mechanisms, and the additional sensitivity of AFM to sample preparation. Later studies (Yamashina and Shigeno, 1995) confirmed the advantage of the use of a diamond knife but did not support suggestions (Yamamoto and Tashiro, 1994) that embedding in LR white resin was preferable to Epon resin. Tapping mode images of ultrathin sections of tissue samples collected from rats revealed details of cytoplasmic organelles (mitochondria or secretory granules), nuclear components (nucleolus and chromatin), rough endoplasmic recticulum and associated ribosomes, microvilli and cilia, and basal bodies in the tracheal epithelial cell (Yamashino and Shigeno, 1995). Similarly contact mode images of ultrathin sections of wool fibre embedded in LR white resin revealed details of the cellular composition; ortho- and para-cortical cells, cell membrane complexes, macrofibrils and nuclear remnants (Titcombe *et al* 1997). Gross surface roughness will limit use of AFM to the study of thin sections. However, detailed differences in surface roughness related to sample structure may provide novel contrast in the images. This type of effect has already been mentioned in section 7.10. When sections were cut from embedded pieces of wood the orientation of fibres with respect to the cut direction caused changes in surface texture, which manifested themselves in the AFM images (Fig. 7.16) as changes in contrast across sections of cell walls (Hanley and Gray, 1994).

7.11.2. Embedment-free sections

An interesting approach has been the use of AFM to study embedment-free sections of cells and tissue (Ushiki *et al* 1994; 1996). This uses methodology originally introduced by Wolosewick (Wolosewick, 1980; Kondo, 1984). Effectively fixed samples are embedded in polyethylene glycol, solidified with liquid nitrogen, sectioned with a microtome, deposited on poly-L-lysine coated glass slides, dehydrated through a graded ethanol series, and critical point dried. Tapping mode has been used to obtain an impressive range of images of kidney, liver, pancreas and small intestine tissue from mice (Ushiki *et al* 1994). Sections of liver tissue revealed an erythrocyte in a sinusoid, chromatin fibres in the nucleus of a hepatocyte, glycogen granules in hepatocyte cytoplasm, mitochondria in liver hepatocytes and the collecting duct cell of the kidney. Microvilli were clearly resolved in samples of small intestine and higher resolution images suggested globular structures, possibly glycoproteins, on the surface of the microvilli (Fig. 7.18). The embedment free method is regarded as an acceptable preservation technique for electron microscopy and may provide a routine methodology for AFM.

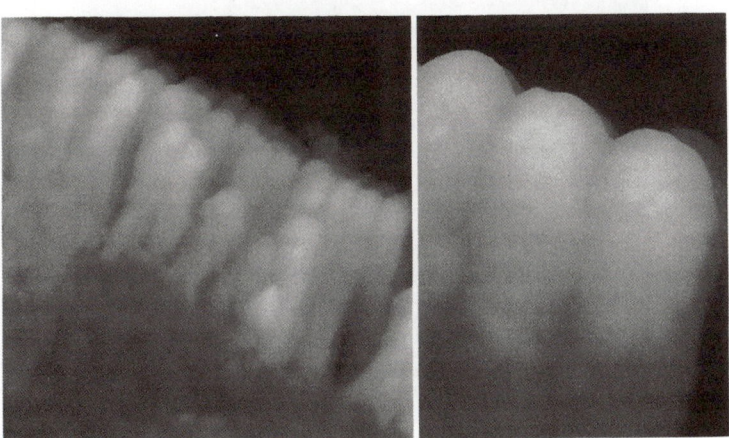

Figure 7.18. AFM images of embedment-free sections of the upper portion of the absorptive cell of the small intestine. (a) Images of microvilli, magnification about 45,000. (b) Higher magnification image showing granular structure which coat the surface of the microvilli, magnification about 192,000. Data reproduced from Ushiki *et al* 1994 with permission.

7.11.3. Hydrated sections

One of the potential benefits of AFM is imaging samples in their native state. An example of the use of freshly cut, wet sections is a study of bovine cornea and sclera (Fullwood *et al* 1994). Sections, 100-200 microns thick, were cut by hand and then glued onto flat steel discs. The samples were imaged in air and, under these conditions, the level of hydration was considered to fall from about 70% to between 20-40%. The major experimental difficulty was locating sufficiently flat regions which remained stable during imaging. However, once such regions were found it proved possible on occasions to resolve features down to 2-3 nm in size. In sections of both the cornea and the sclera the collagen network was visualised, showing details of the periodicity of the collagen fibrils, their surface morphology and high resolution images of cross-bridges believed to involve proteoglycan structures (Fig. 7.19).

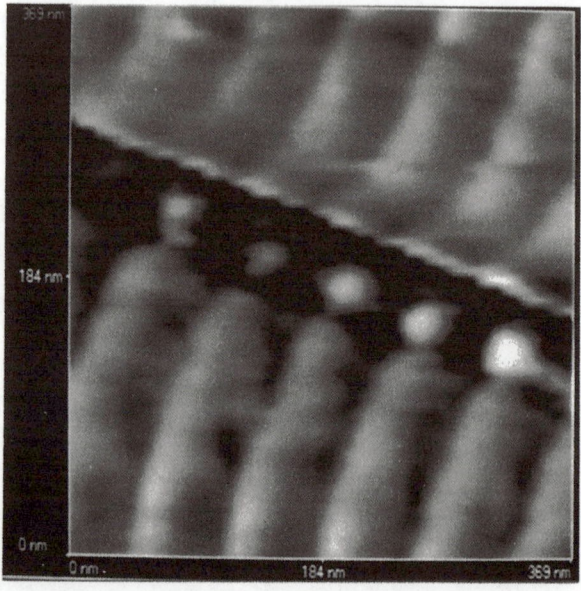

Figure 7.19. AFM error signal mode image of bovine scleral fibres. Cross bridges between the fibrils can clearly be seen. Scan size 369 x 369 nm. Data reproduced from Fullwood *et al* 1994 Current Eye Research 14, 529-535, copyright Swete & Zeitlinger Publishers with permission.

Because the collagen fibres are packed in arrays their measured dimensions are relatively unaffected by probe broadening effects: the measured values correlate better with data from X-ray diffraction rather than TEM, in accordance with evidence from X-ray diffraction that fibre shrinkage only becomes marked when the water content drops below 30%. The resolution achieved by AFM on wet, unstained samples close to their native state approaches that obtainable by conventional TEM studies. Similar high resolution AFM images have been obtained for collagen networks mechanically dissociated from fixed human corneal and scleral tissue (Meller *et al* 1997).

7.11.4. Freeze-fracture replicas

Freeze-fracture replicas have been used successfully to obtain detailed information on cellular and subcellular architecture. Kordylewski and coworkers have reported an AFM study on freeze-fracture replicas of rat atrial tissue (Kordylewski *et al* 1994) which have been well characterised by TEM (Kordylewski *et al* 1993). The AFM can provide direct information on the topography of the replica and both sides of the replica can be imaged. At low magnification the TEM and AFM images reveal similar features. At higher magnification the AFM was able to resolve details not apparent in the TEM data. It was most useful to compare TEM images with error signal mode AFM images, but to use topography images for measurement of dimensions. As with TEM it is possible to generate stereo pairs for viewing the topography of the surface. Features observed by both TEM and AFM included atrial granules, mitochondria with cristae, membrane patches, nuclei and both open and closed pores on the surface of nuclei (Kordylewski *et al* 1994). The major advantages of the use of AFM were direct visualisation of surface geometry, and the ease with which the dimensions of surface features could be determined, although the latter needs to be weighed against the difficulties in correcting for probe broadening effects.

7.11.5. Immunolabelling

A major problem with AFM of cells and tissue is recognition or identification of structural features. Combined AFM and optical microscopy, coupled with specific staining or fluorescent labelling, is a powerful solution to this problem and examples of this approach have already been described. Immunolabelling techniques are well established in both light and electron microscopy. Antibodies have been used to label specific antigens on individual molecules (e.g. DNA section 4.2.4), macromolecular complexes such as chromosomes (section 4.2.5), bacterial biofilms (section 6.2.1), cells (Putman *et al* 1993; Takeuchi *et al* 1998;

Neagu *et al* 1994; Eppell *et al* 1995) or tissue (Saoudi *et al* 1994; Yamashina and Shigeno, 1995). If the surface features are individual molecules, or for flat layered structures, such as bacterial S layers (Ohnesorge *et al* 1992), it may be possible to recognise the antibody-antigen complex directly, although care must be taken to discriminate between specifically bound and passively adsorbed labels. When the surface is rough enhancement of the labels is necessary. Gold labelled antibodies may be used alone or to locate primary antibody-antigen complexes (Mulhern *et al* 1992; Putman *et al* 1993; Saoudi *et al* 1994). The procedure may not be as straightforward for AFM as it is for EM studies. Gold labels may be confused with surface protrusions, or even compressed into the surface by the probe, making them difficult to spot (Mulhern *et al* 1992). Whereas large gold labels may be easy to identify, smaller labels can be difficult to locate by AFM alone (Eppell *et al* 1995). Location of labels can be improved by generating larger particulate deposits: examples used with AFM include silver enhancement (Neagu *et al* 1994; Putman *et al* 1993), peroxidase labelled antibodies and their reaction with DAB (sections 4.2.5 & 4.3), or even fluorescently labelled complexes (McMaster *et al* 1996b,c) (section 4.2.5). With AFM it is possible to modify the label or tip in order to improve sensitivity. Magnetic labels can be detected with magnetic tips and should give enhanced sensitivity. There is at least one report of the use of superparamagnetic beads as labels (Neagu *et al* 1994). Yamashina and Shingo (Yamashina and Shingo, 1995) used Kelvin force probe microscopy to enhance imaging of gold labels: the surface potential is raised in the vicinity of the gold label. Antibody coated tips could be used to locate antigens on sample surfaces. Immunolabelling methods can be used with AFM and there is scope for improving the sensitivity of the technique and in the use of controls.

7.12. Biominerals

Biominerals are probably the biological materials which have been least studied by AFM. These materials are generally complex but the mineral component may mean that they are rigid, possibly well ordered, and hence can be imaged by AFM. A few examples of AFM studies on such materials are collected below.

7.12.1. Bone, tendon and cartilage

The use of AFM to study the structure and assembly of collagen has been discussed in section 4.5.3. In bones the final structure is based on an interaction between collagen and deposited apatite. Combined AFM and TEM studies on mineralised and demineralised tendon have been used to refine composite models for turkey leg tendon. TEM and AFM studies complement each other, with the TEM revealing structure within the specimen, and the AFM showing surface structure (Lees *et al* 1994). AFM Tapping mode studies of *in vitro* and physiologically calcified tendon collagen have suggested that the surface structure of the fibril induces nucleation of apatite crystals, and that their subsequent growth does not markedly alter the fibril structure (Bigi *et al* 1997). Tapping mode AFM on dried collagen fibrils from rat tail tendon have allowed visualisation of proteoglycan bound to the collagen surfaces (Raspanti *et al* 1997). The distribution of the proteoglycans was determined by comparing images obtained for the native structure, samples treated with chondroitinase, and samples incubated with Cupromeronic blue, a dye specifically designed to stabilise the anionic glycosaminoglycan chains. Such studies help build a picture of the ultrastructure of these complex materials. Articular cartilage acts as a low friction bearing in synovial joints. AFM has provided a method for characterising surface and subsurface structures of freshly excised articular tissue in physiologically relevant medium (Jurvelin *et al* 1996). Cartilage discs together with a thin layer of bone were glued at the bony surface to glass coverslips and then imaged under phosphate buffered saline. AFM studies showed the articular surface to be amorphous: blurring of the images was attributed to a high surface viscosity possibly related to its lubrication properties. Surface irregularities, seen previously by SEM, were absent in the AFM images, suggesting that they are preparation artifacts. Small pits were observed but it is not clear whether these are intrinsic features of the native surface or are induced during isolation of the cartilage. Enzymatic digestion of proteoglycan revealed the fibrous substructure. The AFM allowed imaging in physiologically conditions, mechanical measurements to be made on normal and enzymatic digested material, plus measurements of the dimensions and periodicity of the collagen fibrous network. Low resolution AFM topographic and elastic images of bone are similar to those obtained by light and electron microscopy but at higher resolution the AFM revealed dramatic fluctuations of elastic properties over small (*circa* 50 nm) distances (Tao *et al* 1992).

7.12.2. Teeth

A number of studies have demonstrated the value of using AFM to study dentin (Kinney *et al* 1993; 1996; Cassinelli and Morra, 1994; Marshall Jnr *et al* 1993; 1995) and tooth enamel (Sollbohmer *et al* 1995). The dentin body of teeth is protected from abrasive corrosion during eating, and from chemical attack from acidic foods or bacterial metabolites, by an outer coating of enamel composed of about 86% hydroxyl apatite crystals. In order to image the enamel, crowns were extracted, embedded in Epon, and polished to reveal the enamel surface. Using a specially designed cell the enamel could be rapidly exposed to acidic drinks and the AFM used to follow and quantify the etching, or demineralisation as a function of time (Sollbohmer *et al* 1995). Dentine is situated between the pulp and the enamel of the human tooth. It is mainly composed of hydroxyapatite crystals and collagen and has a microtubular structure. Current methods of dentin bonding depend on a demineralisation of the dentin to create a microporous structure which can be penetrated by bonding agents. AFM has been used to observe demineralisation, drying and bonding processes (Kinney *et al* 1993; Cassinelli and Morra, 1994; Marshall Jnr *et al* 1993; 1995). These types of studies are illustrated by investigations of the changes occurring on exposure to a range of conditioning agents (Marshall Jnr *et al* 1995). Discs were cut from sterilised dentin perpendicular to the long axis of the tooth, thus orienting tubules perpendicular to the exposed surface. The surfaces were polished and a gold grid evaporated onto the surface as a reference for monitoring the demineralisation process. Changes in the peritubular and intertubular regions were followed as a function of the exposure time to the conditioning agent (Fig. 7.20).

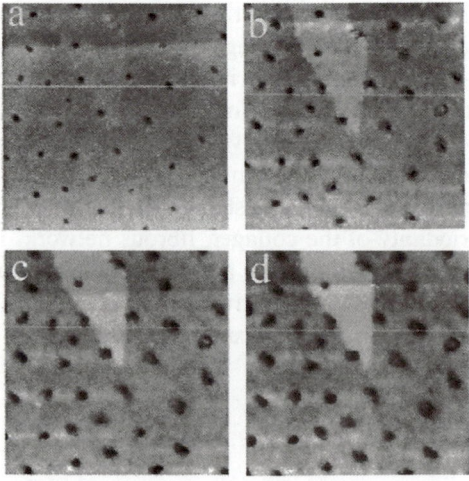

Figure 7.20. AFM studies showing the effect of demineralisation of dentine in dilute phosphoric acid (3 mM). Scan size 40 x 40 μm. Imaged after polishing in deionised water at time intervals (a) 0 s, (b) 30 s, (c) 60 s and (d) 80 s. The upper portion of each image shows a gold reference island. Data reproduced from Marshall Jnr *et al* 1995 AFM of conditioning agents on dentine, J. Biomed. Mat. Res. copyright John Wiley 1995 with permission of the authors and John Wiley & Sons, Inc.

By replacing the normal silicon nitride tip-cantilever assembly with a diamond tip and stainless steel cantilever it has been possible to produce topographic and elasticity maps across the peritubular and intertubular regions of cut faces of dentin specimens (Kinney *et al* 1996). Transport of ions through the dentine tubules is believed to be an important mechanism for nerve stimulation in dentine hypersensitivity. Nughes and Denuault (Nughes and Denuault, 1996) have used a scanning electrochemical microscope to image ion fluxes through the tubules and to identify and monitor the effects of blocked pores. It is suggested that such investigations could help in the assessment of treatments for dentin hypersensitivity (Nughes and Denuault, 1996).

7.12.3. Shells

Shells are one of the few examples of biological materials on which it is possible to resolve detail at the atomic level (Friedbacker *et al* 1991; Manne *et al* 1994). Atomic scale resolution was achieved on pismo (*Tivela stultorum*) and sea urchin (*Stronglyocentrotus purpuratus*) shells, after powdering them and compressing them into KBr discs before imaging with AFM (Friedbacker *et al* 1991). Diatoms are unicellular algae which are frequently used to check the resolution of objectives for light microscopy. The shells are silicon based and coated with layers of polysaccharides, proteins and lipids. An investigation of a range of species of diatom surfaces by AFM showed characteristic pores and ripples, and the raphe and their central nodes were identified (Linder *et al* 1992). AFM has been used to image the matrix proteins of oyster (*Crassostrea virginica*) shells (Donachy *et al* 1992) and to study the distribution of different sized cuticular pigment granules in relation to coloured patches on the surface of Quail egg shells (Makita *et al* 1993). Nacre, the pearly layer of mollusc shells, consists of layers of crystalline argonite tablets. AFM has been used to examine the different distributions of aragonite tablets in a bivalve (*Atrina* sp.) and a gastropod (*Haliotis rufescens*). A feature of the studies was that by imaging in liquids it was possible to follow the dissolution or demineralisation of the nacre: by stripping the structure away layer by layer the structure of individual tablets and their inter-relationship with vertically adjacent tablets was revealed (Manne *et al* 1994).

References: selected reviews

Butt, H-J, Wolff, E.K, Gould, S.A.C, Northern, B.D, Peterson, C.M. and Hansma, P.K. (1990). Imaging cells with the atomic force microscope. *J. Structural Biol.* **105**, 54-61.

Firtel, M. and Beveridge, T.J. (1995). Scanning probe microscopy in microbiology. *Micron* **26**, 347-362.

Hansma, H.G, Kim, K.J, Laney, D.E, Garcia, R.A, Argaman, M, Allen, M.J. and Parsons, S.M. (1997). Properties of biomolecules measured from atomic force microscopic images: a review. *J. Structural Biol.* **119**, 99-108.

Henderson, E. (1994). Imaging of living cells by atomic force microscopy. *Progr. Surface Sci.* **46**, 39-60.

Ohnesorge, F.M, Horber, J.K.H, Haberle, W, Czerny, C-P, Smith, D.P.E. and Binnig, G. (1997). AFM review study on pox viruses and living cells. *Biophys. J.* **73**, 2183-2194.

Shao, Z, Mou, J, Czajkowsky, D.M, Yang, J. and Yuan, J-Y. (1996). Biological atomic force microscopy: what is achieved and what is needed. *Advances Phys.* **45**, 1-86.

Selected references

A-Hassan, E, Heinz, W.F, Antonio, M.D, D'Costa, N.P, Nageswaran, S, Schoenenberger, C-A. and Hoh, J.H. (1998). Relative microelastic mapping of living cells by atomic force microscopy. *Biophys. J.* 74, 1564-1578.

Allen, M.J, Lee, C, Lee, J.D, Pogany, G.C, Balloch, M, Siekhaus, W.J. and Balhorn, R. (1993). Atomic force microscopy of mammalian sperm chromatin. *Chromossoma* **102**, 623-630.

Allen, M.J, Bradbury, E.M. and Balhorn, R. (1995). The natural subcellular surface structure of the bovine sperm cell. *J. Structural Biol.* **114**, 197-208.

Allen, M.J, Lee IV, J.D, Lee, C. and Balhorn, R. (1996). Extent of sperm chromatin hydration determined by atomic force microscopy. *Mol. Reproduction & Development* **45**, 87-92.

Almqvist, N, Backman, L. and Fredriksson, S. (1994). Imaging human erythrocyte spectrin with atomic force microscopy. *Micron* **25**, 227-232.

Amako, K, Takade, A, Umeda, A. and Yoshida, M. (1993). Imaging of the surface structures of Epon thin sections created with a glass knife and a diamond knife by the atomic force microscope. *J. Electron Microscopy* **42**, 121-123.

Antonik, M.D, D'Costa, N.P. and Hoh, J.H. (1997). A biosensor based on micromechanical interrogation of living cells. *IEEE Eng. Medicine Biol.* **March/April**, 66-72.

Barbee, K.A, Davies, P.F. and Lal, R. (1994). Shear stress-induced reorganisation of the surface topography of living endothelial cells imaged by atomic force microscopy. *Circulation Res.* **74**, 163-171.

Barbee, K.A. (1995). Changes in surface topography in endothelial monolayers with time at confluence: influence on subcellular shear stress distribution due to flow. *Biochem. Cell Biol.* **73**, 501-505.

Beckman, M, Kolb, H-A. and Lang, F. (1994). Atomic force microscopy of peritoneal macrophages after particle phagocytosis. *J. Membrane Biol.* **140**, 197-204.

Beech, I.B. (1996). The potential use of atomic force microscopy for studying corrosion of metals in the presence of bacterial biofilms- an overview. *International Biodeterioration & Biodegradation* **37**, 141-149.

Beech, I.B, Cheung, C.W.S, Johnson, D.B. and Smith, J.R. (1996). Comparative studies of bacterial biofilms on steel surfaces using atomic force microscopy and environmental scanning electron microscopy. *Biofouling* **10**, 65-77.

Bigi, A, Gandolfi, M, Roveri, N. and Valdre, G. (1997). *In vitro* calcified tendon collagen: an atomic force and scanning electron microscopy investigation. *Biomaterials* **18**, 657-665.

Braet, F, Kalle, W.H.J, de Zanger, R.B, de Grooth, B.G, Raap, A.K, Tanke, H.J. and Wisse, E. (1996a). Comparative atomic force and scanning electron microscopy: an investigation on fenestrated endothelial cells in vitro. *J. Microscopy* **181**, 10-17.

Braet, F, de Zanger, R, Kalle, W, Raap, A, Tanke, H. and Wisse, E. (1996b). Comparative scanning, transmission and atomic force microscopy of the microtubular cytoskeleton in fenestrated liver endothelial cells. *Scanning Microscopy Supplement* **10**, 225-236.

Braet, F, de Zanger, R. and Wisse, E. (1997a). Drying cells for SEM, AFM and TEM by hexamethyldisilazane: a study on hepatic endothelial cells. *J. Microscopy* **186**, 84-87.

Braet, F, de Zanger, R, Kammer, S. and Wisse, E. (1997b). Noncontact versus contact imaging: an atomic force microscopic study on hepatic endothelial cells in vitro. *Internal. J. Imaging Syst. Technol.* **8**, 162-167.

Braet, F, Seynaeve, C, de Zanger, R. and Wisse, E. (1998a). Imaging surface and submembranous structures with the atomic force microscope: a study on living cancer cells, fibroblasts and macrophages. *J. Microscopy* **190**, 328-338.

Braet, F, Rotsch, C, Wisse, E. and Radmacher, M. (1998b). Comparison of fixed and living liver endothelial cells by atomic force microscopy. *Appl. Phys. A* **66**, S575-S578.

Braga, P.C. and Ricci, D. (1998). Atomic force microscopy: application to investigation of *Escherichia coli* morphology before and after exposure to cefodizime. *Antimicrobial Agents & Chemotherapy* **42**, 18-22.

Bremer, P.J, Geesey, G.G. and Drake, B. (1992). Atomic force microscopy examination of the topography of a hydrated bacterial biofilm on a copper surface. *Current Microbiology* **24**, 223-230.

Braunstein, D. and Spudich, A. (1994). Structure and activation dynamics of RBL-2H3 cells observed with scanning force microscopy. *Biophys. J.* **66**, 1717-1725.

Brown, H.G. and Hoh, J.H. (1997). Entropic exclusion by neurofilament sidearms: a mechanism for maintaining neurofilament spacing. *Biochemistry* **36**, 15035-15040.

Butt, H-J, Wolff, E.K, Gould, S.A.C, Northern, B.D, Peterson, C.M. and Hansma, P.K. (1990). Imaging cells with the atomic force microscope. *J. Structural Biol.* **105**, 54-61.

Butt, H-J. (1992). Measuring local surface charge densities in electrolyte solutions with a scanning force microscope. *Biophys. J.* **63**, 578-582.

Canet, D, Rohr, R, Chamel, A. and Guillian, F. (1996). Atomic force microscopy study of isolated ivy leaf cuticles observed directly and after embedding in Epon. *New Phytol.* **134**, 571-577.

Cassinelli, C. and Morra, M. (1994). Atomic force microscopy studies of the interaction of a dentin adhesive with tooth hard tissue. *J. Biomed. Mater. Res.* **28**, 1427-1431.

Chang, L, Kious, T, Yorgancioglu, M, Keller, D. and Pfeiffer, J. (1993). Cytoskeleton of living, unstained cells imaged by scanning force microscopy. *Biophys. J.* **64**, 1282-1286.

Da Silva, L.B, Trebes, J.E, Balhorn, R, Mrowka, S, Anderson, E, Attwood, D.T, Barbee Jnr, T.W, Brase, J, Corzett, M, Gray, J, Koch, J.A, Lee, C, Kern, D, London, R.A, MacGowan, B.J, Matthews, D.L. and Stone, G. X. (1992). X-ray laser microscopy of rat sperm nuclei. *Science* **258**, 269-271.

Demanet, C.M. and Sankar, K.V. (1996). Atomic force microscopy images of a pollen grain: a preliminary study. *S. African. J. Bot.* **62**, 221-223.

Domke, J, Parak, W.J, George, M, Gaub, H.E. and Radmacher, M. (1999). Mapping the mechanical pulse of single cardiomyocytes with the atomic force microscope. *European Biophys. J. Biophys. Letts.* **28**, 179-186.

Donachy, J.E, Drake, B. and Sikes, C.S. (1992). Sequence and atomic-force microscopy analysis of a matrix protein from the shell of the oyster *Crassostrea virginica*. *Marine Biol.* **114**, 423-428.

Eppell, S.J, Simmons, S.R, Albrecht, R.M. and Marchant, R.E. (1995). Cell-surface receptors and proteins on platelet membranes imaged by scanning force microscopy using immunogold contrast enhancement. *Biophys. J.* **68**, 671-680.

Friedbacher, G, Hansma, P.K, Ramli, E. and Stucky, G.D. (1991), Imaging powders with the atomic force microscope: from biominerals to commercial materials. *Science* **253**, 1261-1263.

Fritz, M, Radmacher, M. and Gaub, H.E. (1993). *In vitro* activation of human platelets triggered and probed by atomic force microscopy. *Experimental Cell Res.* **205**, 187-190.

Fritz, M, Radmacher, M. and Gaub, H.E. (1994). Granule motion and membrane spreading during activation of human platelets imaged by atomic force microscopy. *Biophys. J.* **66**, 1328-1334.

Fullwood, N.J, Hammiche, A, Pollock, H.M, Hourston, D.J. and Song, M. (1994). Atomic force microscopy of the cornea and sclera. *Current Eye Res.* **14**, 529-535.

Gab, M. and Ikai, A. (1996). Method for immobilising microbial cells on gel surface for dynamic AFM studies. *Biophys. J.* **69**, 2226-2233.

Gad, M, Itoh, A. and Ikai, A. (1997). Mapping cell wall polysaccharides of living microbial cells using atomic force microscopy. *Cell Biol. Internat.* **21**, 697-706.

Garcia, C.R.S, Takeuschi, M, Yoshioka, K. and Miyamoto, H. (1997). Imaging *Plasmodium falciparum* -infected ghost and parasite by atomic force microscopy. *J. Structural Biol.* **119**, 92-98.

Geesey, G.G, Mittelman, M.W, Iwaoka, T. and Griffiths, P.R. (1986). Role of bacterial exopolymers in the deterioration of metallic copper surfaces. *Materials Perform.* **25**, 37-40.

Giebel, K-F, Bechinger, C, Herminghaus, S, Riedel, M, Leiderer, P, Weiland, U. and Bastmeyer, M. (1999). Imaging of cell/substrate contacts of living cells with surface plasmon resonance microscopy. *Biophys. J.* **76**, 509-516.

Gould, S.A.C, Drake, B, Prater, C.B, Weisenhorn, A.L, Manne, S, Hansma, H.G, Hansma, P.K, Massie, J, Longmire, M, Elings, V, Northern, B.D, Mukergee, B, Peterson, C.M, Stoeckenius, W, Albrecht, T.R. and Quate, C.F. (1990) From atoms to integrated circuit chips, blood cells, and bacteria with the atomic force microscope. *J. Vac. Sci. & Technol. A* **8**, 369-373.

Gunning, P.A, Kirby, A.R, Parker, M.L, Gunning, A.P. and Morris, V.J. (1996). Comparative imaging of *Pseudomonas putida* bacterial biofilms by Scanning Electron Microscopy and both dc contact and ac non-contact Atomic Force Microscopy. *J. Appl. Bact.* **81**, 276-282.

Haberle, W, Horber, J.K.H. and Binnig, G. (1991). Force microscopy on living cells. *J. Vac. Sci. & Technol. B* **9**, 1210-1213.

Haberle, W, Horber, J.K.H, Ohnesorge, F, Smith, D.P.E. and Binnig, G. (1992). In situ investigations of single living cells infected by viruses. *Ultramicroscopy* **42-44**, 1161-1167.

Han, W, Mou, J, Sheng, J, Yang, J. and Shao, Z. (1995). Cryo atomic force microscopy: a new approach for biological imaging at high resolution. *Biochemistry* **34**, 8215-8220.

Hanley, S.J. and Gray, D.G. (1994). Atomic force microscope images of black spruce wood sections and pulp fibres. *Horzforschung* **48**, 29-34.

Hansma, H.G, Kim, K.J, Laney, D.E, Garcia, R.A, Argaman, M, Allen, M.J. and Parsons, S.M. (1997). Properties of biomolecules measured from atomic force microscopic images: a review. *J. Structural Biol.* **119**, 99-108.

Haydon, P.G, Henderson, E. and Stanley, E.F. (1994). Localization of individual calcium channels at the release face of a presynaptic nerve terminal. *Neuron* **13**, 1275-1280.

Haydon, P.G, Lartius, R, Parpura, V. and Marchese-Ragona, S.P. (1996). Membrane deformation of living glial cells using atomic force microscopy. *J. Microscopy* **182**, 114-120.

Heinz, W.F. and Hoh, J.H. (1999a). Spatially resolved force spectroscopy of biological surfaces using the atomic force microscope. *Trends Biotechnol.* **17**, 143-150.

Heinz, W.F. and Hoh, J.H. (1999b). Relative surface charge mapping with the atomic force microscope. *Biophys. J.* **76**, 528-538.

Henderson, E, Haydon, P.G. and Sakaguchi, D.S. (1992). Actin dynamics in living glial cells imaged by atomic force microscopy. *Science* **257**, 1944-1946.

Henderson, E. (1994). Imaging of living cells by atomic force microscopy. *Progr. Surface Sci.* **46**, 39-60.

Henderson, R.M, Schneider, S, Li, Q, Hornby, D, White, S.J. and Oberleithner, H. (1996a). Atomic force microscopy used to image the inwardly-rectifying ATP-sensitive potassium channel protein, ROMK 1. *Kidney Internal.* **50**, 1780.

Henderson, R.M, Schneider, S, Li, Q, Hornby, D, White, S.J. and Oberleithner, H. (1996b). Imaging ROMK 1 inwardly rectifying ATP-sensitive K+ channel protein using atomic force microscopy. *Proc. Natl. Acad. Sci. USA* **93**, 8756-8760.

Hertz, H. (1882). Uber die Beruhrung fester elastischer Korper. *J. Reine Angew Math.* **92**, 156-171.

Hofmann, U.G, Rotsch, C, Parak, W.J. and Radmacher, M. (1997). Investigating the cytoskeleton of chicken cardiocytes with the atomic force microscope. *J. Structural Biol.* **119**, 84-91.

Hoh, J.H. and Schoenenberger, C-A. (1994). Surface morphology and mechanical properties of MDCK monolayers by atomic force microscopy. *J. Cell Sci.* **107**, 1105-1114.

Hoh, J.H, Heinz, W.F. and A-Hassan, E. (1997). Force volume. *Digital Instruments Support Note No. 240*, Digital Instruments, 112 Robin Hill Road, Santa Barbara, CA 93117, California, USA.

Holstein, T.W, Benoit, M, von Herder, G, Wanner, G, David, C.N. and Gaub, H.E. (1994). Fibrous mini-collagens in hydra nematocysts. *Science* **265**, 402-404.

Horber, J.K.H, Haberle, W, Ohnesorge, F, Binnig, G, Liebich, H.G, Czerny, C.P, Mahnel, H. and Mayr, A. (1992). Investigations of living cells in the nanometer regime with the atomic force microscope. *Scanning Microscopy* **6**, 919-930.

Jaschke, M, Butt, H-J. and Wolff, E.K. (1994). Imaging flageila of Halobacteria by atomic force microscopy. *Analyst* **119**, 1943-1946.

Jolley, J.G, Geesey, G.G, Hankins, M.R, Wright, R.B. and Wichlacz, P.L. (1988). Auger electron spectroscopy and X-ray photoelectron spectroscopy of the biocorrosion of copper by gum arabic, BCS and *Pseudomonas atlantica* exopolymer. *J. Surface & Interface Anal.* **11**, 371-376.

Jurvelin, J.S, Muller, D.J, Wong, M, Studer, D, Engel, A. and Hunziker, E.B. (1996). Surface and subsurface morphology of bovine humeral articular cartilage as assessed by atomic force and transmission microscopy. *J. Structural Biol.* **117**, 45-54.

Kasas, S, Gotzos, V,. and Celio, M.R. (1993). Observation of living cells using atomic force microscopy. *Biophys. J.* **64**, 539-544.

Kasas, S, Fellay, B. and Cargnello, R. (1994). Observation of the action of penicillin on Bacillus-subtilis using atomic force microscopy-techniques for the preparation of bacteria. *Surface & Interface Anal.* **21**, 400-401.

Kasas, S. and Ikai, A. (1996). A method for anchoring round shaped cells for atomic force microscope imaging. *Biophys. J.* **68**, 1678-1680.

Kinney, J.H, Balooch, M, Marshall Jnr., G.W. and Marshall S.J. (1993). Atomic force microscope study of dimensional changes in dentine during drying. *Arch. Oral Biol.* **38**, 1003-1007.

Kinney, J.H, Balooch, M, Marshall, S.J, Marshall Jnr, G.W. and Weihs, T.P. (1996). Atomic force microscope measurements of the hardness and elasticity of peritubular and intertubular human dentin. *J. Biomechanical Eng.* 118, *133-135.*

Kirby, A.R, Fyfe, D.J, Parker, M.L, Gunning, A.P, Gunning, P.A. and Morris, V.J. (1998). Structural studies on human coleretal adenocarcinoma HT29 cells by atomic force microscopy, transmission electron microscopy and scanning electron microscopy. *Probe Microscopy* **1**, 153-162.

Kondo, H. (1984). Polyethylene glycol (PEG) embedding and subsequent de-embedding as a method for the structural and immunocytochemical examination of biological specimens by electron microscopy. *J. Electron Microsc. Tech.* **1**, 227-241.

Kordylewski, L, Goings, G. E. and Page, E. Rat atrial myocyte plasmalemmal caveolae in situ: reversible experimental increases in caveolar size and in surface density of caveolar necks. *Circ. Res.* **73**, 135-146.

Kordylewski, L, Saner, D. and Lal, R. (1994). Atomic force microscopy of freeze-fracture replicas of rat atrial tissue. *J. Microscopy* **173**, 173-181.

Laney, D.E, Garcia, R.A, Parsons, S.M. and Hansma, H.G. (1997). Changes in the elastic properties of cholinergic synaptic vesicles as measured by atomic force microscopy. *Biophys. J.* **72**, 806-813.

Le Grimellec, C, Lesniewska, E, Cachia, C, Schreiber, J.P, de Fornel, F. and Goudonnet, J.P. (1994). Imaging of the membrane surface of MDCK cells by atomic force microscopy. *Biophys. J.* **67**, 36-41.

Le Grimellac, C, Lesniewska, E, Giocondi, M-C, Cachia, C, Schreiber, J.P. and Goudonnet, J.P. (1995). Imaging the cytoplasmic leaflet of the plasma membrane by atomic force microscopy. *Scanning Microscopy* **9**, 401-411.

Le Grimellac, C, Lesniewska, E, Giocondi, M-C, Finot, E. and Goudonnet, J.P. (1997). Simultaneous imaging of the surface and submembraneous cytoskeleton in living cells by tapping mode atomic force microscopy. *C.R. Acad. Sci. Paris, Life Sci.* **320**, 637-643.

Lees, S, Prostak, K.S, Ingle, V.K. and Kjoller, K. (1994). The loci of mineral in turkey leg tendon as seen by atomic force microscope and electron microscopy. *Calcif. Tissue Intern.* **55**, 180-189.

Lesniewska, E, Giocondi, M-C, Vie, V, Finot, E, Goudonnet, J-P. and Le Grimellec, C. (1998). Atomic force microscopy of renal cells: limits and prospects. *Kidney Internal.* **53**, (suppl. 65), S42-S48.

Linder, A, Colchero, J, Apell, H-J, Marti, O. and Mlynek, J. (1992). Scanning force microscopy of diatom shells. *Ultramicroscopy* **42-44**, 329-332.

Ludwig, M, Dettmann, W. and Gaub, H.E. (1997). Atomic force microscope imaging contrast based on molecular recognition. *Biophys. J.* **72**, 445-448.

Makita, T, Ohoue, M, Yamoto, T. and Hakoi, K. (1993). Atomic force microscopy (AFM) of the cuticular pigment globules of the Quail egg shell. *J. Electron Microscopy* **42**, 189-192.

Manne, S, Zaremba, C.M, Giles, R, Huggins, L, Walters, D.A, Becher, A, Morse, D.E, Stucky, G.D, Didymus, J.M, Mann, S. and Hansma, P.K. (1994). Atomic force microscopy of the nacreous layer in mollusc shells. *Proc. Royal Soc. London B* **256**, 17-23.

Margulis, L, Ashen, J.B, Sole, M. and Guerrrero, R. (1993). "Composite, large spirochetes from microbial mats-spirochete structure review". *Proc. Natl. Acad. Sci. USA* **90**, 6966-6970.

Marshall Jnr., G.W, Balooch, M, Tench, R, Kinney, J.H. and Marshall, S.J. (1993). Atomic force microscopy of acid effects on dentin. *Dent. Mater.* **9**, 265-268.

Marshall Jnr., G.W, Balooch, M, Kinney, J.H. and Marshall, S.J. (1995). Atomic force microscopy of conditioning agents on dentin. *J. Biomedical Materials Res.* **29**, 1381-1387.

Maurice, P, Forsythe, J, Hersman, L. and Sposito, G. (1996). Application of atomic-force microscopy to studies of microbial interactions with hydrous Fe(III)-oxides. *Chemical Geology* **132**, 33-43.

McMaster, T.J, Miles, M.J. and Walsby, A.E. (1996a). Direct observation of protein secondary structure in gas vesicles by atomic force microscopy. *Biophys. J.* 70, 2432-2436.

McMaster, T.J, Winfield, M.O, Baker, A.A, Karp, A. and Miles, M.J. (1996b). Chromosome classification by atomic force microscopy volume measurement. *J. Vac. Sci. & Technol. B* **14**, 1438-1443.

McMaster, T.J, Winfield, M.O, Karp, A. and Miles, M.J. (1996c). Analysis of cereal chromosomes by atomic force microscopy. *Genome* **39**, 439-444.

Mulhern, P.J, Blackford, B.L, Jericho, M.H, Southam. G. and Beveridge, T.J. (1992). AFM and STM studies of the interaction of antibodies with S-layer sheath of the archaeobacterium *Methanospiririllum hungatei*. *Ultramicroscopy* **42**, 1214-1224.

Meller, D, Peters, K. and Meller, K. (1997). Human cornea and sclera studied by atomic force microscopy. *Cell Tissue Res.* **288**, 111-118.

Müller, D.J, Schabert, F.A, Buldt, G. and Engel, A. (1995). Imaging purple membranes in aqueous solutions at subnanometer resolution by atomic force microscopy. *Biophys. J.* **68**, 1681-1686.

Neagu, C, van der Werf, K.O, Putman, C.A. J, Kraan, Y.M, de Grooth, B.G, van Hulst, N.F. and Greve, J. (1994). Analysis of immunolabeled cells by atomic force microscopy, optical microscopy, and flow cytometry. *J. Structural Biol.* **112**, 32-40.

Nughes, S. and Denuault, G. (1996). Scanning electrochemical microscopy: amperometric probing of diffusional ion fluxes through porous membranes and human dentine. *J. Electroanaytical Chem.* **408**, 125-140.

Oberleithner, H, Giebisch, G. and Geibel, J. (1993). Imaging the lamellipodium of migrating epithelial cells in vivo by atomic force microscopy. *Pflugers Arch.* **425**, 506-510.

Oberleithner, H, Schwab, A, Wang, W, Giebisch, G, Hume, F. and Geibel, J. (1994). Living renal epithelial cells imaged by atomic force microscopy. *Nephron* **66**, 8-13.

Oberleithner, H, Brinkmann, E, Giebisch, G. and Geibel, J. (1995). Visualizing life on biomembranes by atomic force microscopy. *Kidney International* **48**, 923-929.

Oberleithner, H, Schneider, S, Larmer, J. and Henderson, R.M. (1996). Viewing the renal epithelium with the atomic force microscope. *Kidney & Blood Pressure Res.* **19**, 142-147.

Oberleithner, H, Geibel, J, Guggino, W, Henderson, R.M, Hunter, M, Schneider, S.W, Schwab, A. and Wang, W.H. (1997a). Life on biomembranes viewed with the atomic force microscope. *Wiener Klinische Wochenschrift* **109**, 419-423.

Oberleithner, H, Schneider, S.W. and Henderson, R.M. (1997b). Structural activity of a cloned potassium channel (ROMK 1) monitored with the atomic force microscope: the "molecular sandwich" technique. *Proc. Natl. Acad. Sci. USA* **94**, 14144-14149.

Ohnesorge, F, Heckl, W.M, Harberle, W, Pum, D, Sara, M, SchindlerH, Schilcher, K, Kiener, A, Smith, D.P.E, Sleytr, U.B. and Binnig, G. (1992). Scanning force microscopy studies of the S-layers from *Bacillus coagulans* E38-66, *Bacillus sphaericus* CCM2177 and of an antibody binding process. *Ultramicroscopy* **42-44**, 1236-1242.

Ohnesorge, F.M, Horber, J.K.H, Haberle, W, Czerny, C-P, Smith, D.P.E. and Binnig, G. (1997). AFM review study on pox viruses and living cells. *Biophys. J.* **73**, 2183-2194.

Parker, M.L, Brocklehurst, T.F, Gunning, P.A, Coleman, H.P. and Robins, M.M. (1995). Growth of food-borne pathogenic bacteria in oil-water emulsions: I Methods for investigating the form of growth. *J. Appl. Bacteriol.* **78**, 601-608.

Parpura, V, Haydon, P.G. and Henderson, E. (1993a). Three-dimensional imaging of living neurons and glial with the atomic force microscope. *J. Cell Sci.* **104**, 427-432.

Parpura, V, Haydon, P.G, Sakaguchi, D.S. and Henderson, E. (1993b). Atomic force microscopy and manipulation of living glial cells. *J. Vac. Sci. & Technol. A* **11**, 773-775.

Papura, V, Doyle, R.T, Basarsky, T.A, Henderson, E. and Haydon, P.G. (1995). Dynamic imaging of purified individual synaptic vesicles. *Neuroimag.* **2**, 3-7.

Pietrasanta, L.I, Schaer, A. and Jovin, T.M. (1994). Imaging subcellular structures of rat mammary carcinoma cells by scanning force microscopy. *J. Cell Sci.* **107**, 2427-2437.

Prater, C.B, Weisenhorn, A.L, Northern, B.D, Peterson, C.M, Gould, S.A.C. and Hansma, P.K. (1990). Imaging molecules and cells with the atomic force microscope. *Proc. XIIth International Congress for Electron Microscopy,* San Francisco, USA, San Francisco Press, Inc., pp 254-255.

Putman, C.A.J, van der Werf, K.O, de Grooth, B.G, van Hulst, N.F, Segerink, F.B. and Greve, J. (1992). Atomic force microscope with integrated optical microscope for biological applications. *Rev. Sci. Instrum.* **63**, 1914-1917.

Putman, C.A. J, de Grooth, B.G, Hansma, P.K. and van de Hulst, N.F. (1993). Immunogold labels: cell surface markers in atomic force microscopy, *Ultramicroscopy* **48**, 177-182.

Putman, C.A.J, van der Werf, K.O, de Grooth, B.G, van Hulst, N.F. and Greve, J. (1994). Viscoelasticity of living cells allows high resolution imaging by tapping mode atomic force microscopy. *Biophys. J.* **67**, 1749-1753.

Radmacher, M, Fritz, M. and Hansma, P.K. (1995). Imaging soft samples with the atomic force microscope: gelatin in water and propanol. *Biophys. J.* **69**, 264-270.

Radmacher, M, Fritz, M, Kacher, C.M, Cleveland, J.P. and Hansma, P.K. (1996). Measuring the viscoelastic properties of human platelets with the atomic force microscope. *Biophys. J.* **70**, 556-567.

Raspanti, M, Alessandrini, A, Ottani, V. and Ruggeri, A. (1997). Direct visualisation of collagen-bound proteoglycans by Tapping-mode atomic force microscopy. *J. Structural Biol.* **119**, 118-122.

Rotsch, C. and Radmacher, M. (1997). Mapping local electrostatic forces with the atomic force microscope. *Langmuir* **13**, 2825-2832.

Rowley, J.R, Flynn, J.J. and Takahashi, M. (1995). Atomic force microscopic information on pollen exine substructure. *Nuphar. Bot. Acta* **108**, 300-308.

Rajyaguru, J.M, Kado, M, Richardson, M.C. and Musznski, M.J. (1997). X-ray micrography and imaging of *Escherichia coli* cell shape using laser plasma pulsed point X-ray sources. *Biophys. J.* **72**, 1521-1526.

Schaus, S.S. and Henderson, E.R. (1997). Cell viability and probe-cell interactions of XR1 glial cells imaged by atomic force microscopy. *Biophys. J.* **73**, 1205-1214.

Schoenenberger, C-A. and Hoh, J.H. (1994). Slow cellular dynamics in MDCK and R5 cells monitored by time-lapse atomic force microscopy. *Biophys. J.* **67**, 929-936.

Siedlecki, C.A. and Marchant, R.E. (1998). Atomic force microscopy for characterization of the biomaterial interface. *Biomaterials* **19**, 441-454.

Schabert, F.A, Hefti, A, Goldie, K, Stemmer, A, Engel, A, Meyer, E, Overney, R. and Guntherdodt, H.J. (1992). Ambient pressure scanning probe microscopy of 2D regular protein arrays. *Ultramicroscopy* **42-44**, 1118-1124.

de Souza Pereira, R, Parizotto, N.A. and Baranauskas, V. (1994). Observation of Baker's Yeast strains used in biotransformation by atomic force microscopy. *Appl. Biochem. Biotechnol.* **59**, 135-143.

Schabert, F.A. and Engel, A. (1994). Reproducible acquisition of *Escherichia coli* porin surface topographs by atomic force microscopy. *Biophys. J.* **67**, 2394-2403.

Schabert, F.A, Henn, C. and Engel, A. (1995). Native *Escherichia coli* ompF porin surfaces probed by atomic force microscopy. *Science* **268**, 92-94.

Shroff, S.G, Saner, D.R. and Lal, R. (1995). Dynamic micromechanical properties of cultured rat atrial myocytes measured by atomic force microscopy. *Am. J. Physiol.* **269** (*Cell Physiol.* **38**) C286-C292.

Sollbohmer, O, May, K-P. and Anders, M. (1995). Force microscopical investigation of human teeth in liquids. *Thin Solid Films* **264**, 176-183.

Soon, L.L.L, Bottema, C. and Breed, W.G. (1997). Atomic force microscopy and cytochemistry of chromatin from marsupial spermatoza with special reference to *Sminthopsis crassicaudata*. *Mol. Reproduction & Development* **48**, 367-374.

Southam, F, Firtel, M, Blackford, B.L, Jericho, M.H, Xu, W, Mulhern, P.J. and Beveridge, T.J. (1993). Transmission electron microscopy, scanning tunneling microscopy, and atomic force microscopy of the cell-envelope layers of the archaeobacterium. *Methanospirillium hungatei* GP1. *J. Bacteriol.* **175**, 1946-1955.

Steele, A, Goddard, D.T. and Beech, I.B. (1994). An atomic force microscopy study of the biodeterioration of stainless steel in the presence of bacterial biofilms. *International Biodeterioration & Biodegradation* **341**, 34-46.

Surman, S.B, Walker, J.T, Goddard, D.T, Morton, L.H.G, Keevil, C.W, Weaver, W, Skinner, A, Hanson, K, Caldwell, D. and Kurtz, J. (1996). Comparison of microscope techniques for the examination of biofilms. *J. Microbiol. Methods* **25**, 57-70.

Takeout, M, Miyamoto, H, Sako, Y, Komizu, H. and Kusumi, A. (1998). Structure of the erythrocyte membrane skeleton as observed by atomic force microscopy. *Biophys. J.* **74**, 2171-2183.

Tao, N.J, Lindsay, S.M. and Lees, S. (1992). Measuring the microelastic properties of biological material. *Biophys. J.* **63**, 1165-1169.

Titcombe, L.A, Huson, M.G. and Turner, P.S. (1997). Imaging the internal cellular structure of Merino wool fibres using atomic force microscopy. *Micron* **28**, 69-71.

Ushiki, T, Shigeno, M. and Abe, K. (1994). Atomic force microscopy of embedment-free sections of cells and tissue. *Arch. Histol. Cytol.* **57**, 427-432.

Ushiki, T, Hitomi, J, Ogura, S, Umemoto, T. and Shigeno, M. (1996). Atomic force microscopy in histology and cytology. *Arch. Histol. Cytol.* **59**, 421-431.

Vinckier, A, Heyvaert, I, D'Hoore, A, van Haesendonck, C, Engelborghs, Y. and Hellemans, L. (1995). Immobilising and imaging microtubules by atomic force microscopy. *Ultramicroscopy* **57**, 337-343.

Van der Wel, N.H, Putman, C.A.J, Van Noort, S.J.T, de Grooth, B.G. and Emons, A.M.C. (1996). Atomic force microscopy of pollen grains, cellulose microfibrils, and protoplasts. *Protoplasma* **194**, 29-39.

Willemson, O.H, Snel, M.M.E, van der Werf, K.O, de Grooth, B.G, Greve, J, Hinterdorfer, P, Gruber, H.J, Schindler, H, van Kooyk, Y. and Figdor, C.G. (1998). Simultaneous height and adhesion imaging of antibody-antigen interactions by atomic force microscopy. *Biophys. J.* **75**, 2220-2228.

Wolosewick, J.J. (1980). The application of polyethylene glycol (PEG) to electron microscopy. *J. Cell Biol.* **86**, 675-681.

Xu, W, Mulhern, P.J, Blackford, B.L, Jericho, M.H, Firtel, M. and Beveridge, T.J. (1996). Modeling and measuring the elastic properties of an archaeal surface, the sheath of *Methanospirillum hungatei*, and the implication for methane production. *J. Bacteriol.* **178**, 3106-3112.

Yamashina, S. and Shigeno, M. (1995). Application of atomic force microscopy to ultrastructural and histochemical studies of fixed and embedded cells. *J. Electron Microscopy* **44**, 462-466.

Yamamoto, A. and Tashiro, Y. (1994). Visualisation by an atomic force microscope of the surface of ultrathin sections of rat kidney and liver cells embedded in LR white. *J. Histochem. Cytochem.* **42**, 1463-1470.

Yuan, Y. and Lenhoff, A.M. (1999). Characterisation of phase separation in mixed surfactant films by liquid tapping mode atomic force microscopy. *Langmuir* **15**, 3021-3025.

Zachee, P, Boogaerts, M, Hellemans, L. and Snauwaert, J. (1992). Adverse role of the spleen in hereditary spherocytosis: evidence by the use of the atomic force microscope. *Brit. J. Haematology* **80**, 264-265.

Zachee, P, Boogaerts, M, Snauwaert, J. and Hellemans, L. (1994). Imaging uremic red blood cells with the atomic force microscope. *Amer. J. Nephrology* **14**, 197-200.

Zachee, P, Snauwaert, J, Vandenberghe, P, Hellemans, L. and Boogaerts. (1996). *Brit. J. Haematology* **95**, 472-481.

Zhang, P-C, Bai, C, Huang, Y-M, Zhao, H, Fang, Y, Wang, N-X. and Li, Q. (1995). Atomic force microscopy study of fine structures of the entire surface of red blood cells. *Scanning Microscopy* **9**, 981-988.

Zhang P-C, Bai, C, Cheng, Y, Fang, Y, Liming, F. and Pan, H. (1996). Direct observation of uncoated spectrin with the atomic force microscope. *Science in China Ser. B* **39**, 378-385.

Zhang, Y, Sheng, S.J. and Shao, Z. (1996). Imaging biological structures with the cryo atomic force microscope. *Biophys. J.* **71**, 2168-2176.

CHAPTER 8

OTHER PROBE MICROSCOPES

8.1. Overview

There is a wide variety of other types of probe microscope. They principally differ in the type of surface interaction detected, and therefore the type of sensor used. In most other respects the instruments are similar - for example, they all use piezoelectric devices to initiate tip/sample movement and consequently all utilise high voltage amplifiers. Even the instrument software contains many common elements. The following list gives a flavour of the instruments available, although there are almost certainly others, particularly as prototypes.

Scanning Tunnelling Microscope (STM)
Photon Scanning Tunnelling Microscope (PSTM)
Scanning Capacitance Microscope (SCAM)
Atomic Force Microscope (AFM)
Magnetic Force Microscope (MFM)
Photonic Force Microscope (PFM)
Scanning Near-Field Optical Microscope (SNOM)
Scanning Near-Field Infrared Microscope (SNIM)
Scanning Near-Field Microwave Microscope (SNMM)
Scanning Near-Field Acoustic Microscope (SNAM)
Scanning Near-Field Thermal Microscope (SNTM)
Scanning Plasmon Near-Field Microscope (SPNM)
Scanning Ion Conductance Microscope (SICM)
Scanning Electrochemical Microscope (SECM)
Scanning Thermal Microscope (SThM)

Only some of the instrument types are relevant to biological experiments, with the more promising ones discussed below.

8.2. Scanning tunnelling microscope (STM)

Although this book deals almost exclusively with AFM, it was not the first type of probe microscope. That accolade goes to the STM, which was designed to examine conducting surfaces, particularly semiconductor materials such as silicon under ultra high vacuum (UHV), Fig 8.1. It was such a significant achievement (Binnig *et al*, 1982) that it won its creators, Gerd Binnig and Heinrich Rohrer, the Nobel Prize in 1986. The STM uses a sharp conducting tip which is either mechanically cut or electrochemically etched from metals such as gold, tungsten, or platinum-iridium alloy.

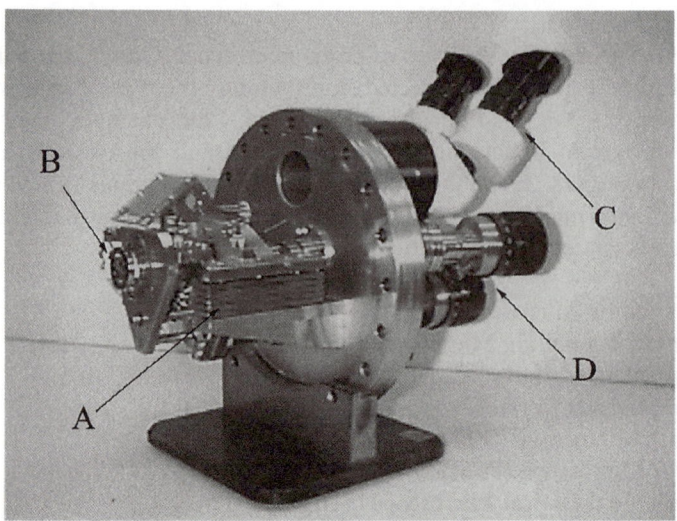

Figure 8.1. A UHV STM head manufactured by WA Technology Ltd, Cambridge, UK. The circular flange enables the head to be bolted onto other UHV apparatus. The stack of horizontal plates (A) is used to provide vibration isolation where the damping is provided by viton rubber beads positioned between each plate. The sample holder (B) secures the sample vertically, with the piezoelectric tube sited horizontally. Alignment of the tip and sample can be observed by the optical stage (C) - it is not, as some visitors to our lab believe, "......for observing the atoms"! Tip/sample separation is effected by the coarse and fine micrometers (D).

A small 'bias' voltage (V_b) of about 1 V is applied between the tip and sample, with the sample usually, but not exclusively, being at a positive potential, Fig 8.2. When the separation between the tip and sample is reduced to about 1 nm, a quantum mechanical effect occurs and a 'tunnelling' current (I_t) is generated and flows between them. This tunnelling current is extremely small - only a few nanoamps, yet despite its size, it is easily detected using a low noise amplifier. The interesting feature of the tunnelling current, is that it decays exponentially with increasing distance between the tip and the sample i.e. its magnitude changes hugely with variation in tip-sample separation. In fact, if the tip-sample separation is reduced by a mere 1 Å, the tunnel current can increase by about an order of magnitude. This makes it ideal for detecting small features on a surface, and it is so sensitive that it can easily resolve atomic lattices. This highlights an important difference when compared with AFM, namely the detection mechanism in STM has an exponential dependence, whereas in AFM it is linear. This is why STM is intrinsically more sensitive than AFM. Interestingly, the extreme vertical sensitivity of the STM dictates that only the atom at the very apex of the tip participates significantly in the tunnelling process. Therefore, an STM tip behaves as though it were atomically sharp - quite some improvement over an AFM tip. Although all these factors add up to what seems like a considerable advantage, the STM has actually been rendered virtually obsolete for biological samples because it requires a conducting substrate. Despite this, it is useful to be familiar with the technique because it set the foundations on which all modern probe microscopes are built.

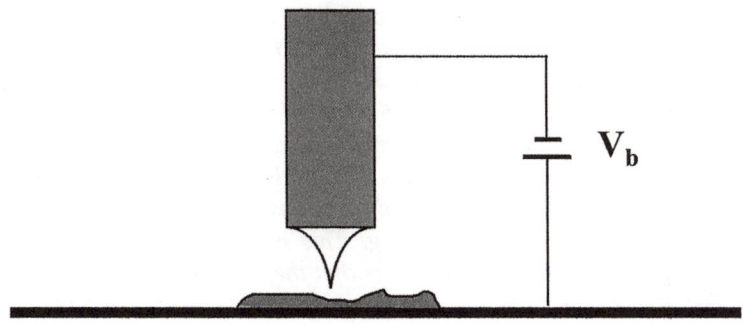

Figure 8.2. Schematic diagram showing the small separation between tip and sample. The bias voltage (V_b) is necessary to initiate the tunnelling current (I_t).

Graphite was a popular choice of substrate in bio-STM, but it is unfortunately hydrophobic and therefore repels any biological samples containing water, leading to poor binding of samples to the substrate. This drastically reduces the probability of locating (biological) samples because they are easily swept aside by the tip, particularly if adhesive forces are present, and as the tip-sample separation in STM is small, there is a significant Van der Waals attractive force between the two. Unfortunately as the control loop monitors the tunnelling current and not the force, it can become surprisingly large, thus leading to movement or damage of the sample. Some of the earliest bio-STM papers contain what are now recognised as artifactual images that actually show details of flaws in the substrate. Another limitation with bio-STM is that as the tunnelling current decays extremely rapidly away from the sample, it quickly becomes difficult to detect - the signal gets lost in the electronic noise. This sets a ridiculously low maximum height on any potential sample, and realistically restricts bio-STM to flat, film-like materials or individual molecules.

Perhaps the most common mode of operating the STM is called 'constant current' as the control loop moves the conducting tip up and down, relative to the sample, in order to maintain a constant tunnelling current. This is analogous to dc mode in AFM where the sample is imaged under a constant force.

Q: *"If the STM maps the variation in tunnel current over the sample, just what does the final image represent? Is it topographical data?"*

A: *The final image is only an approximation of the surface topography. More precisely, the STM records the number of full or empty electron states, commonly known as the 'local density of states' (LDOS). The size of the bias voltage determines exactly which states participate in the tunnelling process, and therefore influences the image contrast. Whilst for metallic conductors the final image is closely related to the topography, semiconductors, and particularly biological materials, may generate an image that is substantially unrelated to the topography. Additionally, the presence of delocalised electrons, such as those found in benzene rings, can promote enhanced tunnelling. This results in the area of the image appearing brighter, which falsely suggests that the area is taller than it really is.*

The tunnelling phenomenon that occurs between an STM tip and a conducting or semiconducting sample is reasonably well understood. However, biological samples are generally very poor conductors, so exactly why these materials should allow the passage of a tunnelling current is unclear. Conducting and semiconducting samples can be imaged reasonably routinely with STM. However, biological samples give a much poorer success rate which could be due to either sample displacement, as outlined earlier, or that the samples will not always sustain a measurable tunnel current. Whatever the reason, users of bio-STM will no doubt be familiar with the "Death or Glory" scenario, where the sample images extremely well or not at all!

8.3. Scanning near-field optical microscope (SNOM)

The performance of conventional optical microscopes is limited by lens aberrations and diffraction effects. The 'Abbé diffraction limit', states that the maximum achievable resolution is approximately equal to half the wavelength of the light source. Green light, in the middle of the visible spectrum, has a wavelength of 550 nm. Therefore, the theoretical spatial resolution is 275 nm (0.275 µm), however this assumes the optics are faultless and that 100 % of the light is perfectly in focus, which is clearly unrealistic. The practical resolution limit for a high quality instrument is in fact closer to 400 nm (0.4 µm). With SNOM (also known as NSOM), a laser acts as a single wavelength light source, which is transmitted through a fine optical fibre onto the sample. To circumvent diffraction problems, the SNOM uses a piezoelectric tube to position the optical fibre in the 'near-field', i.e. so close to the sample that it is illuminated before diffraction causes the light to spread out. As a rule of thumb, the separation between the sample and the optical fibre should be less than about one third of the diameter of the fibre. This ensures that you really are operating in the near-field (Fig. 8.3.).

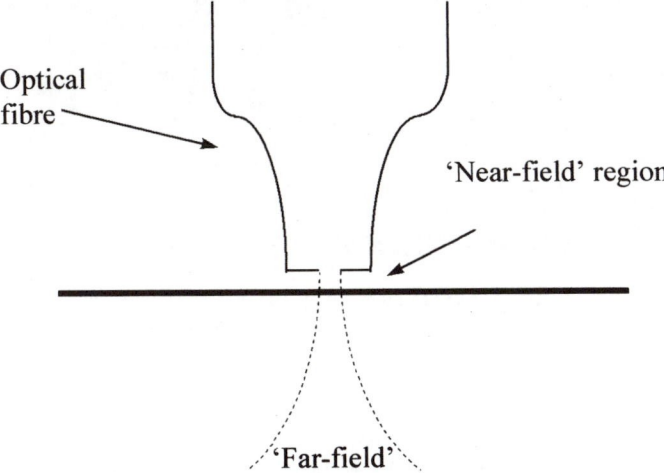

Figure 8.3. By positioning the optical fibre in the 'near-field' region the light interacts with the sample before it diffracts. The resolution of the SNOM is therefore no longer diffraction limited but depends on just how small the aperture at the end of the optical fibre can be manufactured

The optical fibres are effectively sharpened by heating then drawing-out to a point of around 50 nm in diameter. Unfortunately this is some 10 times less than the wavelength of the light source, which causes the laser light to exit not just from the very tip of the optical fibre, but also from the sides. In order to avoid this, the exterior of the fibre is coated with a thin layer of metal, usually aluminium, and consequently light only escapes through the very end of the fibre. When operating in the near field regime the resolution of the instrument is approximately equal to the aperture diameter, and is largely independent of wavelength of the illumination.

This equates to an improvement in resolution by an order of magnitude over a conventional high quality optical instrument. In the future, and with new developments, the aperture size of the optical fibre could be reduced in an effort to improve the resolution even further.

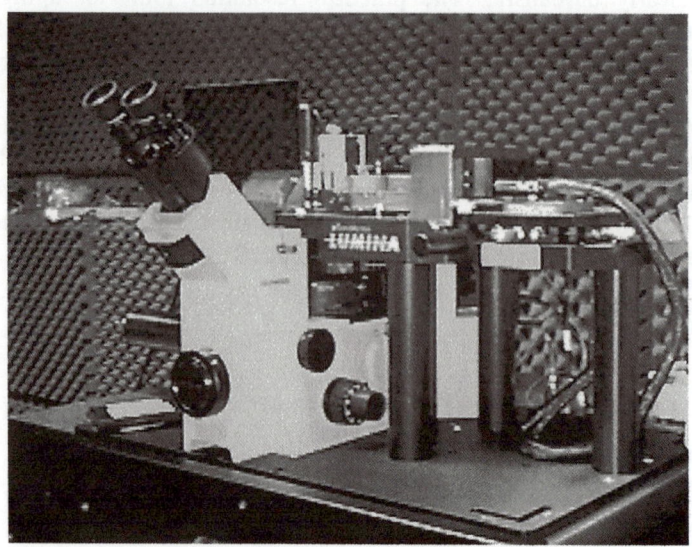

Figure 8.4. A combined SNOM/AFM/STM produced by Thermomicroscopes, California, USA. The SPM stage is mounted onto an inverted microscope which permits both sample observation and the use of fluorescent markers.

The laser light illuminates the sample via the optical fibre, and is consequently scattered then collected by an objective lens, in what is known as the 'far-field'. The objective lens can either be in the reflection or the transmission position, with the latter generally being preferred for semi-transparent biological samples. The light intensity collected by the objective lens provides the information used to construct the optical image. Obviously it is necessary, as with other probe microscopes, to utilise a control loop to keep the optical fibre at the correct height above the sample. In order to provide the control loop with a suitable input signal, a technique known as 'shear-force feedback' is often used. This is similar to non-contact ac mode in AFM, except with SNOM the vertical optical fibre is oscillated laterally - as it approaches the sample the amplitude of oscillation is reduced due to the influence of the attractive Van der Waals force. This information can be used to record the topography of the surface, so that optical and topographical images of the same region can be compared. Since the optical fibre is not in contact with the sample it is free to resonate to any disturbance, therefore a quiet operating environment, with some degree of acoustic isolation is essential. It is also possible to label areas of interest on the sample with fluorescent markers, which are then excited by the laser light and recorded as bright areas. Of course, the laser must emit light of the correct wavelength in order for the marker to fluoresce.

8.4. Scanning ion conductance microscope (SICM)

This instrument, with a sensor rather like a micro-pH probe, was developed by Hansma and co-workers (1989) in order to image insulating materials that were immersed under various electrolytes. A micropipette containing an electrode and filled with electrolyte, is positioned close to the sample, Fig. 8.5. This restricts the flow of ions through the aperture at the end of the micropipette. Consequently, the electrical conductance between the electrode in the micropipette, and a reference electrode situated in the bulk of the electrolyte, decreases.

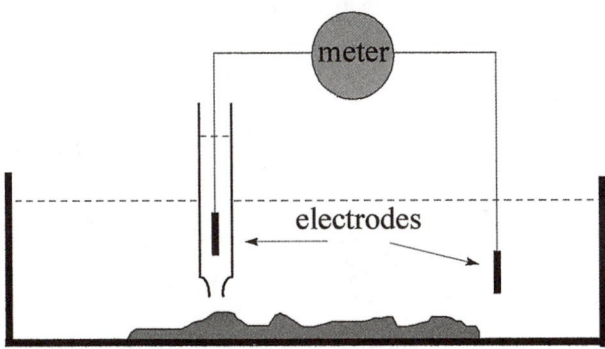

Figure 8.5. The electrical conductance between the two electrodes decreases as the micropipette closely approaches the sample. This is because the flow of ions through the tiny aperture is restricted.

To obtain topographical information, the micropipette is scanned over the sample using a piezoelectric tube, whilst the control loop adjusts its vertical position in order to maintain a constant electrical conductance. Alternatively, it is possible to record local variations in ion-flow (i.e. ionic current) by scanning the micropipette over the sample at a constant height. The ionic current can be monitored using a patch-clamp amplifier. This can be used to provide electrophysiological data on biological systems, for example the ionic currents through pores in the surfaces of membranes. Perhaps unsurprisingly, the achievable resolution is governed by just how small the aperture at the end of the pipette can be manufactured - typical values are similar to those for SNOM optical fibres at about 50 nm in diameter.

8.5. Scanning thermal microscope (SThM)

This type of instrument allows the user to visualise a sample in terms of its thermal properties in addition to its topography. The thermal sensor, first described by Dinwiddie and co-workers (Dinwiddie *et al*, 1994), appears similar to an AFM cantilever but is in fact manufactured from 'Wollaston wire' - fine platinum wire encased in a sheath of silver. At the very end of the sensor the silver is etched away with acid to reveal the platinum core, and this portion is angled downwards to form a tip. The tip of the thermal sensor (Fig. 8.6.) is in contact with the sample during a scan and the topography can be recorded with the aid of a reflective surface attached to legs of the sensor. The sensor acts as both a local resistive heater and as a thermometer. Since the platinum core is only around 5 μm in diameter, the bulk of the probe's electrical resistance is at its tip - the silver coated part has a diameter of about 75 μm.

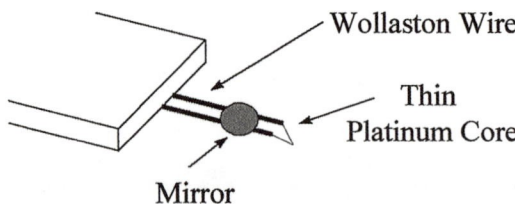

Figure 8.6. The Wollaston wire is etched away to reveal its core of platinum, which then forms the tip of the thermal sensor. The platinum core is extremely fine so the tip is the only part of the sensor with a significant electrical resistance.

This gives the advantages of providing a small localised heat source, and also minimises any thermal drift caused by temperature changes in the flexible legs of the probe. Despite their relatively crude appearance, sensors can be manufactured with force constants of about 1 nN. It is also possible to fabricate silicon nitride AFM cantilevers where a thermistor is integrated into the pyramidal stylus. With the application of a small dc electrical signal, the tip of the sensor heats up and some of this heat is transferred to the sample. The imaging technique is based on keeping the sensor tip at a constant temperature by altering the applied electrical signal as necessary during a scan. It is then possible to determine the electrical power required to maintain a certain temperature across different regions of the sample, and therefore obtain an image illustrating the relative variation in thermal conductivity. In addition, it is possible to add an ac component to the electrical signal so that a short pulse of heat is applied to the surface, usually at a frequency of a few kHz. This provides information on the 'thermal diffusivity' i.e. how the heat penetrates into the sample at various depths. The low mass of the probe ensures a rapid response to changes in thermal properties.

The beauty of this technique is that all three types of data, topography, thermal conductivity and thermal diffusivity, can be acquired simultaneously over the same area. If the tip is held in a fixed x-y position, it is possible to explore the effects of localised sample heating by recording the sensor's z-displacement. This can provide thermomechanical information, such as melting and expansion.

8.6. Optical tweezers and the photonic force microscope (PFM)

A remarkable device (Ashkin *et al*, 1985; 1986; 1987a,b.) lies at the heart of this instrument which can be known by a variety of names; either as an 'optical trap', 'optical tweezers', or even 'laser tweezers'. It is so important and powerful in it's own right that it worthwhile understanding how it operates.

Biological materials are almost invariably low density and therefore mainly semi-transparent. In the presence of light, small particulate biomaterials such as cells and bacteria can act as miniature lenses, causing the light to 'refract' i.e. to bend or deflect. If the light is particularly intense, say from a laser, it can lead to the generation of a sideways force on the illuminated object. This seems rather bizarre as there is no obvious physical interaction to cause this force, much less a force that is actually strong enough to move the object around. To understand how this force arises first refer to Fig. 8.7.

As the laser beam is refracted by the lens-like behaviour of the cell, it gains a little momentum in the positive x-direction. As a fundamental physical quantity, momentum is said to be 'conserved' i.e. the total momentum of the whole system always remains constant. The only way that this can remain true is if the cell gains a equal amount of momentum in the opposite (negative x) direction - hence the cell moves. For simplicity Fig. 8.7. only shows one possible light path. For a broad and unfocussed laser beam, wider than the target cell, it seems reasonable to expect that the individual forces arising from all the various light paths would cancel out so the cell would remain stationary. However, in reality even a laser beam is not homogeneous since the intensity at the centre of the beam is somewhat higher than that at its edge. Fig. 8.8. shows two light paths A and B through a cell. The variation in laser intensity with distance from the optical axis is illustrated by the graduated shading, and the beam is shown travelling from left to right. Path A involves lower intensity light than path B, therefore the momentum gain and subsequent force on the cell is lower for A than for B. This means that the cell will always move to where the light intensity is highest, in this case towards the optical axis. Once at the optical axis the opposing forces arising from the two light paths exactly cancel each other out, and the cell remains trapped at the centre of the beam. The result of a seemingly continuous steam of photons striking the left hand side of the cell, is the generation of a force gently pushing the cell to the right. This force, known as 'radiation pressure', is only significant for intense

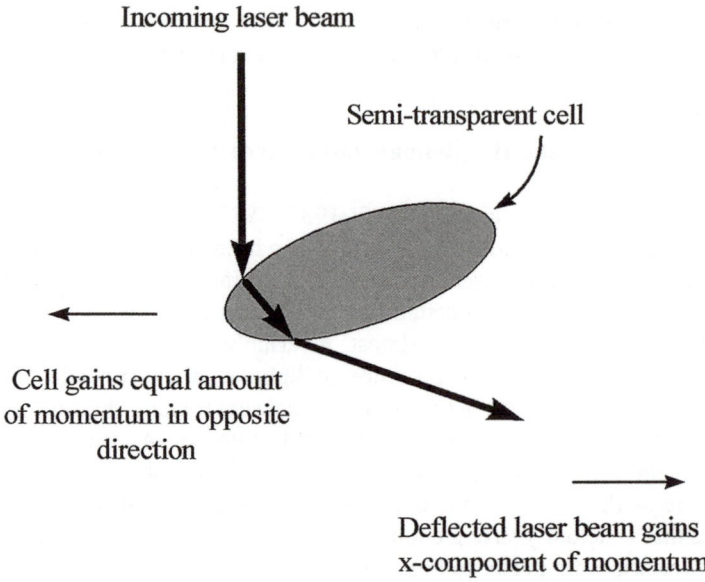

Figure 8.7. The law of conservation of momentum dictates that any momentum gained from the deflection of the laser beam must be exactly balanced by the cell gaining momentum in the opposite direction.

sources of illumination. However when a lens is put into a parallel beam, an extremely concentrated region of light is generated at the focal point. Any small object within a few microns of the focal point will be drawn towards it since that is where the light intensity is highest. With well focussed light the radiation pressure is overcome by the force pushing the object towards the focal point. This means that, depending on its starting position, the object can even travel towards the light source! Once the object arrives at the focal point it remains trapped there and can be positioned by moving the laser beam.

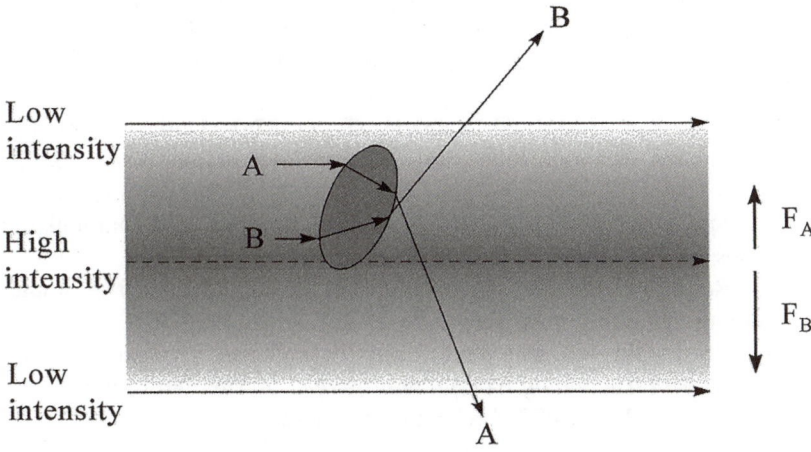

Figure 8.8. The very centre of the laser beam is more intense than at its edge. Therefore, the refracted beam (A) gains less momentum than refracted beam (B). Consequently the balancing force (F_A) on the cell as a result of beam (A) is smaller than the force (F_B) due to beam (B), and the net force pushes the cell towards the optical axis.

The first probe microscope based on optical tweezers was described by Ghislain and Webb (Ghislain and Webb, 1993). A glass shard was scanned over the sample, with scattered laser light being used to determine the displacement of the shard due to sample topography. More recently Florin and co-workers (Florin *et al*, 1997) developed another variant of the instrument also using optical tweezers. In this case it is a fluorescently labelled latex bead that is captured at the focus with optical tweezers and employed as the imaging tip. However, the laser also doubles as an excitation source for the fluorescent label. Consequently, when the latex bead is displaced by features on the surface of the sample, it moves relative to the optical focus and the fluorescence intensity decreases - providing an elegant means of recording the sample topography.

References

Ashkin, A. and Dziedzic, J.M. (1985). Observation of radiation-pressure of particles by alternating light beams. *Phys. Rev. Letts.* **54**, 1245-1248.

Ashkin, A. Dziedzic, J.M. Bjorkholm, J.E. and Chu, S. (1986). Observation of a single-beam gradient force optical trap for dielectric particles. *Optics Letts.* **11**, 288-290.

Ashkin, A. Dziedzic, J.M. and Yamane, T. (1987a). Optical trapping and manipulation of single cells using infrared laser beams. *Nature* **330**, 769-771.

Ashkin, A. and Dziedzic, J.M. (1987b). Optical trapping and manipulation of viruses and bacteria. *Science* **235**, 1517-1520.

Binnig, G. Rohrer, H. Gerber, C. and Weibel, E. (1982). Surface studies by scanning tunnelling microscopy. *Phys. Rev. Letts.* **49**, 57-61.

Dinwiddie, R.B. Pylkki, R.J. and West, P.E. (1994). *Thermal conductivity contrast imaging with a scanning thermal microscope.* In Thermal Conductivity (ed. T.W. Tong,) **22**, 668-677. Technomics: Lancaster PA.

Florin, E-L. Pralle, A. Hörber, J.K.H. and Stelzer, E.H.K. (1997). Photonic force microscope based on optical tweezers and two-photon excitation for biological applications. *J. Structural Biol.* **119**, 202-211.

Ghislain, L.P. and Webb, W.W. (1993). Scanning-force microscope based on an optical trap. *Optics Letts.* **18**, 1678-1680.

Hansma, P.K. Drake, B. Marti, O. Gould. A.C. and Prater, C.B. (1989). The scanning ion-conductance microscope. *Science* **243**, 641-643.

SPM BOOKS

Bai. C. (1995). *Scanning Tunnelling Microscopy and its application.* Springer-Verlag.

Behm, R.J. *et al.* (1990). *Scanning tunnelling microscopy and related methods; proceedings.* Kluwer Academic Publishers.

Bonnell. D.A. (1993) *Scanning tunnelling microscopy and spectroscopy: Theory, techniques and applications.* VCH.

Chen. C.J. (1993). *Introduction to scanning tunnelling microscopy.* Oxford University Press (NY).

Cohen, S.H. *et al.* (1995). *Atomic Force Microscopy/Scanning Tunnelling Microscopy; Proceedings.* Plenum.

Colton, R.J. *et al.* (1998). *Procedures in scanning probe microscopies.* Wiley

Guntherodt, H-J., (1995). Forces in scanning probe methods: Proceedings of the NATO advanced study. Kluwer Academic Publishers.

Guntherodt, H-J., and Weisendanger, R. (1995). Scanning tunnelling microscopy; Further applications and related scanning techniques. Springer-Verlag

Guntherodt, H-J., and Weisendanger, R. (1994). Scanning tunnelling microscopy; General principles and applications to adsorbate covered surfaces. Springer-Verlag

Guntherodt, H-J., and Weisendanger, R. (1993). Scanning tunnelling microscopy; Theory of STM and related scanning probe methods. Springer-Verlag.

Marti, O., and Amrein, M. (1993). *STM and SFM in biology.* Academic Press.

Magnov, S.N., Whangbo, M-H. (1995). *Surface Analysis with STM and AFM: Experimental and Theoretical Aspects of image analysis.* VCH.

Minne, S.C. *et al.* (1999). *Bringing scanning probe microscopy up to speed.* Kluwer Academic Publishers.

Neddermeyer, H. (1993). *Scanning tunnelling microscopy.* Kluwer Academic Publishers.

Sarid. D. (1994). *Scanning Force Microscopy with applications to electric, magnetic, and atomic forces.* Oxford University Press (NY).

Stroscio. J.A., Kaiser. W.J. (1994). *Scanning Tunnelling Microscopy.* Academic Press

Weisendanger. R. (1994). *Scanning probe microscopy and spectroscopy: methods and applications.* Cambridge University Press.

INDEX